UNITEXT

La Matematica per il 3+2

Volume 171

Editor-in-Chief

Alfio Quarteroni, Politecnico di Milano, Milan, Italy
  École Polytechnique Fédérale de Lausanne (EPFL), Lausanne, Switzerland

Series Editors

Luigi Ambrosio, Scuola Normale Superiore, Pisa, Italy

Paolo Biscari, Politecnico di Milano, Milan, Italy

Ciro Ciliberto, Università di Roma "Tor Vergata", Rome, Italy

Camillo De Lellis, Institute for Advanced Study, Princeton, USA

Victor Panaretos, Institute of Mathematics, École Polytechnique Fédérale de Lausanne (EPFL), Lausanne, Switzerland

Lorenzo Rosasco, DIBRIS, Università degli Studi di Genova, Genova, Italy
  Center for Brains Mind and Machines, Massachusetts Institute of Technology, Cambridge, Massachusetts, US
  Istituto Italiano di Tecnologia, Genova, Italy

The **UNITEXT - La Matematica per il 3+2** series is designed for undergraduate and graduate academic courses, and also includes books addressed to PhD students in mathematics, presented at a sufficiently general and advanced level so that the student or scholar interested in a more specific theme would get the necessary background to explore it.

Originally released in Italian, the series now publishes textbooks in English addressed to students in mathematics worldwide.

Some of the most successful books in the series have evolved through several editions, adapting to the evolution of teaching curricula.

Submissions must include at least 3 sample chapters, a table of contents, and a preface outlining the aims and scope of the book, how the book fits in with the current literature, and which courses the book is suitable for.

For any further information, please contact the Editor at Springer: francesca.bonadei@springer.com

THE SERIES IS INDEXED IN SCOPUS

Davide Lombardo

L-Functions

An Elementary Introduction

 Springer

Davide Lombardo
Dipartimento di Matematica
Università di Pisa
Pisa, Italy

ISSN 2038-5714 ISSN 2532-3318 (electronic)
UNITEXT
ISSN 2038-5722 ISSN 2038-5757 (electronic)
La Matematica per il 3+2
ISBN 978-3-031-85144-5 ISBN 978-3-031-85145-2 (eBook)
https://doi.org/10.1007/978-3-031-85145-2

© The Editor(s) (if applicable) and The Author(s), under exclusive license to Springer Nature Switzerland AG 2025

This work is subject to copyright. All rights are solely and exclusively licensed by the Publisher, whether the whole or part of the material is concerned, specifically the rights of translation, reprinting, reuse of illustrations, recitation, broadcasting, reproduction on microfilms or in any other physical way, and transmission or information storage and retrieval, electronic adaptation, computer software, or by similar or dissimilar methodology now known or hereafter developed.
The use of general descriptive names, registered names, trademarks, service marks, etc. in this publication does not imply, even in the absence of a specific statement, that such names are exempt from the relevant protective laws and regulations and therefore free for general use.
The publisher, the authors and the editors are safe to assume that the advice and information in this book are believed to be true and accurate at the date of publication. Neither the publisher nor the authors or the editors give a warranty, expressed or implied, with respect to the material contained herein or for any errors or omissions that may have been made. The publisher remains neutral with regard to jurisdictional claims in published maps and institutional affiliations.

This Springer imprint is published by the registered company Springer Nature Switzerland AG
The registered company address is: Gewerbestrasse 11, 6330 Cham, Switzerland

If disposing of this product, please recycle the paper.

A mia mamma

Preface

The understanding of L-functions is central to modern number theory, where they play a fundamental role in the study of various arithmetic phenomena. There are ways of attaching an L-function to various mathematical objects, and in particular to many that are of interest in number theory, such as algebraic number fields, Dirichlet characters, and elliptic curves. These L-functions encode a wealth of information about the corresponding objects, allowing us to analyse their properties and also to make connections between seemingly unrelated mathematical entities. A famous example of this is given by the celebrated *modularity theorem* due to Breuil-Conrad-Diamond-Taylor [1], building on the groundbreaking work of Wiles [5] and Taylor-Wiles [4]: roughly speaking, this result asserts that the set of L-functions of elliptic curves over $\mathbb{Q}$ coincides with the set of L-functions of certain modular forms.

The aim of this book is much more modest: starting from scratch, I develop enough of the theory of L-functions to be able to use them to deduce interesting arithmetic consequences, such as the Prime Number Theorem and Dirichlet's theorem on primes in arithmetic progressions. The latter is one of the most significant unconditional applications of L-functions, and historically the first: the very name L-functions originally appeared in the work of Dirichlet, and all the known proofs rely on them to some extent.

A major step forward in the theory of L-functions came with Tate's doctoral thesis, in which he systematically used tools from harmonic analysis to reprove in a unified way many results that had previously been shown by complicated and ad-hoc arguments. One of the main objectives of this book is to acquaint the reader with Tate's beautiful ideas, which have been the starting point for many modern developments. The style of Tate's thesis is famously concise, and I hope to help the reader appreciate its content by fleshing out many arguments that are only sketched in the original.

Contents In Part I of the book I focus on the derivation of number-theoretic results from analytic properties of L-functions. I mostly take the analytic statements for granted, postponing their discussion until the third part of the book, where they are eventually proved in full. I try to give a unifying framework to understand many different constructions by discussing Artin's general definition of L-functions, but I focus mostly on the so-called *abelian* ones (namely, in the language of Artin, those that correspond to abelian extensions of number fields). This already covers a huge class of L-functions, including all zeta functions of number fields, as well as the classical functions studied by Dirichlet. I give a proof of the more sophisticated theorem of Chebotarev, concerning the distribution of Frobenius automorphisms in Galois groups, and show how the argument generalise in a transparent manner the proof of Dirichlet's theorem. This proof of Chebotarev's theorem, while known to experts, to the best of my knowledge does not appear in print. I also offer a direct algebraic proof, closer in spirit to Chebotarev's original approach, to show how the use of L-functions streamlines the hardest parts of the argument.

Overall, in this first part I try to highlight the role of representation theory and discrete Fourier analysis in motivating and proving identities of L-functions. Here the underlying groups acting are Galois groups of *finite* extensions of number fields. One of Tate's great insights was that this machinery—at least in the abelian setting—can be pushed much further by replacing the Galois group of a finite extension of number fields L/k with a more complicated (infinite) object, the idèle group I_k of the field k. In hindsight, this is hardly surprising: class field theory gives a canonical surjection from I_k to the Galois group $\mathrm{Gal}(k^{\mathrm{ab}}/k)$, where k^{ab} is the maximal abelian extension of k. Thus, studying the representation-theoretic properties of I_k amounts to understanding all finite *abelian* extensions L/k at the same time. This also explains the immense success of Tate's method in recovering all previously known results for abelian L-functions.

Naturally, the shift from finite Galois groups to the infinite idèle group comes with significant technical complications: while for finite groups both the topological and measure-theoretic aspects are trivial (every finite group carries the discrete topology and the counting measure), for the idèle group they become crucial. In particular, by its very definition, the idèle group is a locally compact topological group that arises as a so-called *restricted product* of groups that are related to the (infinitely many) metric completions of the field k. For this reason, Part II of this book is dedicated to establishing the technical prerequisites for Tate's thesis. I begin by constructing the Haar measure for an arbitrary locally compact topological group, which is later used to discuss integration on the idèle group. I then review some important tools of abstract Fourier analysis, describe the main properties of local fields, and discuss many aspects of restricted products—group theoretic, topological, measure-theoretic, and Fourier-analytic.

Finally, Part III focuses on the proof of the main analytic properties of abelian L-functions, such as functional equations, analytic continuation, and the analytic class number formula. I follow closely the method of Tate's thesis, but I provide many more details and also explicitly derive all the classical results introduced in Part I, a task which is mostly left implicit in Tate's original work. In the last chapter, I construct the L-function of a specific elliptic curve with complex multiplication: this turns out to be a Hecke L-function, thus providing a bridge between the purely arithmetic L-functions studied in the rest of the book and the world of geometric L-functions.

About the Book This book grew out of notes written for a course aimed at master's and PhD students that I taught at the University of Pisa in the spring of 2023. This is one of the reasons why I have included several exercises throughout the text; while solutions are not generally provided, I have given extensive hints on the most important problems.

Although mathematically the book does not contain much original material, I hope that the choice of topics and their presentation are somewhat novel. For instance, I begin with concrete number-theoretic applications to motivate readers to delve into the proofs of the underlying analytic statements. In addition, the definition of Artin's L-functions appears early on, and their functorial properties serve as a unifying framework to understand many constructions that might otherwise seem ad hoc.

The target audience for this book also differentiates it from others on similar topics. Experience suggests that a one-semester class based on this book can be taught to master's students with minimal background in algebraic number theory. In order to make the book as self-contained as possible, in Chap. 3 I give a comprehensive review of some basic results about number fields, which should allow even readers with almost no background in number theory to appreciate and understand the rest of the results in the book. In Chap. 4 I also give an almost self-contained account of the basic theorems in the representation theory of finite groups. Moreover, apart from brief forays into the theory of curves over finite and number fields—used solely to motivate the terminology *arithmetic Riemann-Roch theorem* and the example in the last chapter of the book—I also avoid any use of algebraic geometry. These choices are intended to make the presentation accessible to almost any graduate student.

Thus, for example, the prerequisites of this text are different from those of Kahn's book on L-functions of varieties [2], which requires a substantial background in algebraic geometry. The aims are also quite distinct from those of the monograph by Ramakrishnan and Valenza [3] which, while developing from scratch all the harmonic analysis necessary for Tate's thesis in greater detail than I do, does not emphasise the derivation of concrete number-theoretic results from the general formalism of L-functions. It is my hope that this text can be read in parallel to [3], to give additional motivation, and that it can serve as a gentle introduction to [2].

The subject of L-functions is vast, and while of course I mention only very briefly many important topics (L-functions of geometric origin, automorphic forms, the Langlands programme), I hope that what *is* there can be useful to students, providing a different take on this very classical and important area of number theory.

Acknowledgments I would like to thank Lucrezia Bertoletti, Sebastiano Boscardin, Davide Colpo, Lorenzo Furio, Andrea Gallese, Livia Grammatica, Francesco Moroniti, Luca Speciale, Mirko Torresani, and Cristofer Villani for pointing out mistakes in a preliminary version of the notes that eventually became this book. I am especially grateful to Davide Colpo and Livia Grammatica for suggesting a simplification of the proof of Theorem 8.3. I also thank Alessandra Caraceni, Lorenzo Furio, Andrea Gallese, and Luca Mauri for their help with a second thorough reading. I am grateful to Alberto Perelli for an interesting discussion about functional equations, for some insightful comments on the definition of the Selberg class, and for sharing with me the notes of one of his courses. I thank Andrea Maffei for many interesting mathematical discussions, especially on the definition of automorphic L-functions.

Pisa, Italy
December 2024

Davide Lombardo

Competing Interests The author has no competing interests to declare that are relevant to the content of this manuscript.

References

1. Breuil, C., Conrad, B., Diamond, F., Taylor, R.: On the modularity of elliptic curves over **Q**: wild 3-adic exercises. J. Am. Math. Soc. **14**(4), 843–939 (2001). https://doi.org/10.1090/S0894-0347-01-00370-8
2. Kahn, B.: Zeta and L-functions of varieties and motives. London Mathematical Society Lecture Note Series, vol. 462. Cambridge University Press, Cambridge (2020). https://doi.org/10.1017/9781108691536. Translated from the 2018 French original [3839285]
3. Ramakrishnan, D., Valenza, R.J.: Fourier analysis on number fields. Graduate Texts in Mathematics, vol. 186. Springer-Verlag, New York (1999). https://doi.org/10.1007/978-1-4757-3085-2
4. Taylor, R., Wiles, A.: Ring-theoretic properties of certain Hecke algebras. Ann. Math. (2) **141**(3), 553–572 (1995). https://doi.org/10.2307/2118560
5. Wiles, A.: Modular elliptic curves and Fermat's last theorem. Ann. Math. (2) **141**(3), 443–551 (1995). https://doi.org/10.2307/2118559

Contents

Part I Classical L-functions and Applications

1 What Is an L-function? 3
   1.1 The Riemann ζ Function 4
   1.2 Dedekind ζ Functions 6
   1.3 Dirichlet L-functions 6
   1.4 The Selberg Class 9
   1.5 Automorphic Versus Galois: A Tale of Two (Classes of) L-functions 10
   Problems 14
   References 16

2 The Prime Number Theorem 17
   2.1 The Riemann–von Mangoldt Exact Formula 26
   Problems 28
   References 28

3 Review of Algebraic Number Theory 29
   3.1 Structure of the Ring of Integers 29
   3.2 Unique Factorisation of Ideals 30
   3.3 Splitting of Primes 30
   3.4 Galois Action on the Primes 32
   3.5 Frobenius Elements 33
   3.6 Dirichlet's Unit Theorem and the Regulator 35
   3.7 The Class Group 38
   3.8 Completed ζ Functions: The Local Factors at Infinity 39
   Problems 40
   Reference 40

4 A Primer of Representation Theory 41
   4.1 Basic Definitions 41
   4.2 Constructions on Representations 43
      4.2.1 Direct Sum 43

		4.2.2 Dual Representation	44
		4.2.3 Tensor Products and Homomorphisms	44
	4.3	Complete Reducibility	45
	4.4	Character Theory	47
		4.4.1 Further Orthogonality Relations	53
	4.5	Further Operations on Representations	54
	Problems		59
	Reference		60
5	**The L-function of a Complex Galois Representation**		**61**
	5.1	Independence of the Choice of $\mathfrak{P}$	62
	5.2	Convergence of the Euler Product	64
	5.3	The Riemann and Dedekind ζ Functions, and Dirichlet's L-functions, Are Artin L-functions	65
	5.4	The Formalism of Artin L-functions	68
	5.5	Artin's Conjecture on Analytic Continuation	73
	5.6	Factorisation of the Dedekind ζ-function	75
	Problems		77
	References		79
6	**Dirichlet's Theorem on Arithmetic Progressions**		**81**
	6.1	Pontryagin Duality: Finite Case	81
	6.2	Densities	85
	6.3	Factorisation of the Cyclotomic Dedekind ζ Function	87
	6.4	Infinitely Many Primes in Arithmetic Progressions	88
	6.5	The Philosophy of Special Values	91
		6.5.1 Further Special Values	92
	Problems		93
	References		94
7	**The Chebotarev Density Theorem**		**95**
	7.1	Analytic Proof	97
	7.2	Algebraic Proof	103
	Problems		111
	References		112

Part II Prerequisites for Tate's Thesis

8	**The Haar Measure**		**115**
	8.1	Preliminaries	115
	8.2	Haar Measure: Existence	117
	8.3	Haar Measure: Uniqueness (up to Constants)	124
	Problems		126
	References		127
9	**Abstract Fourier Analysis**		**129**
	9.1	Pontryagin Duality: General Case	129

	9.2	The Abstract Fourier Transform	132
	References		134
10	**Review of Local Fields**		135
	Problems		140
	References		141
11	**Restricted Direct Products**		143
	11.1	Abstract Group Theory	143
	11.2	Topological Groups	144
	11.3	(Quasi-)Characters of a Restricted Product	146
	11.4	Measure Theory	149
	11.5	Fourier Analysis	153
	Problems		154

Part III Tate's Thesis

12	**The Local Theory**			157
	12.1	The Additive Group		160
		12.1.1	Choice of Haar Measure	162
	12.2	The Multiplicative Group		165
		12.2.1	Choice of Haar Measure	168
	12.3	Local Zeta Functions I: The General Functional Equation		169
	12.4	Local Zeta Functions II: Computation of the Local Factors		174
		12.4.1	Real Case	174
		12.4.2	Complex Case	178
		12.4.3	p-Adic Case	181
		12.4.4	Non-vanishing of the Standard Local ζ Functions	187
	Problems			187
	References			189
13	**The Global Theory**			191
	13.1	The Additive Group: The Adèles		191
		13.1.1	The Field as a Subring of the Adèles	193
		13.1.2	The Poisson Formula	198
		13.1.3	Analogy with the Geometric Riemann-Roch Theorem	204
	13.2	The Multiplicative Group: The Idèles		208
		13.2.1	Multiplicative Fundamental Domain	210
		13.2.2	The Quasi-Characters of $I_k/k^\times$	215
	13.3	Global Zeta Functions		216
	Problems			222
	Reference			222
14	**Hecke L-functions**			223
	14.1	Characters of the Idèles		225
	Problems			229
	Reference			230

15 Recovering the Classical Theory ... 231
- 15.1 The Riemann Zeta Function ... 231
- 15.2 Dedekind ζ Functions ... 232
- 15.3 The General Case: L-functions of Characters ... 235
  - 15.3.1 Idèlic Character ... 235
  - 15.3.2 Choice of the Adèlic Function ... 235
  - 15.3.3 Fourier Transform ... 236
  - 15.3.4 The ζ-Function ... 236
  - 15.3.5 Conclusion: Analytic Continuation and Functional Equation ... 238
- Problems ... 241

16 An Extended Example: The L-function of a CM Elliptic Curve ... 243
- 16.1 The L-function of an Elliptic Curve ... 243
- 16.2 Interpretation as a Hecke L-function ... 252
- 16.3 Interpretation as the L-function of a Representation ... 255
- Problems ... 258
- References ... 258

Index ... 259

Part I
Classical L-functions and Applications

Chapter 1
What Is an L-function?

Abstract In this chapter we meet the main abelian L-functions that were introduced, for different reasons, in the nineteenth century: Riemann's ζ function and its generalisation to arbitrary number fields by Dedekind, and Dirichlet's L-functions. We also state their main properties: proving these deep facts will occupy most of parts II and III of the book.

Given their importance in number theory, it is surprisingly hard to define L-functions. While it is possible to conjecturally describe the class of all L-functions by means of the so-called *axioms of the Selberg class* (see below), we believe it is much more natural to form one's idea of L-functions by looking at examples. It is also certainly closer to the historical development of the notion. For this reason, we postpone the definition of *Artin L-functions* (which encompasses all the main examples we will meet in this course) until Sect. 5, and begin instead with some important examples. To help contextualise the discussion, we start by mentioning that L-function are meromorphic functions of a complex variable s, often defined by means of a *Dirichlet series*:

Definition 1.1 (Dirichlet Series) Let $(a_n)_{n \geq 1}$ be a sequence of complex numbers. The associated **Dirichlet series** is

$$\sum_{n \geq 1} \frac{a_n}{n^s},$$

seen as a function of the complex variable s (if the sum converges). The **abscissa of absolute convergence** is

$$\sigma_0 := \inf \left\{ \sigma \in \mathbb{R} : \Re s > \sigma \Rightarrow \sum_{n \geq 1} \frac{a_n}{n^s} \text{ converges absolutely} \right\}.$$

The function $s \mapsto \sum_{n \geq 1} \frac{a_n}{n^s}$ is holomorphic for $s \in \{\Re s > \sigma_0\}$.

When the sequence a_n has some arithmetic significance, the resulting function $\sum_{n\geq 1} \frac{a_n}{n^s}$ often has special properties. We now look at some important examples where the sequence a_n is either constantly equal to 1 (this will give Riemann's ζ function), is built out of the arithmetic of a number field (Dedekind's ζ functions of fields), or is related to a Dirichlet character (Dirichlet's original L-functions).

1.1 The Riemann ζ Function

The single most important example of L-function is given by Riemann's ζ function.

Definition 1.2 (Riemann ζ Function) The **Riemann ζ function** is given by

$$\zeta(s) = \sum_{n\geq 1} \frac{1}{n^s}$$

for all $s \in \mathbb{C}$ with $\Re s > 1$.

By standard results, $\zeta(s)$ is well-defined (since $\sum_{n\geq 1} n^{-s}$ converges for all real numbers $s > 1$) and defines a holomorphic function on $\{\Re s > 1\}$. We will later show:

Theorem 1.1 *The function $\zeta(s)$ extends to a meromorphic function on the whole of $\mathbb{C}$, with a single simple pole at $s = 1$ with residue 1. In particular, $\zeta(s) = \frac{1}{s-1} + O(1)$ as $s \to 1$.*

This is a consequence of the famous *functional equation* for $\zeta(s)$. In order to discuss it, we need to recall Euler's Γ function:

Definition 1.3 We define

$$\Gamma(s) = \int_0^\infty t^s e^{-t} \frac{dt}{t}$$

for $\Re s > 0$.

Remark 1.1 One way to remember the definition of $\Gamma(s)$ (and the reason we write it in this way, instead of the more usual $\int_0^\infty t^{s-1} e^{-t}\, dt$) is to notice that it is the Mellin transform (=abstract Fourier transform for the group $(\mathbb{R}_{>0}, \cdot)$) of the function e^{-t}, see Remark 12.9.

Having introduced the Γ function, we can define a further auxiliary function (which will eventually turn out to be somewhat more natural than $\zeta(s)$):

Definition 1.4 (Landau's ξ Function) We set $\xi(s) = \frac{1}{2} s(s-1)\pi^{-s/2} \Gamma\left(\frac{s}{2}\right) \zeta(s)$.

In terms of $\xi(s)$, one has:

1.1 The Riemann ζ Function

Theorem 1.2 (Functional Equation for $\xi(s)$) *The ξ function is holomorphic on the whole complex plane and satisfies*

$$\xi(s) = \xi(1-s).$$

Remark 1.2 The factor $s(s-1)$ in the definition of $\xi(s)$ is invariant under the transformation $s \mapsto 1-s$. It follows from the functional equation that the simple function $f(s) = \pi^{-s/2} \Gamma\left(\frac{s}{2}\right) \zeta(s)$ satisfies the functional equation $f(s) = f(1-s)$. From the point of view we will later take, this function f is probably the 'most natural version' of the Riemann ζ function, see Sect. 15.1.

The last basic property of $\zeta(s)$ we want to recall is its representation as an Euler product. More generally, we recall the following result:

Theorem 1.3 (Euler Product) *Let $f : \mathbb{N} \to \mathbb{C}$ be a multiplicative[1] function and let $F(s) = \sum_{n \geq 1} \frac{f(n)}{n^s}$ be the associated Dirichlet series. Denote by σ_0 the abscissa of absolute convergence. There is an equality of holomorphic functions*

$$F(s) := \sum_{n \geq 1} \frac{f(n)}{n^s} = \prod_{p \text{ prime}} \left(\sum_{n \geq 0} \frac{f(p^n)}{p^{ns}} \right),$$

valid over $\{\Re s > \sigma_0\}$. If furthermore f is completely multiplicative,[2] one has $\sum_{n \geq 0} \frac{f(p^n)}{p^{ns}} = \sum_{n \geq 0} \left(\frac{f(p)}{p^s} \right)^n = \frac{1}{1 - f(p)p^{-s}}$, and hence

$$\sum_{n \geq 1} \frac{f(n)}{n^s} = \prod_{p \text{ prime}} \left(1 - f(p)p^{-s}\right)^{-1}.$$

In particular, taking $f(n) = 1$ for all $n \geq 1$ we get

$$\zeta(s) = \prod_{p \text{ prime}} (1 - p^{-s})^{-1}.$$

Remark 1.3 One of the objectives of this course will be to give an interpretation of the function $\pi^{-s/2} \Gamma\left(\frac{s}{2}\right) \zeta(s) = \pi^{-s/2} \Gamma\left(\frac{s}{2}\right) \prod_{p \text{ prime}} (1 - p^{-s})^{-1}$ from Remark 1.2 as an extended Euler product, where the additional factor $\pi^{-s/2} \Gamma\left(\frac{s}{2}\right)$ comes from the infinite place of $\mathbb{Q}$ (cf. Definition 10.1 for the notion of *place*).

---

[1] That is, $(m, n) = 1$ implies $f(mn) = f(m)f(n)$.
[2] That is, $f(mn) = f(m)f(n)$ for all positive integers m, n.

1.2 Dedekind ζ Functions

Our next family of L-functions is given by the so-called (Dedekind) ζ functions of number fields. Before defining them, we quickly recall the notion of *ring of integers* of a number field:

Definition 1.5 (Ring of Integers) Let K be a number field, that is, a finite extension of $\mathbb{Q}$. The **ring of integers of** K, denoted by O_K, is the subring

$$\{\alpha \in K : \mu_\alpha(x) \in \mathbb{Q}[x] \text{ has integral coefficients}\}$$

of K. Here $\mu_\alpha(x)$ is the unique monic minimal polynomial of α over $\mathbb{Q}$.

We will also need the notion of *norm* of an ideal:

Definition 1.6 The **norm** of an ideal $I \triangleleft O_K$ is the cardinality of the quotient O_K/I. We will denote it by $N(I)$.

Problem 1.3 shows that the following definition makes sense:

Definition 1.7 (Dedekind ζ Function) Let K be a number field. The **Dedekind ζ function of** K is

$$\zeta_K(s) := \sum_{\substack{I \triangleleft O_K \\ I \neq (0)}} \frac{1}{N(I)^s} = \sum_{n \geq 1} \frac{\#\{I \triangleleft O_K : N(I) = n\}}{n^s}.$$

These functions satisfy properties similar to those of the Riemann ζ function. In particular, we will later establish the following (see Sect. 15.2):

Theorem 1.4 (Analytic Continuation of $\zeta_K(s)$) *The function $\zeta_K(s)$ extends to a meromorphic function on the entire complex plane, with a single simple pole at $s = 1$.*

In fact, the residue of $\zeta_K(s)$ at $s = 1$ carries interesting arithmetic information: we will discuss this in Theorem 6.4, which we will eventually prove in Sect. 15.2.

1.3 Dirichlet L-functions

The very name 'L-function' comes from a class of functions introduced by Dirichlet in his study of primes in arithmetic progressions. In order to define them, we need to first introduce the notion of *Dirichlet character*.

Definition 1.8 (Dirichlet Characters) Let m be a positive integer. The **group of characters modulo** m, denoted by $\mathbb{D}_m$, is the group $\text{Hom}((\mathbb{Z}/m\mathbb{Z})^\times, \mathbb{S}^1)$, where $\mathbb{S}^1$ is the multiplicative group of complex numbers of norm 1. Given an element

1.3 Dirichlet L-functions

$\chi \in \mathbb{D}_m$, we extend χ to a function (again denoted by χ)

$$\chi : \mathbb{Z} \to \mathbb{C}$$

by setting

$$\chi(n) = \begin{cases} \chi(n \bmod m), & \text{if } (m, n) = 1; \\ 0, & \text{if } (m, n) > 1. \end{cases}$$

This extended function is called a **Dirichlet character modulo** m.

Remark 1.4 Let $m \geq 2$ be an integer and let $\chi \in \mathbb{D}_m$ be the trivial element, that is, the homomorphism sending every element of $(\mathbb{Z}/m\mathbb{Z})^\times$ to 1. The corresponding Dirichlet character $\chi : \mathbb{Z} \to \mathbb{C}$ *depends on m*, because

$$\chi(n) = \begin{cases} 1, & \text{if } (m, n) = 1 \\ 0, & \text{if } (m, n) > 1. \end{cases}$$

All such characters are called the **trivial** (or **principal**) **character** (or more precisely, the **trivial character mod** m), and one should be aware that there are infinitely many of them!

See Problem 1.5 for more properties of Dirichlet characters, including in particular the existence of an isomorphism $\mathbb{D}_m \cong (\mathbb{Z}/m\mathbb{Z})^\times$.

Remark 1.5 The isomorphism $\mathbb{D}_m \cong (\mathbb{Z}/m\mathbb{Z})^\times$ is not canonical, and as such, it is better to distinguish the groups $(\mathbb{Z}/m\mathbb{Z})^\times$ and $\mathbb{D}_m$. We will later call these two groups *dual to each other in the sense of Pontryagin*, see Proposition 6.1, Remark 6.3 and Theorem 9.1.

To each Dirichlet character we now attach a corresponding L-function:

Definition 1.9 (Dirichlet L-functions) Let χ be a Dirichlet character modulo m. We set

$$L(s, \chi) := \sum_{n \geq 1} \frac{\chi(n)}{n^s}.$$

From Problems 1.6 and 1.5 and Theorem 1.3 it follows that for $s \in \{\Re s > 1\}$ one has

$$L(s, \chi) = \prod_{p \text{ prime}} \left(1 - \chi(p) p^{-s}\right)^{-1}. \tag{1.1}$$

Problem 1.7 and the properties of the Riemann ζ function take care of the principal character. For all other characters, we will later show the following:

Theorem 1.5 (Analyticity of Dirichlet L-functions of Non-trivial Characters)
Let χ be a non-principal character modulo m. The function $L(s, \chi)$ extends to an entire function (that is, a holomorphic function on the whole complex plane).

For later use, we also quickly review the notion of **primitivity** for Dirichlet characters.

Definition 1.10 (Primitive Character, Conductor) Let m be a positive integer. A character $\chi : (\mathbb{Z}/m\mathbb{Z})^\times \to \mathbb{S}^1$ is said to be **imprimitive** if there exist a divisor d of m, with $d < m$, and a character $\tilde{\chi} : (\mathbb{Z}/d\mathbb{Z})^\times \to \mathbb{S}^1$ such that $\chi = \tilde{\chi} \circ \pi$, where $\pi : (\mathbb{Z}/m\mathbb{Z})^\times \to (\mathbb{Z}/d\mathbb{Z})^\times$ is the canonical projection. A character is **primitive** if it is not imprimitive.

Any d such that χ factors as above is called a **modulus** for the character χ, while the *minimal* such d is called the **conductor**. The character $\tilde{\chi} : (\mathbb{Z}/d\mathbb{Z})^\times \to \mathbb{S}^1$ which induces χ (where d is the conductor) is called the **primitive character inducing** χ.

Finally, a Dirichlet character is called primitive if its modulus coincides with the conductor of the multiplicative character that induces it.

Example 1.1 Let $\chi : \mathbb{Z} \to \{0, \pm 1\}$ be the function given by

$$\chi(n) = \begin{cases} 0, & \text{if } (6, n) > 1 \\ 1, & \text{if } (6, n) = 1 \text{ and } n \equiv \pm 1 \pmod{8} \\ -1, & \text{if } (6, n) = 1 \text{ and } n \equiv \pm 3 \pmod{8} \end{cases}$$

We also identify χ to the character $\chi : (\mathbb{Z}/24\mathbb{Z})^\times \to \{\pm 1\}$ given essentially by the same rule. It is clear that χ is not primitive, since it is induced by the homomorphism $\tilde{\chi} : (\mathbb{Z}/8\mathbb{Z})^\times \to \{\pm 1\}$ given by

$$\tilde{\chi}(n) = \begin{cases} 1, & \text{if } n \equiv \pm 1 \pmod{8} \\ -1, & \text{if } n \equiv \pm 3 \pmod{8}. \end{cases}$$

One can check easily that $\tilde{\chi}$ is primitive, so that the conductor of χ is 8. Finally, letting $K = \mathbb{Q}(\sqrt{2})$, it is not hard to show that $\zeta_K(s) = \zeta(s) L(s, \tilde{\chi})$.

We have now met the main characters of this course. We will later introduce the Hecke L-functions, which generalise Dirichlet L-functions to arbitrary number fields, but the examples we have seen so far are already enough to discuss two important theorems in analytic number theory: the prime number theorem and Dirichlet's theorem on arithmetic progressions. We will prove them in the next chapters, but before doing so, we give the promised definition of the Selberg class.

1.4 The Selberg Class

We briefly discuss an idea of Selberg's to define axiomatically a class of functions that (conjecturally) consists precisely of those we want to call 'L-functions'. However, at present, we are unable to prove that many functions we *do* want to call L-functions actually belong to this class. For this reason, we will make no use of the notion of the Selberg class. The interested reader can consult Selberg's original paper [6] and the survey by Kaczorowski and Perelli [4], as well as the later work of the same two authors.

The formal definition of the class S is the set of all Dirichlet series

$$F(s) = \sum_{n=1}^{\infty} \frac{a_n}{n^s}$$

absolutely convergent for $\Re s > 1$ that satisfy the following four conditions:

1. Analyticity: $F(s)$ has a meromorphic continuation to the entire complex plane, with the only possible pole (if any) when s equals 1. More precisely, there exists an integer $m \geq 0$ such that $(s-1)^m F(s)$ has analytic continuation to an entire function *of finite order*;[3]
2. Ramanujan conjecture: $a_1 = 1$ and $a_n \ll_\varepsilon n^\varepsilon$ for any $\varepsilon > 0$;
3. Functional equation: there is a gamma factor of the form

$$\gamma(s) = Q^s \prod_{i=1}^{k} \Gamma(\omega_i s + \mu_i)$$

where Q is real and positive, the ω_i are real and positive, and the μ_i are complex with non-negative real part, as well as a so-called root number $\alpha \in \mathbb{C}$, $|\alpha| = 1$, such that the function

$$\Phi(s) = \gamma(s) F(s)$$

satisfies

$$\Phi(s) = \alpha \overline{\Phi(1 - \bar{s})};$$

---

[3] At the present state of knowledge it is not clear whether this condition can be removed, nor whether it follows from the other axioms.

4. Euler product: for $\Re s > 1$, $F(s)$ can be written as a product over primes,
$$F(s) = \prod_p F_p(s),$$
with $F_p(s) = \exp\left(\sum_{n=1}^{\infty} \frac{b_{p^n}}{p^{ns}}\right)$ and, for some $\vartheta < \frac{1}{2}$, $b_{p^n} = O(p^{n\vartheta})$.

1.5 Automorphic Versus Galois: A Tale of Two (Classes of) L-functions

While not logically necessary for the path we follow in this book, it seems important to point out that mathematicians are aware of *two* main sources of L-functions of number-theoretic importance: automorphic representations and Galois representations. (This does not include Selberg's approach, since we do not know how to prove that functions in the Selberg class really are L-functions.) In this book we mostly focus on the Galois side, which we will discuss at length, starting with Artin's definition of the L-function associated with a Galois representation (Definition 5.1). In this section, meant mostly as an advertisement of some beautiful mathematics not covered in this book, we briefly describe the automorphic side of the story. We will freely reference facts from later chapters, so this discussion may be more accessible after the reader has some acquaintance with the rest of the book. In fact, a reader with no previous exposure to the theory of L-functions may wish to skip this section on the first pass.

Even just defining automorphic representations takes considerable effort, so we only give a sketch, focusing on the case where the ground field is $\mathbb{Q}$. See for example the book by Bump [2] for a detailed treatment of the case of GL_2 and many information on the general case; a precise definition of automorphic form for general groups is given in [1].

As a rough first approximation, let G be a reductive algebraic group defined over $\mathbb{Q}$, let Z be the largest $\mathbb{Q}$-split subgroup of the centre of G, and let $\mathbb{A}_\mathbb{Q}$ be the adèle ring of $\mathbb{Q}$ (see Definition 11.3). The reader who is not familiar with algebraic groups may want to keep in mind the case where $G = GL_N$ and Z is the subgroup given by the multiples of the identity. Since $\mathbb{A}_\mathbb{Q}$ is a $\mathbb{Q}$-algebra, we have a well-defined group of adèlic points of G, denoted by $G(\mathbb{A}_\mathbb{Q})$. In the case of GL_N, this is the group of $N \times N$ invertible matrices with coefficients in the adèle ring. There is a factorisation $\mathbb{A}_\mathbb{Q} \cong \mathbb{R} \times \mathbb{A}_\mathbb{Q}^f$, where $\mathbb{A}_\mathbb{Q}^f$, the ring of finite adèles, is given by $\prod_{p\text{ prime}} \mathbb{Z}_p \otimes_\mathbb{Z} \mathbb{Q}$ (equivalently, it is the restricted product of the p-adic fields $\mathbb{Q}_p$ with respect to the subrings $\mathbb{Z}_p$, see Definition 11.3). We have a corresponding decomposition $G(\mathbb{A}_\mathbb{Q}) \cong G(\mathbb{R}) \times G(\mathbb{A}_\mathbb{Q}^f)$. There is a natural embedding $G(\mathbb{Q}) \to G(\mathbb{A}_\mathbb{Q})$ induced by the diagonal embedding $\mathbb{Q} \to \mathbb{A}_\mathbb{Q}$. Let

1.5 Automorphic Versus Galois: A Tale of Two (Classes of) L-functions

$\chi : Z(\mathbb{Q}) \backslash G(\mathbb{A}_{\mathbb{Q}}) \to \mathbb{S}^1$ be a homomorphism (called a *character* in this context) and K be a maximal compact subgroup of $G(\mathbb{A}_{\mathbb{Q}})$.

Automorphic forms on G (with respect to the subgroup K) are smooth[4] functions $f : G(\mathbb{A}_{\mathbb{Q}}) \cong G(\mathbb{R}) \times G(\mathbb{A}_{\mathbb{Q}}^f) \to \mathbb{C}$ satisfying certain special conditions, among which the most relevant for the present discussion are expressed by the functional equations

$$f(\gamma z) = f(z) \quad \forall \gamma \in G(\mathbb{Q}),$$

$$f(gz) = \chi(g) f(z) \quad \forall g \in Z(\mathbb{A}_{\mathbb{Q}}), \tag{1.2}$$

together with the fact that the right-translates $f(zk)$, for $k \in K$, span a finite-dimensional vector space.

To complete the definition of automorphic form one also needs a *slow growth* condition, and the fact that f satisfies certain natural differential equations, see [1, 4.2(c)]. An automorphic form is further called *cuspidal* if the integral of any of its translates over certain subgroups of $G(\mathbb{A})$ vanishes (specifically, one needs to consider the unipotent radicals of parabolic subgroups).

Example 1.2 We make the cuspidality condition concrete in the case $G = \mathrm{GL}_N$. Fix a partition $N = n_1 + \cdots + n_r$ of N (one can show that these partitions parametrise the conjugacy classes of parabolic subgroups of GL_N). The associated parabolic subgroup is

$$P_{n_1,\ldots,n_r} := \left\{ \begin{pmatrix} g_1 & A_{11} & \cdots & A_{1r} \\ 0 & g_2 & \cdots & A_{2r} \\ \vdots & & \ddots & \vdots \\ 0 & 0 & \cdots & g_r \end{pmatrix} \;\Big|\; g_i \in \mathrm{GL}_{n_i}, A_{ij} \in \mathrm{Mat}_{n_i \times n_j} \right\}.$$

The group $P_{n_1,\ldots,n_r}$ is the product of the Levi subgroup $\mathrm{GL}_{n_1} \times \cdots \times \mathrm{GL}_{n_r}$ (embedded in the obvious way as block-diagonal matrices) and of the unipotent radical

$$U_{n_1,\ldots,n_r} := \left\{ \begin{pmatrix} \mathrm{Id}_{n_1} & A_{11} & \cdots & A_{1r} \\ 0 & \mathrm{Id}_{n_2} & \cdots & A_{2r} \\ \vdots & & \ddots & \vdots \\ 0 & 0 & \cdots & \mathrm{Id}_{n_r} \end{pmatrix} \;\Big|\; A_{ij} \in \mathrm{Mat}_{n_i \times n_j} \right\}.$$

---

[4] Write $G(\mathbb{A}_{\mathbb{Q}})$ as the product of the real Lie group $G(\mathbb{R})$ and $G(\mathbb{A}_{\mathbb{Q}}^f)$. The smoothness condition means that f is continuous and, seen as a function of $(x, y) \in G(\mathbb{R}) \times G(\mathbb{A}_{\mathbb{Q}}^f)$, is smooth in x for fixed y and locally constant in y for fixed x.

An automorphic form f for GL_N is cuspidal if

$$\int_{U_{n_1,\ldots,n_r}(\mathbb{Q})\backslash U_{n_1,\ldots,n_r}(\mathbb{A})} f(ug)\, du = 0$$

for all $g \in \mathrm{GL}_N(\mathbb{A})$ and all partitions $N = n_1 + \ldots + n_r$. In fact, it suffices to consider parabolic subgroups P_{n_1,n_2} corresponding to non-trivial partitions $N = n_1 + n_2$ with exactly two parts.

If we let V be the complex vector space of adèlic automorphic forms as defined above, the group $G(\mathbb{A}_{\mathbb{Q}}^f)$ acts on V by right translation on the variable z: for $\varphi \in V$ and $g \in G(\mathbb{A}_{\mathbb{Q}}^f)$ we set $(g \cdot \varphi)(z) = \varphi(zg)$. The irreducible subquotients of V which satisfy an additional property called *smoothness* are essentially the desired automorphic representations (more precisely, V is endowed with three actions: right translation by elements of $G(\mathbb{A}_{\mathbb{Q}}^f)$, right translation by elements of K, and the action of certain differential operators; all three should be taken into account). The subspace of cusp forms is preserved by the natural actions on V, and the (smooth) subquotients of this subspace give rise to cuspidal automorphic representations. Finally, the whole discussion can be further generalised by replacing $\mathbb{Q}$ with an arbitrary number field F.

In favorable cases, to an automorphic representation (together with a representation of the Langlands dual of G) one can attach an L-function, which—since the objects we started with are analytic in nature—is *sometimes* known to have good analytic properties. This result is known for some arithmetically very relevant cases: for example, Godement and Jacquet [3] proved analytic continuation for the L-functions associated with the automorphic forms of $G = \mathrm{GL}_N$ and the standard representation of GL_N, interpreted here as the dual Langlands group of G ('standard L-functions'). For reasons we will explain below, this includes the case of Hecke L-functions. The work of Godement and Jacquet generalises Tate's approach, which is the focus of the last part of this book.

While this is a very complicated and rich story, the moral is simple: there is an *analytic* source of L-functions, namely, automorphic forms (suitably defined) on reductive groups. The analytic properties of these L-functions, while still very hard to establish, are better understood than the corresponding properties for the Galois L-functions we will consider.

There is a special case where most of the complexity of the above description disappears: the group $G = \mathrm{GL}_1 = \mathbb{G}_m$ over $\mathbb{Q}$. In this case we have $Z = G$, and the functional Eq. (1.2) shows that an automorphic form is essentially a character

$$\chi : Z(\mathbb{Q})\backslash G(\mathbb{A}_{\mathbb{Q}}) \to \mathbb{S}^1,$$

where the source group is isomorphic to $\mathbb{Q}^\times \backslash I_{\mathbb{Q}} \cong \widehat{\mathbb{Z}}^\times \times \mathbb{R}_{>0}$. As we will explain in Example 14.1, such characters are precisely the idèlic Hecke characters of the field $\mathbb{Q}$ (which in turn are closely related to Dirichlet characters). More generally,

1.5 Automorphic Versus Galois: A Tale of Two (Classes of) L-functions

the automorphic representations for GL_1 over a number field F are the Hecke characters of that field. The standard automorphic L-functions of Hecke characters are then nothing but Hecke's L-functions, as originally defined by Hecke in analytic terms, see Remark 5.4. Historically, the analytic continuation of Hecke L-functions was proved (by Hecke himself) using the analytic definition; the special case of Dirichlet L-functions is elementary, and was already known to Dirichlet. From the automorphic point of view, the aforementioned work by Godement-Jacquet gives analytic continuation for all standard automorphic L-functions of GL_N, hence in particular for Hecke L-functions.

It was Artin who made the connection to the Galois side: using the full strength of class field theory (and his own reciprocity law), he showed that the L-function of a Hecke character with finite image could also be interpreted as the (Artin) L-function of a Galois representation. As Langlands [5] pointed out, a sufficiently precise version of this statement is equivalent in strength to the Artin reciprocity law itself, a very deep theorem in class field theory.

Since in this book we focus mostly on the Galois side, we will first introduce a subclass of the Hecke L-functions from the Galois point of view, see Definition 5.2. Only later, when discussing harmonic analysis on GL_1, will we introduce the full class of Hecke L-functions (Definition 14.2). We stress that this second definition is essentially *on the automorphic side*, and it is only through the power of class field theory that we can relate it back to our previous Galois approach (the details of this comparison are spelled out in Chap. 14).

We conclude this overview section by emphasising that—at least as far as we know at present—good analytic properties tend to flow from the automorphic side to the Galois side, and not vice versa. In particular, Dirichlet L-functions can be shown to admit meromorphic (and even analytic, for the non-trivial characters) continuation by simple, direct analytic means. Our approach may partially obscure this fact, which is why we highlight it here. The route we take interprets Dirichlet L-functions as a special case of our 'Galois-side Hecke L-functions' (themselves a special case of Artin L-functions), whose good analytic properties are later proved by *first* transferring back to the automorphic side via class field theory, and *then* establishing the necessary results on the automorphic side by Tate's harmonic analysis methods. In other words, it is very much *not* the case that Dirichlet or Hecke L-functions are (known to be) analytic because they are special cases of Artin L-functions; rather, it's the other way around: Hecke L-series have good analytic properties because they are automorphic, and then—with some effort—this can be transferred back to the Galois side to prove that some Artin L-functions have analytic continuation, and that all of them have meromorphic continuation (see Theorem 5.3).

Problems

1.1 Check the following properties of $\Gamma(s)$:

1. $\Gamma(s)$ is a holomorphic function of s in the right half-plane $\{\Re s > 0\}$;
2. $\Gamma(s+1) = s\Gamma(s)$ for all $s \in \mathbb{C}$ with $\Re s > 0$;
3. $\Gamma(s)$ extends to a meromorphic function on $\mathbb{C}$, with poles only at the non-positive integers;
4. $\Gamma(\frac{1}{2}) = \sqrt{\pi}$;
5. Legendre's duplication formula:

$$\frac{\Gamma(s)\Gamma(s+\frac{1}{2})}{\Gamma(2s)} = \frac{\Gamma(\frac{1}{2})}{2^{2s-1}} = \frac{\sqrt{\pi}}{2^{2s-1}}.$$

Hint. This is harder than the other parts of the exercise. Here is a possible strategy.

(a) Introduce the Beta function

$$B(m,n) = \int_0^1 u^{m-1}(1-u)^{n-1}\, du$$

and prove that $B(m,n) = \frac{\Gamma(m)\Gamma(n)}{\Gamma(m+n)}$.

(b) Replacing $m = n = z$ and then $u = \frac{1+x}{2}$, obtain

$$\frac{\Gamma(z)^2}{\Gamma(2z)} = 2^{1-2z}\left(2\int_0^1 (1-x^2)^{z-1}\, dx\right).$$

(c) Prove the following identity for the Beta function:

$$B(m,n) = 2\int_0^1 x^{2m-1}(1-x^2)^{n-1}\, dx.$$

(d) Obtain the equality

$$\frac{\Gamma(z)^2}{\Gamma(2z)} = 2^{1-2z} B(1/2, z) = 2^{1-2z}\frac{\Gamma(1/2)\Gamma(z)}{\Gamma(z+1/2)}$$

and conclude.

6. It is useful to also mention **Euler's reflection formula**,

$$\Gamma(z)\Gamma(1-z) = \frac{\pi}{\sin(\pi z)},$$

whose proof lies a bit deeper than the rest of the statements in this exercise.

7. $\Gamma(s)$ has no zeroes in $\mathbb{C}$ (even though it is not necessary, you might want to use Euler's formula to prove this).

1.2 Assuming the functional equation $\xi(s) = \xi(1-s)$ in Theorem 1.2, prove that $\xi(s)$ is everywhere holomorphic.

1.3 Recall that, given an ideal $I \triangleleft O_K$, we have defined $N(I)$ as the size of the quotient O_K/I.

1. Show that $N(I)$ is finite if and only if $I \neq (0)$.
2. Show that for every positive integer m the set $\{I \triangleleft O_K : N(I) = m\}$ is finite.

1.4 (Euler Products)

1. Prove the second equality appearing in Definition 1.7.
2. Show that there is an Euler product representation of the form

$$\zeta_K(s) = \prod_{P \text{ non-zero prime ideal of } O_K} \left(1 - N(P)^{-s}\right)^{-1}.$$

Hint. See Theorem 3.1 if necessary.
3. Show that $\zeta_K(s)$ converges for $\Re s > 1$.

1.5 Check the following statements:

1. $\mathbb{D}_m$ is isomorphic to $(\mathbb{Z}/m\mathbb{Z})^\times$;
2. any Dirichlet character $\chi : \mathbb{Z} \to \mathbb{C}$ is a completely multiplicative function.

1.6 Show that the abscissa of absolute convergence for the Dirichlet series of Definition 1.9 is $\sigma_0 = 1$.

1.7 Let χ be the trivial character modulo m. Is $L(s, \chi)$ the same as the Riemann ζ function? Express $L(s, \chi)$ in terms of $\zeta(s)$ and simple holomorphic functions.

1.8 Prove the last statement in Example 1.1: $\zeta_{\mathbb{Q}(\sqrt{2})}(s) = \zeta(s)L(s, \tilde{\chi})$.
Hint. You can (and should) do this in at least two ways, which are equivalent but offer slightly different points of view:

1. using the development of $\zeta(s)$, $\zeta_{\mathbb{Q}(\sqrt{2})}(s)$ as Euler products;
2. writing $\mathbb{Q}(\sqrt{2}) = \mathbb{Q} \oplus \mathbb{Q} \cdot \sqrt{2}$ as a sum of irreducible representations of $\text{Gal}(\mathbb{Q}(\sqrt{2})/\mathbb{Q})$ and using Theorem 5.2.

1.9 (Characters vs Dirichlet Characters) There are some subtleties concerning the distinction between characters considered as homomorphisms $(\mathbb{Z}/m\mathbb{Z})^\times \to \mathbb{S}^1$ or as Dirichlet characters $\mathbb{Z} \to \mathbb{C}$. The best you can do is think about this yourself; if you want a specific exercise, here is one that should be instructive.

Let $\tilde{\chi} : (\mathbb{Z}/d\mathbb{Z})^\times \to \mathbb{S}^1$ be a primitive character modulo d, let m be a multiple of d, and let $\chi = \tilde{\chi} \circ \pi$ be the character modulo m that is induced by $\tilde{\chi}$. Finally, let $\chi_{\text{Dirichlet}}$ be the Dirichlet character corresponding to χ.

1. Show that the non-zero values of $\chi_{\text{Dirichlet}}$ are periodic of minimal period d.

2. Show that $\chi_{\text{Dirichlet}}$ is periodic of period m.
3. Show that the minimal period of $\chi_{\text{Dirichlet}}$ can be equal to $m > d$.
4. Show that the minimal period of $\chi_{\text{Dirichlet}}$ can be equal to d even when $m > d$.

1.10 Let $q \geq 2$ be an integer and let χ be a primitive character modulo q. Determine the zeroes of the function $L(s, \chi)$ for $s \in \mathbb{R}, s < 0$ (the answer will depend on whether χ is *even* or *odd*, that is, whether $\chi(-1) = 1$ or -1).

1.11 Classify all Dirichlet characters taking values in $\{\pm 1\}$.

References

1. Borel, A., Jacquet, H.: Automorphic forms and automorphic representations. In: Automorphic Forms, Representations and L-Functions (Proc. Sympos. Pure Math., Oregon State Univ., Corvallis, Ore., 1977), Part 1, Proc. Sympos. Pure Math., vol. XXXIII, pp. 189–207. American Mathematics Society, Providence (1979). With a supplement "On the notion of an automorphic representation" by R. P. Langlands
2. Bump, D.: Automorphic forms and representations. Cambridge Studies in Advanced Mathematics, vol. 55. Cambridge University Press, Cambridge (1997). https://doi.org/10.1017/CBO9780511609572
3. Godement, R., Jacquet, H.: Zeta functions of simple algebras, *Lecture Notes in Mathematics*, vol. 260. Springer-Verlag, Berlin-New York (1972)
4. Kaczorowski, J., Perelli, A.: The Selberg class: a survey. In: Number Theory in Progress, (Zakopane-Kościelisko, 1997) vol. 2 , pp. 953–992. de Gruyter, Berlin (1999)
5. Langlands, R.P.: L-functions and automorphic representations. In: Proceedings of the International Congress of Mathematicians (Helsinki, 1978), pp. 165–175. Academiae Scientiarum Fennicae, Helsinki (1980)
6. Selberg, A.: Old and new conjectures and results about a class of Dirichlet series. In: Proceedings of the Amalfi Conference on Analytic Number Theory (Maiori, 1989), pp. 367–385. University of Salerno, Salerno (1992)

Chapter 2
The Prime Number Theorem

Abstract In this essentially self-contained chapter we give a complete proof of the Prime Number Theorem. We follow the method of Newman, but also discuss Riemann's original ideas to relate his ζ function to the distribution of prime numbers.

The purpose of this chapter is to prove the Prime Number Theorem, namely, to show the following:

Theorem 2.1 (Prime Number Theorem) *Let $\pi(x) = \#\{p \text{ prime} : p \leq x\}$ be the prime-counting function. As $x \to \infty$, we have the asymptotic relation*

$$\pi(x) \sim \frac{x}{\log x}.$$

The theorem was originally proven (independently) by Hadamard [2] and de la Vallée Poussin [4, 5] in 1896, based on ideas that are rooted in Riemann's work on his ζ function. We will instead follow the strategy of Newman [3], as streamlined by Zagier [6].

As is traditional, we introduce the auxiliary functions

$$\Phi(s) = \sum_p \frac{\log p}{p^s} \quad \text{and} \quad \vartheta(x) = \sum_{p \leq x} \log p,$$

where every sum indexed by p (here and below) ranges over the prime numbers.

Proposition 2.1 *We have $\vartheta(x) = O(x)$.*

Proof Let N be a positive integer. Notice that every prime p with $N < p \leq 2N$ divides $\binom{2N}{N}$, so

$$\vartheta(2N) - \vartheta(N) = \sum_{N < p \leq 2N} \log p \leq \log \binom{2N}{N} \leq \log 2^{2N} = 2N \log 2.$$

In particular, $\vartheta(2^{k+1}) - \vartheta(2^k) \leq 2^{k+1} \log 2$. Summing over $k = 0, \ldots, n$ we get

$$\vartheta(2^{n+1}) = \vartheta(2^{n+1}) - \vartheta(1) \leq \log 2 \left(2 + 2^2 + 2^3 + \cdots + 2^{n+1}\right) < 2^{n+2} \log 2.$$

For generic $x \geq 1$, we have $2^n \leq x < 2^{n+1}$ for some $n \in \mathbb{N}$, hence

$$\vartheta(x) \leq \vartheta(2^{n+1}) \leq 2^{n+2} \log 2 \leq (2 \log 2)x.$$

$\square$

The key point in the proof of the Prime Number Theorem is the following non-vanishing result:

Theorem 2.2 (Non-vanishing of ζ Along $\mathfrak{R}s = 1$) *The function $\zeta(s)$ does not have any zeroes along the line $\mathfrak{R}s = 1$, and the function $\Phi(s) - \frac{1}{s-1}$ extends to a holomorphic function on the closed[1] right half-plane $\{\mathfrak{R}s \geq 1\}$.*

Proof We start by noticing the following formal identity, valid for $\mathfrak{R}s > 1$:

$$\log \zeta(s) = \log \prod_p (1 - p^{-s})^{-1} = -\sum_p \log(1 - p^{-s}).$$

Taking the derivative of both sides,

$$\frac{\zeta'(s)}{\zeta(s)} = -\sum_p \frac{\log p \cdot p^{-s}}{1 - p^{-s}} = -\sum_p \frac{\log p}{p^s(1 - p^{-s})}$$

$$= -\sum_p \frac{\log p}{p^s} \sum_{k \geq 0} p^{-ks} = -\sum_p \sum_{k \geq 1} \frac{\log p}{p^{ks}} \quad (2.1)$$

$$= -\sum_{n \geq 1} \frac{\Lambda(n)}{n^s},$$

where $\Lambda(n)$ is the **von Mangoldt function**,

$$\Lambda(n) = \begin{cases} \log p, & \text{if } n = p^k \text{ for a prime } p \\ 0, & \text{otherwise.} \end{cases}$$

---

[1] This means that every point in this set has an open neighbourhood on which the function in question is holomorphic. These open neighbourhoods will necessarily contain complex numbers with real part strictly less than 1.

Essentially the same calculation shows

$$-\frac{\zeta'(s)}{\zeta(s)} = \sum_p \frac{\log p}{p^s(1-p^{-s})} = \sum_p \frac{\log p(1-p^{-s}+p^{-s})}{p^s(1-p^{-s})}$$
$$= \sum_p \frac{\log p}{p^s} + \sum_p \frac{\log p}{p^s(p^s-1)} = \Phi(s) + \sum_p \frac{\log p}{p^s(p^s-1)}, \quad (2.2)$$

where the sum $\sum_p \frac{\log p}{p^s(p^s-1)}$ converges (to a holomorphic function) for $\Re s > \frac{1}{2}$. Hence, $\Phi(s) = -\frac{\zeta'(s)}{\zeta(s)} - \sum_p \frac{\log p}{p^s(p^s-1)}$ is holomorphic over $\{\Re s > \frac{1}{2}\}$, except for the poles of $\frac{\zeta'(s)}{\zeta(s)}$, which are $s = 1$ and the zeros of $\zeta(s)$. Indeed, recall that the logarithmic derivative $f'(s)/f(s)$ of an analytic function $f(s)$ is analytic except at the zeroes and poles of f. At each zero (of multiplicity $m > 0$) or pole (of multiplicity $-m > 0$) of $f(s)$, the logarithmic derivative has a simple pole with residue m. Finally, we already know (Theorem 1.2) that $\zeta(s)$ doesn't have any poles apart from $s = 1$, hence that $-\frac{\zeta'(s)}{\zeta(s)} = \frac{1}{s-1} + O(1)$ for s near 1. We then obtain that $\Phi(s) - \frac{1}{s-1}$ extends holomorphically to $\{\Re s \geq 1\}$ if and only if $\zeta(s)$ does not have any zeroes on the line $\{\Re s = 1\}$. We now prove this crucial statement.

Using again the properties of the logarithmic derivative, we obtain that the order of vanishing of $\zeta(s)$ at $1 + it$ is given by

$$\mathrm{ord}_{1+it}\, \zeta = \lim_{\varepsilon \to 0^+} \varepsilon \frac{\zeta'(1+it+\varepsilon)}{\zeta(1+it+\varepsilon)}. \quad (2.3)$$

Notice furthermore that $\overline{\zeta(s)} = \zeta(\bar s)$ since the coefficients of the Dirichlet series defining ζ are real. Let α be a positive real number and denote by $\mu \geq 0$, $\nu \geq 0$ the orders of vanishing of ζ at $1 + i\alpha$ and $1 + 2i\alpha$. Combining Eqs. (2.3) and (2.2), and recalling that $\sum_p \frac{\log p}{p^s(p^s-1)}$ is holomorphic along $\{\Re s = 1\}$, we obtain

$$\lim_{\varepsilon \to 0^+} \varepsilon \Phi(1+\varepsilon) = 1, \quad \lim_{\varepsilon \to 0^+} \varepsilon \Phi(1+\varepsilon \pm i\alpha) = -\mu,$$
$$\lim_{\varepsilon \to 0^+} \varepsilon \Phi(1+\varepsilon \pm 2i\alpha) = -\nu. \quad (2.4)$$

On the other hand, we have the following inequality,

$$\sum_{r=-2}^{2} \binom{4}{r+2} \Phi(1+\varepsilon+ri\alpha) = \sum_p \frac{\log p}{p^{1+\varepsilon}} \left(p^{i\alpha/2}+p^{-i\alpha/2}\right)^4 \geq 0$$

which follows directly from the definitions, the binomial expansion, and the positivity of squares. Multiplying by $\varepsilon > 0$, passing to the limit $\varepsilon \to 0^+$ and

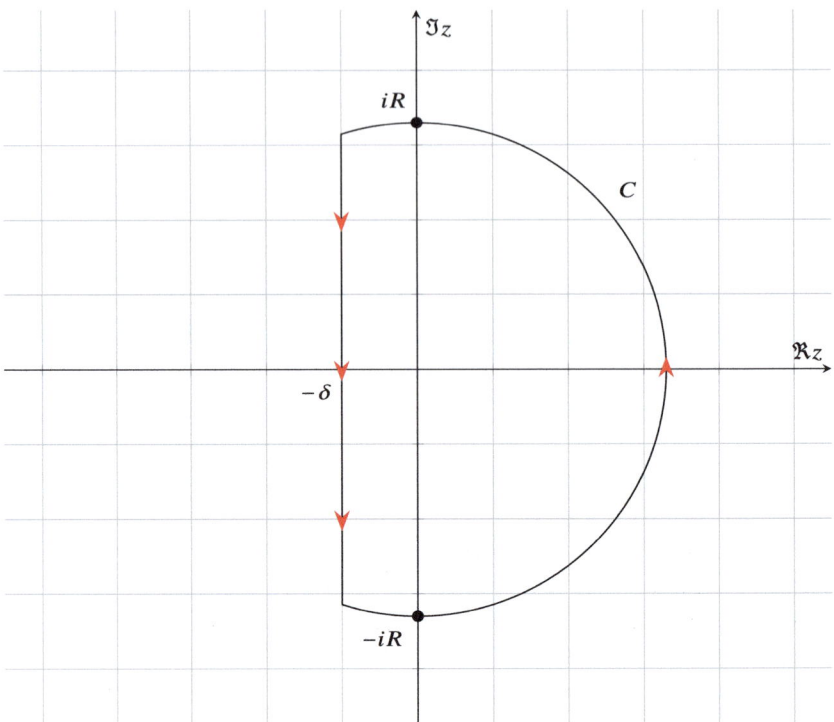

Fig. 2.1 The contour C

replacing the values given by (2.4) we finally get

$$-2\nu - 8\mu + 6 \geq 0,$$

which clearly gives $\mu < 1$, hence $\mu = 0$. By definition of μ, this means $\zeta(1+i\alpha) \neq 0$, as desired. □

Theorem 2.3 (Tauberian Theorem) *Let $f(t) : \mathbb{R}_{\geq 0} \to \mathbb{C}$ be a bounded, locally integrable function. Suppose that the function $g(z) = \int_0^\infty f(t)e^{-zt}\,dt$, which is defined and holomorphic for $\Re z > 0$, extends holomorphically to $\Re z \geq 0$. The integral $\int_0^\infty f(t)\,dt$ exists and equals $g(0)$.*

Proof For $T > 0$ set $g_T(z) = \int_0^T f(t)e^{-zt}\,dt$. This function is holomorphic on the whole complex plane. We will show that $\lim_{T \to \infty} g_T(0) = g(0)$. Let R be large and let C be the boundary of the region $D = \{z \in \mathbb{C} : |z| \leq R, \Re z \geq -\delta\}$ (see Fig. 2.1). Here $\delta > 0$ is chosen as a function of R in such a way that $g_T - g$ is holomorphic inside and on C.

To show that such a δ exists, notice that g_T is everywhere holomorphic, whereas, by assumption, the function $g(z)$ is holomorphic along the segment $I = \{it : -R \leq t \leq R\}$. Since being holomorphic is an open property, for every point z of I there is a small disc centred at z in which $g(z)$ is holomorphic. By compactness of I, a finite union of such discs covers it. We can then take δ to be the minimum of the radii of these finitely many discs.

Let $h_T(z) = (g(z) - g_T(z))e^{zT}\left(1 + \frac{z^2}{R^2}\right)$. Cauchy's integral formula gives

$$h_T(0) = g(0) - g_T(0) = \frac{1}{2\pi i}\int_C (g(z) - g_T(z))\, e^{zT}\left(1 + \frac{z^2}{R^2}\right)\frac{dz}{z}.$$

Our aim is to show that $\lim_{T\to\infty} h_T(0) = 0$. We study Cauchy's integral separately along the arcs

$$C_+ := C \cap \{\Re z > 0\} \quad \text{and} \quad C_- := C \cap \{\Re z < 0\}.$$

Along C_+ we have

$$|g(z) - g_T(z)| = \left|\int_T^\infty f(t) e^{-zt}\, dt\right| \le \int_T^\infty |f(t)||e^{-zt}|\, dt =$$
$$\le \|f\|_\infty \int_T^\infty |e^{-zt}|\, dt = \frac{\|f\|_\infty e^{-\Re zT}}{\Re z}. \tag{2.5}$$

Note also that (by essentially the same calculation as in Eq. (2.5)) when $\Re z$ is negative we have $|g_T(z)| \le \frac{\|f\|_\infty e^{-\Re(z)T}}{|\Re z|}$. Furthermore, all along the circle $|z| = R$ we can estimate

$$\left|e^{zT}\left(1 + \frac{z^2}{R^2}\right)\frac{1}{z}\right| = e^{\Re(z)T}\left|\left(\frac{|z|^2}{z} + z\right)\frac{1}{R^2}\right|$$
$$= \frac{e^{\Re(z)T}}{R^2}|\bar{z} + z| = \frac{2|\Re z|}{R^2}e^{\Re(z)T}. \tag{2.6}$$

Hence, the integral of $h_T(z)$ along C_+ is bounded in absolute value by

$$\|f\|_\infty \cdot \frac{2}{R^2} \cdot (\pi R) = \frac{2\pi \|f\|_\infty}{R}.$$

In particular, we see that the contribution from the integral along C_+ vanishes in the limit $R \to \infty$.

We now consider the integral along C_-, separating the contributions from $g(z)$ and $g_T(z)$. As for $g_T(z)$, which is entire, we can deform the integration contour to the semi-circle $D_- := \{|z| = R, \Re z < 0\}$. Along this semi-circle we can use the estimates $|g_T(z)| \le \frac{\|f\|_\infty e^{-\Re(z)T}}{|\Re z|}$ and (2.6) to obtain as above

$$\left|\int_{D_-} g_T(z) e^{zT}\left(1 + \frac{z^2}{R^2}\right)\frac{1}{z}\, dz\right| \le \frac{2\pi \|f\|_\infty}{R}.$$

This quantity also vanishes in the limit $R \to \infty$, so we are left with considering the integral $\int_{C_-} g(z)e^{zT}\left(1 + \frac{z^2}{R^2}\right)\frac{dz}{z}$. We will show that this integral vanishes in the limit $T \to \infty$ (note that here we take the limit in T, not in R: this contribution tends to 0 also for fixed, finite values of R, provided that T is taken large enough). To handle this integral, note that the function $T \mapsto |g(z)e^{zT}\left(1 + \frac{z^2}{R^2}\right)\frac{1}{z}|$ is decreasing (since $|e^{zT}| = e^{-|\Re(z)|T}$), and it is integrable (even holomorphic) for any fixed value of T. By the dominated convergence theorem, we obtain

$$\lim_{T \to \infty} \int_{C_-} g(z)e^{zT}\left(1 + \frac{z^2}{R^2}\right)\frac{dz}{z} = \int_{C_-} \lim_{T \to \infty} g(z)e^{zT}$$
$$\times \left(1 + \frac{z^2}{R^2}\right)\frac{dz}{z} = \int_{C_-} 0\,dz = 0,$$

where we have used the pointwise convergence of $g(z)e^{zT}\left(1 + \frac{z^2}{R^2}\right)\frac{1}{z}$ to 0 (which again follows from $\Re z < 0$ along C_-). We have thus proved that $\lim_{T \to \infty} |h_T(0)| \leq \frac{4\pi}{R}\|f\|_\infty$. As R is arbitrary, this shows $\lim_{T \to \infty} h_T(0) = 0$, as desired. □

To prove our next result we will need Abel's summation by parts formula:

Theorem 2.4 (Abel's Summation by Parts) *Let $(a_n)_{n \geq 1}$ be a sequence of complex numbers and let $\varphi : [1, \infty) \to \mathbb{R}$ be a C^1 function. For all $x > 1$ we have*

$$\sum_{n \leq x} a_n \varphi(n) = \left(\sum_{n \leq x} a_n\right)\varphi(x) - \int_1^x \left(\sum_{n \leq t} a_n\right)\varphi'(t)\,dt.$$

Proof This is elementary, but we give a proof using distributions, which shows that this is *exactly* (and not just philosophically) the integration-by-parts formula. Consider the function $A(x) = \sum_{n \leq x} a_n$. This function is constant on all intervals of the form $[n, n+1)$. Its derivative (in the distributional sense) is concentrated on the integers, and it is easy to see that it is $\sum_n a_n \delta(x - n)$, where δ is Dirac's delta. Since integration by parts works for distributions, we get (for any $\varepsilon \in (0, 1)$)

$$\sum_{n \leq x} a_n \varphi(n) = \sum_n a_n \int_{1-\varepsilon}^x \delta(t - n)\varphi(t)\,dt = \int_{1-\varepsilon}^x \sum_n a_n \delta(t - n)\varphi(t)\,dt$$
$$= \int_{1-\varepsilon}^x A'(t)\varphi(t)\,dt = [A(t)\varphi(t)]_{1-\varepsilon}^x - \int_{1-\varepsilon}^x A(t)\varphi'(t)\,dt$$
$$= \left(\sum_{n \leq x} a_n\right)\varphi(x) - \int_{1-\varepsilon}^x A(t)\varphi'(t)\,dt,$$

where we have used $A(1-\varepsilon) = 0$ for any $\varepsilon > 0$. Passing to the limit $\varepsilon \to 0^+$ yields the result. □

Proposition 2.2 *The integral*

$$\int_1^\infty \frac{\vartheta(t) - t}{t^2} dt$$

converges.

Proof We fix $s > 1$ and apply Theorem 2.4 to the sequence

$$a_n = \begin{cases} \log n, & \text{if } n \text{ is a prime number}; \\ 0, & \text{otherwise} \end{cases}$$

and to the function $\varphi(x) = x^{-s}$. By definition, the function $\sum_{n \leq x} a_n$ coincides with $\vartheta(x)$. Abel's formula yields

$$\sum_{p \leq x} \frac{\log p}{p^s} = \frac{1}{x^s} \sum_{p \leq x} \log p + s \int_1^x \frac{\vartheta(t)}{t^{s+1}} dt.$$

Letting $x \to \infty$ we obtain

$$\Phi(s) = \lim_{x \to \infty} \frac{1}{x^s} \vartheta(x) + s \int_1^\infty \frac{\vartheta(t)}{t^{s+1}} dt,$$

and by Proposition 2.1 we have $\lim_{x \to \infty} \frac{1}{x^s} \vartheta(x) = 0$ since $s > 1$. Thus, we have

$$\Phi(s) = s \int_1^\infty \frac{\vartheta(t)}{t^{s+1}} dt.$$

The exponential change of variables $t = e^u$ allows us to rewrite this as

$$\Phi(s) = s \int_0^\infty \vartheta(e^u) e^{-us} du. \tag{2.7}$$

Note that we have proved this for $s \in \mathbb{R}_{>1}$, but by analytic continuation the two sides of this equation coincide wherever both are defined and analytic.

We now apply Theorem 2.3 to the functions

$$f(t) = \vartheta(e^t)e^{-t} - 1, \quad g(z) = \frac{\Phi(z+1)}{z+1} - \frac{1}{z}.$$

We check the assumptions:

1. $f(t)$ is bounded and locally integrable: we know that $\vartheta(e^t) = O(e^t)$ by Proposition 2.1, which shows that $f(t)$ is bounded, and $\vartheta(e^t)$, e^{-t} are certainly locally integrable.
2. Next we need to check that $\int_0^\infty f(t)e^{-zt}\,dt = g(z)$ in $\{\Re z > 0\}$, and that $g(z)$ extends holomorphically to $\{\Re z \geq 0\}$. We have

$$\int_0^\infty f(t)e^{-zt}\,dt = \int_0^\infty \left(\vartheta(e^t)e^{-t} - 1\right)e^{-zt}\,dt$$

$$= \int_0^\infty \vartheta(e^t)e^{-(z+1)t}\,dt - \int_0^\infty e^{-zt}\,dt.$$

From Eq. (2.7) we know that $\int_0^\infty \vartheta(e^t)e^{-(z+1)t}\,dt = \frac{\Phi(z+1)}{z+1}$ whenever $\Re(z+1) > 1$, that is, $\Re z > 0$. The integral $\int_0^\infty e^{-zt}\,dt$ is immediate to compute, and evaluates to $\left[\frac{e^{-zt}}{-z}\right]_0^\infty = \frac{1}{z}$. Thus, the functions $\int_0^\infty f(t)e^{-zt}\,dt$ and $g(z)$ coincide for all z with $\Re z > 0$. On the other hand, the crucial Theorem 2.2 implies that $g(z)$ has analytic continuation to $\{\Re z \geq 0\}$: indeed, we know that $\Phi(z) - \frac{1}{z-1}$ is analytic in $\{\Re z \geq 1\}$, hence $\Phi(z+1) - \frac{1}{z}$ is analytic in $\{\Re z \geq 0\}$. Multiplying by $\frac{1}{z+1}$, which is analytic in $\{\Re z \geq 0\}$, we obtain that

$$\frac{\Phi(z+1)}{z+1} - \frac{1}{z(z+1)}$$

is also analytic on the same set. The difference between this function and $\frac{\Phi(z+1)}{z+1} - \frac{1}{z}$ is

$$\frac{1}{z} - \frac{1}{z(z+1)} = \frac{1}{z+1},$$

which is also analytic in $\{\Re z \geq 0\}$.

The conclusion of the theorem is that $\int_0^\infty f(t)\,dt$ exists, that is, the integral

$$\int_0^\infty (\vartheta(e^t)e^{-t} - 1)\,dt$$

is convergent. Substituting back $t = \log u$, we obtain that

$$\int_1^\infty \left(\frac{\vartheta(u)}{u} - 1\right)\frac{du}{u}$$

converges, which (up to renaming u to t) is exactly the statement of the proposition. □

Proposition 2.3 *The function $\vartheta(x)$ is asymptotic to x as $x \to \infty$.*

2 The Prime Number Theorem

Proof This follows from Proposition 2.2. More precisely, suppose by contradiction that there exists $\lambda > 1$ such that $\frac{\vartheta(x_n)}{x_n} > \lambda$ for a sequence x_n going to infinity. Since $\vartheta(x)$ is clearly monotonically increasing, we obtain

$$\int_{x_n}^{\lambda x_n} \frac{\vartheta(t) - t}{t^2} dt \geq \int_{x_n}^{\lambda x_n} \frac{\vartheta(x_n) - t}{t^2} dt \geq \int_{x_n}^{\lambda x_n} \frac{\lambda x_n - t}{t^2} dt$$

$$\stackrel{t=yx_n}{=} \int_1^{\lambda} \frac{\lambda x_n - y x_n}{(y x_n)^2} x_n dy = \int_1^{\lambda} \frac{\lambda - y}{y^2} dy.$$

This is a contradiction: convergence of the integral $\int_1^{\infty} \frac{\vartheta(t)-t}{t^2} dt$ implies that the 'partial tail' $\int_x^{\lambda x} \frac{\vartheta(t)-t}{t^2} dx$ can be made arbitrarily small by choosing x large enough.

Conversely, suppose that for some $\lambda < 1$ there is an unbounded sequence x_n such that $\frac{\vartheta(x_n)}{x_n} < \lambda$. Reasoning as above, we obtain

$$\int_{\lambda x_n}^{x_n} \frac{\vartheta(t) - t}{t^2} dt \leq \int_{\lambda}^1 \frac{\lambda - y}{y^2} dy < 0,$$

which is again a contradiction. $\square$

Proof of Theorem 2.1 On the one hand, we have

$$\vartheta(x) = \sum_{p \leq x} \log p \leq \sum_{p \leq x} \log x = \pi(x) \log(x),$$

while on the other we also have

$$\vartheta(x) \geq \sum_{x^{1-\varepsilon} \leq p \leq x} \log p \geq \sum_{x^{1-\varepsilon} \leq p \leq x} (1 - \varepsilon) \log x$$

$$= (1 - \varepsilon) \log x \left(\sum_{x^{1-\varepsilon} \leq p \leq x} 1 \right) = (1 - \varepsilon) \log x \left(\pi(x) - \pi(x^{1-\varepsilon}) \right).$$

Since clearly $\pi(x^{1-\varepsilon}) \leq x^{1-\varepsilon} = O(x^{1-\varepsilon})$, we have obtained

$$(1 - \varepsilon) \log(x) \left(\pi(x) + O(x^{1-\varepsilon}) \right) \leq \vartheta(x) \leq \pi(x) \log(x).$$

Dividing through by x and using Proposition 2.3 we get

$$(1 - \varepsilon) \left(\frac{\pi(x)}{x / \log x} + o(1) \right) \leq 1 + o(1) \leq \frac{\pi(x)}{x / \log x} \quad \text{as } x \to \infty,$$

which (since ε is arbitrary) implies the theorem. $\square$

2.1 The Riemann–von Mangoldt Exact Formula

Although this is not a book on analytic number theory, we feel it is important to explain more carefully the role of the zeros of the ζ function in controlling the distribution of prime numbers. We will not give full proofs, but hopefully the content of this section will be enough to convince the reader that information about the distribution of the zeros of ζ translates fairly directly into information about the distribution of primes. To make this concrete, we state and sketch a proof of an exact formula for a close relative of the function ϑ:

Theorem 2.5 (Riemann–von Mangoldt Exact Formula) *For every non-integer x we have*

$$\sum_{n \leq x} \Lambda(n) = x - \lim_{T \to \infty} \sum_{\rho : |\operatorname{Imm}(\rho)| \leq T} \frac{x^\rho}{\rho} - \log(2\pi) - \frac{1}{2}\log(1 - x^{-2}),$$

where the sum ranges over the zeroes of ζ in the critical strip $\Re s \in [0, 1]$.

Remark 2.1 Let $\Theta(x) := \sum_{n \leq x} \Lambda(n)$. The difference $\Theta(x) - \vartheta(x)$ is given by

$$\sum_{p \leq x} \sum_{\substack{n \geq 2 \text{ such} \\ \text{that } p^n \leq x}} \log(p) = \sum_{n=2}^{\log_2(x)} \sum_{p \leq x^{1/n}} \log p = \sum_{n=2}^{\log_2(x)} \vartheta(x^{1/n}) \ll x^{1/2} \log(x),$$

where we have used Proposition 2.1. Thus, precise estimates on $\Theta(x)$ lead to precise estimates on the function $\vartheta(x)$, which—as we have seen—is intimately tied to the actual distribution of prime numbers.

The following sketch is very rough (we ignore a number of problems related to the convergence and well-posedness of integrals and sums), but we hope it gives an idea of the inextricably close connection between the distribution of prime numbers (in the form of $\Lambda(n)$) and $\zeta(s)$.

Sketch of Proof of Theorem 2.5 Setting aside the analytic difficulties, the key point lies in an application of Perron's formula (see Problem 2.1). In particular, we start from the equality

$$\sum_{n \geq 1} \frac{\Lambda(n)}{n^s} = -\frac{\zeta'(s)}{\zeta(s)}$$

that we obtained in Eq. (2.1). Setting $g(s) = -\frac{\zeta'(s)}{\zeta(s)}$ in Perron's formula, we get

$$\Theta(x) = \sum_{n \leq x} \Lambda(n) = -\int_{c-i\infty}^{c+i\infty} \frac{\zeta'(z)}{\zeta(z)} \frac{x^z}{z} \, dz.$$

2.1 The Riemann–von Mangoldt Exact Formula

Now, using the residue theorem, we shift the integration contour from $c + i\mathbb{R}$ (where c, in order to use Perron's formula, is taken to be > 1) to $-R + i\mathbb{R}$, and then take the limit $R \to \infty$. In so doing, by the residue theorem, we pick up a correction term every time we cross a pole ρ of $\frac{\zeta'}{\zeta}$; these corrections are of the form $-2\pi i \operatorname{Res}_{z=\rho} \left(\frac{\zeta'(z)}{\zeta(z)} \frac{x^z}{z} \right)$, contribution which then gets divided by the factor $2\pi i$ in Perron's formula.

The poles of ζ'/ζ are precisely the poles of ζ (of which there is one, at $z = 1$) and its zeroes. The term corresponding to the pole gives a residue of 1 for $-\zeta'/\zeta$, which—multiplied by x^z/z at $z = 1$—gives a contribution of x. Each zero ρ, on the other hand, gives a residue of -1 for ζ'/ζ, which—when multiplied by $\frac{x^z}{z}|_{z=\rho} = \frac{x^\rho}{\rho}$—gives the contribution x^ρ/ρ in the Riemann-von Mangoldt exact formula.

If you want to understand the (bounded) terms $\log(2\pi)$ and $\frac{1}{2}\log(1 - x^{-2})$ you should read a complete proof of the theorem, for example in Chapter 17 of [1]. We will simply point out that these additional contributions come from working with the completed ζ function of Definition 1.4 instead of ζ itself. □

Remark 2.2 (Prime Number Theorem Under the Riemann Hypothesis) It is impossible to talk of the Prime Number theorem and not mention the Riemann hypothesis. As the reader presumably already knows, this is the statement that all the zeroes of $\zeta(s)$ in the 'critical strip' $\{0 < \Re(s) < 1\}$ have real part equal to $\frac{1}{2}$. In particular, for every zero ρ of ζ in this strip we have $|x^\rho| = x^{1/2}$. Assuming that the Riemann hypothesis holds, and ignoring again all the analytic difficulties, we then see from Theorem 2.5 that

$$\sum_{n \leq x} \Lambda(n) = x + O\left(\sum_\rho \frac{x^{1/2}}{|\rho|} \right).$$

Since one can show that there aren't too many zeroes of ζ in the critical strip, this leads to

$$\sum_{n \leq x} \Lambda(n) = x + O_\varepsilon(x^{1/2+\varepsilon}).$$

Recalling Remark 2.1 we then get $\vartheta(x) = x + O(x^{1/2+\varepsilon})$, which in turn leads to a strong form of Theorem 2.1, namely, $\pi(x) = \operatorname{Li}(x) + O_\varepsilon\left(x^{1/2+\varepsilon}\right)$ for every $\varepsilon > 0$, where $\operatorname{Li}(x) = \int_2^x \frac{dt}{\log t}$.

Problems

2.1 (Perron's Formula) Let

$$g(s) = \sum_{n=1}^{\infty} \frac{a_n}{n^s}$$

be a Dirichlet series. Assume that it converges uniformly for $\Re(s) > \sigma$, and let $x > 0$ be a real number which is *not* an integer. Also fix $c > \max\{0, \sigma\}$. We have

$$\sum_{n \leq x} a_n = \frac{1}{2\pi i} \int_{c-i\infty}^{c+i\infty} g(z) \frac{x^z}{z} \, dz.$$

References

1. Davenport, H.: Multiplicative number theory. Graduate Texts in Mathematics, vol. 74, 3rd edn. Springer-Verlag, New York (2000). Revised and with a preface by Hugh L. Montgomery
2. Hadamard, J.: Sur la distribution des zéros de la fonction $\zeta(s)$ et ses conséquences arithmétiques. Bull. Soc. Math. Fr. **24**, 199–220 (1896)
3. Newman, D.J.: Simple analytic proof of the prime number theorem. Am. Math. Monthly **87**(9), 693–696 (1980). https://doi.org/10.2307/2321853
4. de la Vallée Poussin, C.J.: Recherches analytiques de la théorie des nombres premiers. Ann. Soc. Sci. Bruxelles (20), 183–256, 281–352, 363–397 (1896)
5. de la Vallée Poussin, C.J.: Recherches analytiques de la théorie des nombres premiers. Ann. Soc. Sci. Bruxelles (21), 351–368 (1896)
6. Zagier, D.: Newman's short proof of the prime number theorem. Am. Math. Monthly **104**(8), 705–708 (1997). https://doi.org/10.2307/2975232

Chapter 3
Review of Algebraic Number Theory

Abstract In this chapter we review without proof many basic facts from algebraic number theory. In contrast to many introductory texts on the subject, we emphasise the importance of Frobenius elements in the case of Galois extensions of number fields: these elements will prove essential for defining Artin's L-functions in the next chapter.

Our next main objective is to prove Dirichlet's theorem on primes in arithmetic progressions. Before doing this, however, we want to give a unified interpretation of all the L-functions we have seen this far in terms of Galois representations. This requires a fair amount of basic algebraic number theory, which we now review, and of representation theory, which we review in the next chapter. All results in this chapter are standard, so we will not provide proofs (for which the reader can refer to any standard textbook on algebraic number theory, such as for example the classical book of Marcus [1]). The reader familiar with the basics of algebraic number theory can safely skip to Chap. 4.

3.1 Structure of the Ring of Integers

We have already met the notion of *ring of integers* of a number field, see Definition 1.5. It is useful to recall that, if K is a number field of degree $n = [K : \mathbb{Q}]$, the ring O_K is isomorphic as an additive group to the free group $\mathbb{Z}^n$. Thus, one can fix a $\mathbb{Z}$-basis $\alpha_1, \ldots, \alpha_n$ of O_K; any two $\mathbb{Z}$-basis are related by a base-change matrix in $\mathrm{GL}_n(\mathbb{Z})$. We denote by $\sigma_1, \ldots, \sigma_n$ the embeddings of K into $\mathbb{C}$.

Definition 3.1 (Discriminant) The **discriminant** of K is

$$d_K := \det\left(\sigma_i(\alpha_j)\right)^2.$$

It is an integer independent of the choice of the basis $\alpha_1, \ldots, \alpha_n$.

Example 3.1 For $K = \mathbb{Q}(\sqrt{2})$ one has $O_K = \mathbb{Z}[\sqrt{2}]$, hence we can take $\alpha_1 = 1$ and $\alpha_2 = \sqrt{2}$. The discriminant is therefore

$$d_K = \det \begin{pmatrix} 1 & \sqrt{2} \\ 1 & -\sqrt{2} \end{pmatrix}^2 = (-2\sqrt{2})^2 = 8.$$

3.2 Unique Factorisation of Ideals

The ring O_K enjoys the following properties:

Theorem 3.1

1. If I is any non-zero ideal of O_K, the quotient O_K/I is finite. The cardinality of O_K/I is called the **norm** of I, see Definition 1.6. The ideal norm is multiplicative: if $I = I_1 I_2$, then $N(I) = N(I_1)N(I_2)$.
2. Non-zero prime ideals of O_K are maximal. Every primary ideal of O_K is the power of a prime ideal. The norm of a prime ideal is of the form p^f, where $p \in \mathbb{Z}$ is prime and f is a positive integer.
3. Every non-zero ideal I of O_K factors uniquely (up to reordering the factors) as a product $I = \mathfrak{p}_1^{e_1} \cdots \mathfrak{p}_r^{e_r}$ of prime ideals $\mathfrak{p}_1, \ldots, \mathfrak{p}_r$.
4. In particular, if $I = \prod_i \mathfrak{p}_i^{e_i}$ and $N(\mathfrak{p}_i) = p_i^{f_i}$ for every i, then

$$N(I) = N\left(\prod_i \mathfrak{p}_i^{e_i}\right) = \prod_i p_i^{e_i f_i}.$$

3.3 Splitting of Primes

Let $\mathfrak{p}$ be a non-zero prime ideal of O_K. The contraction $\mathfrak{p} \cap \mathbb{Z}$ is a non-zero prime ideal of $\mathbb{Z}$, so it is of the form (p). We say that $\mathfrak{p}$ lies over p, or equivalently, that p lies under $\mathfrak{p}$ (the terminology is justified, at least *a posteriori*, by the scheme-theoretic interpretation: there is a natural map $\operatorname{Spec} O_K \to \operatorname{Spec} \mathbb{Z}$, which we can think of as a ramified cover, and the point (p) is the image of the point $\mathfrak{p}$ for this topological map, which is usually drawn with $\operatorname{Spec} O_K$ lying above $\operatorname{Spec} \mathbb{Z}$).

Conversely, starting from a non-zero prime (p) of $\mathbb{Z}$, one can factor the ideal $(p)O_K$ using Theorem 3.1 to obtain an expression of the form

$$(p)O_K = \mathfrak{p}_1^{e_1} \cdots \mathfrak{p}_r^{e_r}.$$

We say that $\mathfrak{p}_1, \ldots, \mathfrak{p}_r$ are the **primes of O_K lying over** (p), that e_i is the **ramification index of $\mathfrak{p}_i$ over** p, and that the exponent f_i defined by $N(\mathfrak{p}_i) = p^{f_i}$

3.3 Splitting of Primes

is the **inertia degree** of $\mathfrak{p}_i$ over p. One can easily show that f_i is also the degree of the field extension $\frac{O_K}{\mathfrak{p}_i}/\mathbb{F}_p$. The field $\frac{O_K}{\mathfrak{p}_i}$ is called the **residue field of (or at)** $\mathfrak{p}_i$. Borrowing the standard notation from scheme theory, we will denote it by $\kappa(\mathfrak{p}_i)$.

There is also a fundamental formula, which is ultimately a consequence of the flatness[1] of O_K over $\mathbb{Z}$, relating the invariants e_i, f_i with the degree $[K : \mathbb{Q}]$.

Theorem 3.2 *Let p be a prime of $\mathbb{Z}$ and write $p = \prod_{i=1}^{r} \mathfrak{p}_i^{e_i}$ for the factorisation of p in O_K. Letting f_i be the inertia degree of $\mathfrak{p}_i$ over p, we have*

$$\sum_{i=1}^{r} e_i f_i = [K : \mathbb{Q}].$$

Consider now the relative setting of an extension L/K, and let $\mathfrak{p}$ be a prime of O_K. As above, one may factor $\mathfrak{p}O_L$ as $\prod_{i=1}^{r} \mathfrak{P}_i^{e_i}$, and we say that the $\mathfrak{P}_i$ are the primes of O_L (or, more informally, of L) lying over $\mathfrak{p}$. We call e_i the **ramification index** of $\mathfrak{P}_i$ over $\mathfrak{p}$, and define f_i as

$$f_i = [\kappa(\mathfrak{P}_i) : \kappa(\mathfrak{p})], \tag{3.1}$$

that is, the degree of the extension between the residue fields at $\mathfrak{P}_i$ and at $\mathfrak{p}$. The analogue of Theorem 3.2 in this setting is as follows.

Theorem 3.3 *With the above notation we have*

$$\sum_{i=1}^{r} e_i f_i = [L : K].$$

We say that a prime ideal $\mathfrak{p}$ of O_K **ramifies** in L if in the factorisation $\mathfrak{p}O_L = \prod_i \mathfrak{P}_i^{e_i}$ at least one exponent e_i is strictly greater than 1. When this is the case, we say more precisely that the prime $\mathfrak{P}_i$ is ramified in the extension L/K.

Remark 3.1 It would be more precise to always speak of the extension O_L/O_K. However, it is both traditional and quite practical to talk about the extension L/K (implicitly meaning the corresponding extension of rings of integers), just like it is common use to write *primes of L* instead of *primes of O_L*.

Finally, a fundamental fact is that only finitely many primes ramify in any given (finite) extension L/K:

Theorem 3.4 *Let L/K be an extension of number fields.[2] The set of primes $\mathfrak{p}$ of O_L that are ramified in L/K is finite.*

---

[1] Since the local rings of $\mathbb{Z}$ are all PIDs, flatness is equivalent to torsion-freeness, which is obvious.

[2] By definition, a number field is a finite extension of $\mathbb{Q}$. As a consequence, any extension of number fields is automatically finite.

The following theorem is also often useful:

Theorem 3.5 (Minkowski) *The only number field K such that no prime of $\mathbb{Q}$ ramifies in K is $K = \mathbb{Q}$ itself.*

3.4 Galois Action on the Primes

We now specialise to the case of L/K being Galois, with group G. In this case, there is an obvious action of G on O_L: indeed, it is clear from the definition that if $\alpha \in L$ is an algebraic integer and σ is any element of G, then $\sigma(\alpha)$ is still an algebraic integer.

Let $\mathfrak{p}$ be a prime of O_K and let $\mathfrak{P}$ be a prime of O_L lying over $\mathfrak{p}$. For every $\sigma \in G$ we have

$$\sigma(\mathfrak{P}) \cap O_K = \sigma(\mathfrak{P} \cap O_K) = \sigma(\mathfrak{p}) = \mathfrak{p},$$

so $\sigma(\mathfrak{P})$ is another prime ideal of O_L lying over $\mathfrak{p}$: the Galois action permutes the primes of L over $\mathfrak{p}$. This action has many nice properties:

Theorem 3.6 (Galois Action on the Primes) *Let $\mathfrak{p}$ be a prime of O_K and denote by $X = \{\mathfrak{P}_1, \ldots, \mathfrak{P}_r\}$ the set of primes of O_L lying over $\mathfrak{p}$. The following hold:*

1. *G acts transitively on X.*
2. *Let $D_i = D(\mathfrak{P}_i \mid \mathfrak{p})$ be the stabiliser of $\mathfrak{P}_i$ for this action. The groups D_i are all conjugate to each other, and $r = [G : D_i]$ for every i. We call D_i the **decomposition group** of $\mathfrak{P}_i$ over $\mathfrak{p}$.*
3. *Let $I_i = \{\sigma \in G : \sigma(x) \equiv x \pmod{\mathfrak{P}_i} \; \forall x \in O_L\}$. The group I_i, called the **inertia group** of $\mathfrak{P}_i$, is normal in D_i, and there is a canonical isomorphism*

$$\frac{D_i}{I_i} \cong \mathrm{Gal}\left(\kappa(\mathfrak{P}_i) \big/ \kappa(\mathfrak{p})\right).$$

   We will denote I_i by $I(\mathfrak{P}_i \mid \mathfrak{p})$.
4. *The group I_i is trivial whenever $\mathfrak{P}_i$ is unramified over $\mathfrak{p}$.*

Remark 3.2 Even though we will not prove this theorem, we note that part (2) is an obvious consequence of (1) and standard facts about group actions.

It is not hard to see that the transitivity of the action (Theorem 3.6 (1)) implies that all the ramification indices e_i are equal to each other and all the inertia degrees f_i are equal to each other. Writing e, f for their common values, the formula of Theorem 3.3 takes the simple form

$$[L : K] = r \cdot e \cdot f. \tag{3.2}$$

3.5 Frobenius Elements

We now come to the real protagonists of this section, namely, Frobenius elements in Galois groups. We keep the setup of the previous section, namely, we let L/K be a Galois extension with group G. Let $\mathfrak{p}$ be a prime of K that is unramified in L and let $\mathfrak{P}$ be a prime of L lying over it. Let f be the inertia degree of $\mathfrak{P}$ over $\mathfrak{p}$ (note that $e = 1$ since $\mathfrak{P}$ is unramified).

The group $\mathrm{Gal}(\kappa(\mathfrak{P})/\kappa(\mathfrak{p}))$ is cyclic of order f, generated by the Frobenius automorphism $\mathrm{Frob} : x \mapsto x^{N(\mathfrak{p})}$. Using the assumption $e = 1$ and Theorem 3.6 (4), we see that the inertia group $I := I(\mathfrak{P} \mid \mathfrak{p})$ is trivial. The isomorphism of Theorem 3.6 (3) then shows that $\mathrm{Gal}(\kappa(\mathfrak{P})/\kappa(\mathfrak{p}))$ is canonically isomorphic to the decomposition group $D := D(\mathfrak{P} \mid \mathfrak{p})$. In particular, $\mathrm{Frob} \in \mathrm{Gal}(\kappa(\mathfrak{P})/\kappa(\mathfrak{p}))$ corresponds to a unique element $\mathrm{Frob}_\mathfrak{P} \in D \subseteq G$. The following all-important definition will be crucial for us.

Definition 3.2 (Artin Symbol) Let $\mathfrak{p}$ be a prime of O_K and let $\mathfrak{P}$ be a prime of O_L lying over it. Suppose that $\mathfrak{P}$ is unramified over $\mathfrak{p}$. We let

$$\left(\frac{L/K}{\mathfrak{P}}\right)$$

denote the element $\mathrm{Frob}_\mathfrak{P} \in G$ constructed above. The element $\left(\frac{L/K}{\mathfrak{P}}\right)$ is called the **Artin symbol** of $\mathfrak{P}$ in the extension L/K, and is often also called the **Frobenius (element) at** $\mathfrak{P}$. More generally, when $\mathfrak{p}$ is ramified and $\mathfrak{P}$ is a prime lying over $\mathfrak{p}$, we say that $g \in G$ is **a Frobenius element at** $\mathfrak{P}$ if it lies in D and its image in $D/I \cong \mathrm{Gal}(\kappa(\mathfrak{P})/\kappa(\mathfrak{p}))$ is the Frobenius automorphism of the residue field. We will also denote by $\left(\frac{L/K}{\mathfrak{P}}\right)$ any such Frobenius element.

Remark 3.3 Unwinding the definitions, and assuming that $\mathfrak{P}$ is unramified in L/K, we see that $\left(\frac{L/K}{\mathfrak{P}}\right)$ is the unique element $\sigma \in G$ that satisfies

$$\sigma(x) \equiv x^{N(\mathfrak{p})} \pmod{\mathfrak{P}} \tag{3.3}$$

for all $x \in O_L$.

Remark 3.4 When $\mathfrak{P}$ is ramified over $\mathfrak{p}$, there are several choices for a Frobenius element at $\mathfrak{P}$. However, by definition, they all lie in D and they all have the same image in D/I, and therefore, given any two choices g, g' of elements of G that are Frobenii at $\mathfrak{P}$, there exists an $h \in I$ such that $g = g'h$.

Remark 3.5 Suppose $\mathfrak{P}'$ is a different prime of O_L lying over $\mathfrak{p}$. By Theorem 3.6 (1), there is an element $\sigma \in G$ such that $\mathfrak{P}' = \sigma \mathfrak{P}$. It is then easy to check that

$$\left(\frac{L/K}{\mathfrak{P}'}\right) = \sigma \left(\frac{L/K}{\mathfrak{P}}\right) \sigma^{-1} :$$

indeed, this follows easily from the characterisation given in Eq. (3.3).

Remark 3.5 allows us to define an 'Artin symbol' that only depends on $\mathfrak{p}$ and not on $\mathfrak{P}$:

Definition 3.3 (Conjugacy Class of Frobenius) Let L/K be a Galois extension of number fields with group G. Let $\mathfrak{p}$ be a prime of O_K unramified in L. We define

$$\left(\frac{L/K}{\mathfrak{p}}\right)$$

as the *conjugacy class* in G of the Artin symbol $\left(\frac{L/K}{\mathfrak{P}}\right)$, where $\mathfrak{P}$ is any prime of L lying over $\mathfrak{p}$. We will often write $\mathrm{Frob}_\mathfrak{p}$ for this conjugacy class, or for any element in it, if the choice of the element does not make any difference.

Remark 3.6 The following special case is particularly interesting: if L/K is an *abelian* extension, that is, it is a Galois extension whose group is commutative, then conjugacy classes consist of a single element. *In this special case*, the Artin symbol of Definition 3.3 can be identified with a specific element of the Galois group.

Example 3.2 (Frobenius Elements for the Cyclotomic Extensions) Consider the Galois extension $\mathbb{Q}(\zeta_n)/\mathbb{Q}$, with group $G \cong (\mathbb{Z}/n\mathbb{Z})^\times$. Recall that, under this isomorphism, the residue class $a \in (\mathbb{Z}/n\mathbb{Z})^\times$ corresponds to the unique automorphism σ_a of $\mathbb{Q}(\zeta_n)$ that sends ζ_n to ζ_n^a.

The primes that ramify in $\mathbb{Q}(\zeta_n)$ are precisely those that divide n. Let p be any other prime, and let $\mathfrak{p}$ be a prime of $\mathbb{Q}(\zeta_n)$ lying over p. By Eq. (3.3), the Artin symbol $\left(\frac{\mathbb{Q}(\zeta_n)/\mathbb{Q}}{\mathfrak{p}}\right)$ is the unique $\sigma \in G$ such that

$$\sigma(x) \equiv x^p \pmod{\mathfrak{p}}$$

for all $x \in O_{\mathbb{Q}(\zeta_n)}$. It is well-known that $O_{\mathbb{Q}(\zeta_n)} = \mathbb{Z}[\zeta_n]$, so we require

$$\sigma\left(\sum c_i \zeta_n^i\right) \equiv \left(\sum c_i \zeta_n^i\right)^p \equiv \sum c_i^p \zeta_n^{pi} \equiv \sum c_i \zeta_n^{pi} \pmod{\mathfrak{p}},$$

where we have used the fact that in the residue field $\kappa(\mathfrak{p})$ (of characteristic p) we have $(x+y)^p = x^p + y^p$ (freshman's dream) and $c_i^p = c_i$ since $c_i \in \mathbb{Z}$. Now, the σ we are looking for must be of the form σ_a for some $a \in (\mathbb{Z}/n\mathbb{Z})^\times$, and it is clear that taking $a = p \bmod n$ works. Since there exists at most one element in the Galois

3.6 Dirichlet's Unit Theorem and the Regulator

group that satisfies (3.3) (this follows from Theorem 3.6 (3)), we conclude that

$$\left(\frac{\mathbb{Q}(\zeta_n)/\mathbb{Q}}{\mathfrak{p}}\right) = \sigma_p,$$

and that, moreover, the conjugacy class $\left(\frac{\mathbb{Q}(\zeta_n)/\mathbb{Q}}{p}\right)$ consists of the single element σ_p.

We conclude this section by mentioning a fundamental theorem (which we will prove in Sect. 7) that gives some motivation as to why Frobenius elements and Artin symbols are so important:

Theorem 3.7 (Chebotarev, First Approximate Form) *Let L/K be a Galois extension of number fields with group G. For every element σ of G, there exist infinitely many primes $\mathfrak{P}$ of O_L such that $\left(\frac{L/K}{\mathfrak{P}}\right) = \sigma$ and $N(\mathfrak{P})$ is a prime number.*

While this may not look like much, we point out right away that Theorem 3.7 contains Dirichlet's theorem on primes in arithmetic progressions as an extremely special case.

Theorem 3.8 (Dirichlet's Theorem on Primes in Arithmetic Progresions) *Let a, m be integers with $(a, m) = 1$. There exist infinitely many primes p that satisfy $p \equiv a \pmod{m}$.*

Proof of Theorem 3.8 Assuming Theorem 3.7 Apply Theorem 3.7 to the Galois extension $\mathbb{Q}(\zeta_m)/\mathbb{Q}$ and to the element σ_a of the Galois group $(\mathbb{Z}/m\mathbb{Z})^\times$. It yields infinitely many primes $\mathfrak{p}$ of $\mathbb{Q}(\zeta_m)$ such that $\left(\frac{\mathbb{Q}(\zeta_m)/\mathbb{Q}}{\mathfrak{p}}\right) = \sigma_a$ and $N(\mathfrak{p})$ is a prime number p. We have shown in Example 3.2 that $\left(\frac{\mathbb{Q}(\zeta_m)/\mathbb{Q}}{\mathfrak{p}}\right) = \sigma_{N(\mathfrak{p})} = \sigma_p$, so we obtain $\sigma_p = \sigma_a$, that is, $p \equiv a \pmod{m}$. Since Chebotarev's theorem guarantees the existence of infinitely many such p, we are done. □

3.6 Dirichlet's Unit Theorem and the Regulator

Before returning to our main topic of L-functions, we review two more facts from algebraic number theory: the structure of the group of units of a number ring and the finiteness of the class group.

Definition 3.4 The **signature** of a number field K of degree n is the pair (r_1, r_2), where r_1 is the number of distinct embeddings of K into $\mathbb{R}$, and r_2 is the number of pairs of complex conjugate embeddings of K into $\mathbb{C}$ whose image is not contained in $\mathbb{R}$. One has $r_1 + 2r_2 = n$.

Theorem 3.9 (Dirichlet's Unit Theorem) *Let K be a number field of signature (r_1, r_2). The group $O_K^\times$ is isomorphic to $T \times \mathbb{Z}^{r_1+r_2-1}$, where T—the torsion subgroup—is precisely given by the set of roots of unity in $K^\times$.*

The standard proof of Theorem 3.9 uses in a fundamental way the so-called **logarithmic embedding**. We now recall this map.

Let $\sigma_1, \ldots, \sigma_{r_1}, \tau_1, \overline{\tau_1}, \ldots, \tau_{r_2}, \overline{\tau_{r_2}}$ be the set of embeddings $K \hookrightarrow \mathbb{C}$, where $\sigma_1, \ldots, \sigma_{r_1}$ are the embeddings with image in $\mathbb{R}$ and $\tau_1, \overline{\tau_1}, \ldots, \tau_{r_2}, \overline{\tau_{r_2}}$ are the r_2 pairs of complex embeddings.

It is also useful to set

$$\rho_1 = \sigma_1, \ldots, \rho_{r_1} = \sigma_{r_1}, \quad \rho_{r_1+1} = \tau_1, \ldots, \rho_{r_1+r_2} = \tau_{r_2}$$

and

$$N_1 = \ldots = N_{r_1} = 1, \quad N_{r_1+1} = \ldots = N_{r_1+r_2} = 2.$$

This will allow for more uniform formulas below.

Definition 3.5 (Logarithmic Embedding) The map

$$\begin{aligned} L : O_K \setminus \{0\} &\to \mathbb{R}^{r_1} \times \mathbb{R}^{r_2} = \mathbb{R}^{r_1+r_2} \\ x &\mapsto (\log(|\sigma_i(x)|))_{i=1,\ldots,r_1}, \left(\log |\tau_j(x)|^2\right)_{j=1,\ldots,r_2} \end{aligned}$$

is called the **logarithmic embedding**. It can equivalently be defined as

$$\begin{aligned} L : O_K \setminus \{0\} &\to \mathbb{R}^{r_1+r_2} \\ x &\mapsto (N_i \log(|\rho_i(x)|))_{i=1,\ldots,r_1+r_2}. \end{aligned}$$

Now recall that every element α in $O_K^\times$ satisfies

$$\pm 1 = N_{K/\mathbb{Q}}(\alpha) = \prod_{i=1}^{r_1} \sigma_i(\alpha) \cdot \prod_{j=1}^{r_2} \tau_j(\alpha)\overline{\tau_j(\alpha)},$$

which implies

$$0 = \log|\pm 1| = \sum_{i=1}^{r_1} \log|\sigma_i(\alpha)| + \sum_{j=1}^{r_2} \log(|\tau_j(\alpha)|^2).$$

This shows that $L(O_K^\times)$ is contained in the hyperplane Π of $\mathbb{R}^{r_1+r_2}$ given by the condition that the sum of the coordinates vanish. A more precise version of Dirichlet's unit theorem shows that $L(O_K^\times)$ is a *lattice* in Π, that is, it is a discrete subgroup such that $\Pi/L(O_K^\times)$ is a compact topological space; equivalently, $L(O_K^\times)$ is discrete in Π and spans it.

3.6 Dirichlet's Unit Theorem and the Regulator

Definition 3.6 (Regulator of a Number Field) The **regulator** of K is by definition

$$\text{Reg}_K = \frac{1}{\sqrt{r_1 + r_2}} \text{vol}\left(\Pi/L(O_K^\times)\right).$$

Equivalently, and more practically, Reg_K can be computed as follows.

Let $u_1, \ldots, u_{r_1+r_2-1} \in O_K^\times$ be a lift of a basis of the free group $O_K^\times/\text{torsion}$. Consider the matrix

$$M := \left(N_i \log |\rho_i(u_j)|\right)_{\substack{i=1,\ldots,r_1+r_2 \\ j=1,\ldots,r_1+r_2-1}}.$$

The matrix M has size $(r_1 + r_2) \times (r_1 + r_2 - 1)$, and every line sums to zero. This implies that any minor of size $(r_1+r_2-1) \times (r_1+r_2-1)$ gives the same determinant, up to sign. The common value of the absolute values of these determinants is the regulator of K.

Example 3.3 We compute the regulator of $K = \mathbb{Q}(\sqrt{2})$. Notice that K has 2 real embeddings, so $r_1 = 2, r_2 = 0$, and Dirichlet's unit theorem gives $O_K^\times \cong \langle -1 \rangle \times \mathbb{Z}$.

One has $O_K = \mathbb{Z}[\sqrt{2}]$. Units of O_K are numbers of the form $x + y\sqrt{2} \in O_K$ with $x, y \in \mathbb{Z}$ and $x^2 - 2y^2 = \pm 1$. The theory of Pell equations shows that all solutions to this equation are given by $x + y\sqrt{2} = \pm(1 + \sqrt{2})^n$ for $n \in \mathbb{Z}$. Hence, a generator of the free part of the group of units can be taken to be $1 + \sqrt{2}$. Now consider the matrix M from Definition 3.6. The two embeddings of K into $\mathbb{C}$ send $\sqrt{2}$ to $\pm\sqrt{2}$, and we have $N_1 = N_2 = 1$, so the matrix M is given by

$$\left(\log|1+\sqrt{2}| \;\; \log|1-\sqrt{2}|\right) = \left(\log(1+\sqrt{2}) \;\; \log\left|\frac{-1}{1+\sqrt{2}}\right|\right)$$
$$= \left(\log(1+\sqrt{2}) \;\; -\log(1+\sqrt{2})\right).$$

As already observed, every line of M sums to zero. The regulator is simply the absolute value of any of the two coefficients of M:

$$\text{Reg}_{\mathbb{Q}(\sqrt{2})} = \log(1 + \sqrt{2}).$$

Finally, we discuss more generally the structure of the so-called group of S-**units**.

Theorem 3.10 *Let K be a number field and let S be a finite set of primes of O_K. The group*

$$O_{K,S}^\times = \left\{x \in K^\times : \begin{array}{c} \mathfrak{p} \text{ does not appear in the} \\ \text{factorisation of the principal ideal } (x) \end{array} \;\; \forall \mathfrak{p} \notin S\right\}$$

is a finitely generated abelian group of rank $|S| + (r_1 + r_2 - 1)$. Its torsion part is given by the set of roots of unity in K.

Remark 3.7 This can be stated more elegantly in the following form, by using the notion of *place* (see Definition 10.1). Let S be a finite set of places of K, containing all the infinite ones. The ring of S-integers is

$$O_{K,S} = \{x \in K : \|x\|_v \leq 1 \quad \forall v \notin S\},$$

and the group of S-units $O_{K,S}^\times$ is simply the multiplicative group of this ring. The previous theorem can then be restated simply as: $O_{K,S}^\times$ is a finitely generated abelian group of rank $|S| - 1$.

3.7 The Class Group

To conclude this review of basic algebraic number theory, we recall the definition of the *class group* of a number field. This group can be considered as a measure of the failure of unique factorisation in O_K, but we will not use this interpretation much. In order to define the class group, we begin by introducing the notion of fractional ideal:

Definition 3.7 (Fractional Ideal) A **fractional ideal** $\mathcal{I}$ of K is a finitely generated O_K-submodule of K. Equivalently, it is a subset of K of the form $\mathcal{I} = \frac{1}{d}I = \{\frac{i}{d} : i \in I\}$, where I is a (usual[3]) ideal of O_K and $d \in K^\times$.

Fractional ideals can be multiplied, as in the next definition:

Definition 3.8 (Product of Fractional Ideals) Let $\mathcal{I}_1, \mathcal{I}_2$ be two fractional ideals. The product $\mathcal{I}_1 \mathcal{I}_2$ is the set $\{\sum x_i y_i \mid x_i \in \mathcal{I}_1, y_i \in \mathcal{I}_2\}$. Equivalently, if $\mathcal{I}_1 = \frac{1}{d_1} I_1$ and $\mathcal{I}_2 = \frac{1}{d_2} I_2$ with I_1, I_2 integral ideals, then

$$\mathcal{I}_1 \mathcal{I}_2 = \frac{1}{d_1 d_2} I_1 I_2.$$

The product makes the set of non-zero fractional ideals into a group.

Theorem 3.11 (Group of Fractional Ideals) *The set $\mathcal{F}(K)$ of non-zero fractional ideals of K forms an abelian group with respect to the multiplication of Definition 3.8. In particular, let $\mathcal{I}$ be a non-zero fractional ideal. There exists a fractional ideal $\mathcal{J}$ such that $\mathcal{I}\mathcal{J} = O_K$.*

---

[3] In this context, an ideal of O_K is sometimes called an *integral* ideal.

We are almost ready to introduce the class group, but we need one more definition:

Definition 3.9 (Principal Fractional Ideal) A fractional ideal $\mathcal{I} = \frac{1}{d}I$ is **principal** if the integral ideal I is principal in the usual sense. Equivalently, $\mathcal{I}$ is principal if and only if it is of the form $O_K \cdot \alpha$ for some $\alpha \in K$.

It is clear that the (non-zero) principal fractional ideals $\mathrm{Princ}(K)$ form a subgroup of $\mathcal{F}(K)$, so we can consider the quotient:

Definition 3.10 (Class Group) The quotient $\dfrac{\mathcal{F}(K)}{\mathrm{Princ}(K)}$ is the **class group** of K, usually denoted by $\mathrm{Cl}(K)$. The order of $\mathrm{Cl}(K)$ is called the **class number** and is denoted by h_K.

The next theorem gives two very important properties of $\mathrm{Cl}(K)$.

Theorem 3.12 (Finiteness of the Class Group) *The group $\mathrm{Cl}(K)$ is finite for every number field K. It is trivial if and only if O_K is a unique factorisation domain.*

3.8 Completed ζ Functions: The Local Factors at Infinity

We now have all the tools to state the following generalisation of Theorem 1.2:

Theorem 3.13 (Functional Equation for the Dedekind ζ Function) *Let K be a number field of signature (r_1, r_2) and discriminant d_K. The **completed ζ function***

$$\Lambda_K(s) = \left(\frac{|d_K|}{4^{r_2}\pi^n}\right)^{s/2} \Gamma\left(\frac{s}{2}\right)^{r_1} \Gamma(s)^{r_2} \zeta_K(s)$$

satisfies the functional equation $\Lambda_K(1-s) = \Lambda_K(s)$. The function $\zeta_K(s)$ has meromorphic continuation to $\mathbb{C}$, with a simple pole at $s = 1$.

As is the case for the ξ function (see Theorem 1.2), one should consider that

$$\left(\frac{|d_K|}{4^{r_2}\pi^n}\right)^{s/2} \Gamma\left(\frac{s}{2}\right)^{r_1} \Gamma(s)^{r_2}$$

plays the role of a 'local factor at infinity'. Notice that K has r_1 places at which the completion is $\mathbb{R}$, and r_2 places at which the completion is $\mathbb{C}$ (see Definition 10.1 for the notion of *place* of a number field). Thus, it is very tempting to think that this 'local factor at infinity' factors even further, as an actual product over the infinite places of K. Tate's approach will make it clear that this is indeed the case.

Problems

3.1 For every integer d that is not a square, compute a $\mathbb{Z}$-basis of the ring of integers of the field $\mathbb{Q}(\sqrt{d})$ and deduce the value of the discriminant of this number field.

3.2 Classify all number fields K such that $O_K^\times$ is finite.

3.3 For $n \geq 1$ let ζ_n be a primitive n-th root of unity.

1. Show that, for every integer $n \geq 3$, the ring of integers of $\mathbb{Q}(\zeta_n)$ is $\mathbb{Z}[\zeta_n]$.
   Hint. This is not an easy problem. You should start by considering the case when n is a prime number, then a prime power, and finally you should prove that if $(m, n) = 1$, then the ring of integers of $\mathbb{Q}(\zeta_{mn})$ is generated by the rings of integers of $\mathbb{Q}(\zeta_m)$ and $\mathbb{Q}(\zeta_n)$.
2. Let p be a prime number. Compute the discriminant of the cyclotomic field $\mathbb{Q}(\zeta_p)$.
3. Let again p be a prime number. Prove that any element of the form $\frac{1-\zeta_p^r}{1-\zeta_p}$ with $r \in \mathbb{Z}$ is a unit of the ring $O_{\mathbb{Q}(\zeta_p)}$. Use this to determine a finite-index subgroup of $\mathbb{Z}[\zeta_5]^\times$.

3.4 Compute the regulator of $\mathbb{Q}(\sqrt{d})$ for every positive $d \leq 20$.

3.5 In this problem you will prove that the ring of integers of $K = \mathbb{Q}(\sqrt{-5})$ does not have unique factorisation and that its class group has even order.

1. Check the identity $(1+\sqrt{-5})(1-\sqrt{-5}) = 2 \cdot 3$ and show that $2, 3, 1+\sqrt{-5}, 1-\sqrt{-5}$ are all irreducible elements in the ring $\mathbb{Z}[\sqrt{-5}] = O_K$.
2. Deduce that $h_K > 1$.
3. Prove that the ideal $I := (2, 1 + \sqrt{-5})$ is not principal, but I^2 is. Deduce that $2 \mid h_K$.

Note. Using the result known as *Minkowski's bound* one can easily prove that $h_K = 2$.

3.6 Compute the factorisation of the ideal (3) in the ring of integers of the fields $\mathbb{Q}(\sqrt{2}), \mathbb{Q}(\sqrt{7}), \mathbb{Q}(\sqrt[3]{3})$.

3.7 Let K be the splitting field over $\mathbb{Q}$ of the polynomial $x^3 - 2$.

1. Prove that the prime factorisation of the prime (7) in O_K is $\mathfrak{p}_1\mathfrak{p}_2$, where $\mathfrak{p}_1, \mathfrak{p}_2$ are both primes of inertia degree 3.
2. Describe the Frobenius elements $\left(\frac{K/\mathbb{Q}}{\mathfrak{p}_1}\right)$ and $\left(\frac{K/\mathbb{Q}}{\mathfrak{p}_2}\right)$ by determining their action on the roots of the polynomial $x^3 - 2$.

Reference

1. Marcus, D.A.: Number fields. Universitext, 2nd edn. Springer, Cham (2018). https://doi.org/10.1007/978-3-319-90233-3. With a foreword by Barry Mazur

Chapter 4
A Primer of Representation Theory

Abstract In this chapter we review the theory of finite-dimensional complex representations of finite groups.

In this book we are mainly concerned with L-functions associated with representations of Galois groups. We will therefore need the language of representation theory, which we review in this chapter. We prove essentially all the results we will use in later chapters, but of course this short summary is not intended to replace a complete textbook on the representation theory of finite groups. For a more thorough treatment, the interested reader may wish to consult, for example, [1].

4.1 Basic Definitions

Given a finite group G, a (complex, finite-dimensional) **representation** of G is a group homomorphism

$$\rho : G \to \operatorname{GL}(V),$$

where V is a finite-dimensional complex vector space and $\operatorname{GL}(V)$ denotes the group of invertible linear transformations of V. Equivalently, it is an action of G on V via linear transformations. To lighten the notation, we will sometimes write $g \cdot v$ for $\rho(g)(v)$, as is customary for a group action. When the action of G on V is clear, a representation $\rho : G \to \operatorname{GL}(V)$ as above is often denoted simply by V.

The **dimension** of a representation V is the dimension of the underlying vector space V.

Let $\rho_1 : G \to \operatorname{GL}(V_1)$ and $\rho_2 : G \to \operatorname{GL}(V_2)$ be two representations. A **homomorphism of G-representations** or **G-equivariant homomorphism** between ρ_1 and ρ_2 is by definition a $\mathbb{C}$-linear map $\varphi : V_1 \to V_2$ between the underlying vector spaces that is equivariant with respect to the action of G, that is,

$$\varphi(\rho_1(g)(v_1)) = \rho_2(g)(\varphi(v_1)) \quad \forall g \in G, \forall v_1 \in V_1. \tag{4.1}$$

The space of G-equivariant homomorphisms $V_1 \to V_2$ is denoted by $\operatorname{Hom}_G(V_1, V_2)$; it is a $\mathbb{C}$-vector subspace of $\operatorname{Hom}(V_1, V_2)$. A homomorphism of G-representations is an **isomorphism** (of G-representations) if the underlying linear map is bijective. We will write $V_1 \cong V_2$ to mean that there exists an isomorphism of G-representations $\varphi : V_1 \to V_2$.

A subspace W of V is called G-invariant or G-stable if $\rho(g)(W) \subseteq W$ for all $g \in G$. In this case, W is also called a **subrepresentation**: indeed, the map

$$\rho|_W : G \to \operatorname{GL}(W)$$
$$g \mapsto \rho(g)|_W$$

endows W with the structure of a G-representation.

A representation V is called **irreducible** (or **simple**) if V is not the trivial vector space $\{0\}$ and the only G-subrepresentations (equivalently, G-invariant subspaces) of V are $\{0\}$ and V itself. If V is not irreducible, then it is called **reducible**. The condition of being irreducible (or reducible) is clearly invariant under isomorphism. We will sometimes write **irrep** for *irreducible complex representation*. The **trivial** representation of G is the 1-dimensional vector space $\mathbb{C}$ with G-action given by the trivial homomorphism $G \to \operatorname{GL}(\mathbb{C})$ that sends every element of G to the identity.

Given $\rho : G \to \operatorname{GL}(V)$, the **character** of ρ is the function

$$\chi_\rho : G \to \mathbb{C}$$
$$g \mapsto \operatorname{tr} \rho(g).$$

Note that isomorphic representations clearly have the same character.

Remark 4.1 Let $\rho : G \to \operatorname{GL}(V)$ be a representation. Lagrange's theorem implies that for every $g \in G$ we have $\rho(g)^{|G|} = \operatorname{Id}_V$. Elementary facts in linear algebra imply that $\rho(g)$ is diagonalisable, with eigenvalues that are roots of unity of order dividing $|G|$.

Remark 4.2 (The Value of Characters at the Identity) Let V be a representation with character χ_V. By definition, we have $\chi_V(\operatorname{id}_G) = \operatorname{tr} \rho(\operatorname{id}_G) = \operatorname{tr} \operatorname{Id}_V = \dim V$.

Finally, if V is a G-representation, we let

$$V^G := \{v \in V : g \cdot v = v \quad \forall g \in G\}$$

be the subspace of G-fixed points. It is easy to check that V^G is a subrepresentation of G (in fact, it is by definition the largest subrepresentation of V on which G acts trivially).

Example 4.1 (A Representation of S_3) Let S_3 act on $V = \mathbb{C}^3$ by permuting the coordinates in the natural way. It is easy to check that $W_1 := V^G$ is the 1-dimensional subspace generated by $\begin{pmatrix} 1 \\ 1 \\ 1 \end{pmatrix}$, so V is not irreducible. Another subrepresentation of V is given by $W_2 := \{{}^t(x, y, z) \mid x + y + z = 0\}$: the permutation action on the coordinates preserves the condition $x + y + z = 0$. Note that $V = W_1 \oplus W_2$. An easy calculation shows that the character χ_V is the function that sends every $\sigma \in S_3$ to the number of its fixed points. Thus, for example,

$$\chi_V((1,2)) = 1 \quad \text{and} \quad \chi_V((1,2,3)) = 0.$$

4.2 Constructions on Representations

4.2.1 Direct Sum

The **direct sum** of two representations V_1, V_2, corresponding to homomorphisms $\rho_1 : G \to \mathrm{GL}(V_1)$ and $\rho_2 : G \to \mathrm{GL}(V_2)$, is the vector space $V_1 \oplus V_2$ endowed with the action $\rho_1 \oplus \rho_2$ given by

$$g \cdot (v_1, v_2) = (\rho_1(g)v_1, \rho_2(g)v_2).$$

Write $n_i = \dim(V_i)$ for $i = 1, 2$ and fix bases of V_1, V_2. Using these bases we can identify $\mathrm{GL}(V_i)$ with $\mathrm{GL}_{n_i}(\mathbb{C})$, and each $\rho_i(g)$ with an $n_i \times n_i$ matrix that we still denote by $\rho_i(g)$. In these terms, the vector space $V_1 \oplus V_2 \cong \mathbb{C}^{n_1+n_2}$ is endowed with the G-action corresponding to the homomorphism

$$\begin{aligned} \rho_1 \oplus \rho_2 : G &\to \mathrm{GL}_{n_1+n_2}(\mathbb{C}) \\ g &\mapsto \begin{pmatrix} \rho_1(g) & 0 \\ 0 & \rho_2(g) \end{pmatrix}. \end{aligned}$$

It is then clear that $\chi_{\rho_1 \oplus \rho_2} = \chi_{\rho_1} + \chi_{\rho_2}$.

Example 4.2 Let G be the group $\mathbb{Z}/2\mathbb{Z} = \{\bar{0}, \bar{1}\}$, acting on $V = \mathbb{C}^2$ via the homomorphism

$$\begin{aligned} \rho : G &\to \mathrm{GL}(V) \\ \bar{1} &\mapsto \begin{pmatrix} 0 & 1 \\ 1 & 0 \end{pmatrix}. \end{aligned}$$

The representation V is not irreducible, because it admits the 1-dimensional invariant subspaces $W_1 = \langle \begin{pmatrix} 1 \\ 1 \end{pmatrix} \rangle$ and $W_2 = \langle \begin{pmatrix} 1 \\ -1 \end{pmatrix} \rangle$. The vectors $\begin{pmatrix} 1 \\ 1 \end{pmatrix}, \begin{pmatrix} 1 \\ -1 \end{pmatrix}$ form a basis of V that simultaneously diagonalises all the matrices in the image of ρ. The element $\bar{1} \in G$ acts trivially on W_1 and as $-\mathrm{Id}$ on W_2.

Example 4.3 Let ρ be the representation of Example 4.2. Consider the representations

$$\rho_1 : G \to \mathrm{GL}(W_1) \qquad \text{and} \qquad \rho_2 : G \to \mathrm{GL}(W_2)$$
$$\bar{n} \mapsto \mathrm{Id} \qquad\qquad\qquad \bar{n} \mapsto (-1)^n \mathrm{Id} :$$

it is easy to check that $\rho_1 \oplus \rho_2 \cong \rho$.

4.2.2 Dual Representation

If $\rho : G \to \mathrm{GL}(V)$ is a representation of G, the dual vector space $V^\vee = \mathrm{Hom}(V, \mathbb{C})$ naturally acquires a structure of G-representation, given by the action

$$(g \cdot \psi)(v) := \psi(\rho(g^{-1})(v)) \quad \forall g \in G, v \in V, \psi \in V^\vee.$$

It is called the **dual** or **contragredient** representation of V. Fix a basis $v_1, \ldots, v_n$ of V and let $v_1^*, \ldots, v_n^*$ be the basis of $V^\vee$ that is dual to $v_1, \ldots, v_n$. If we use these bases to identify ρ and its dual to homomorphisms

$$\rho : G \to \mathrm{GL}_n(\mathbb{C}), \quad \rho^\vee : G \to \mathrm{GL}_n(\mathbb{C})$$

with values in matrix groups, then $\rho^\vee(g) = {}^t\rho(g)^{-1}$, where ${}^t\rho(g)^{-1}$ denotes the inverse transpose matrix. Since all the eigenvalues of $\rho(g)$ are roots of unity (see Remark 4.1), and the inverse of a root of unity is its complex conjugate, the eigenvalues of ${}^t\rho(g)^{-1}$ are the complex conjugates of the eigenvalues of $\rho(g)$. It follows that $\chi_{V^\vee} = \overline{\chi_V}$.

4.2.3 Tensor Products and Homomorphisms

Let $\rho_V : G \to \mathrm{GL}(V), \rho_W : G \to \mathrm{GL}(W)$ be two representations. The vector space $V \otimes W$ can be endowed with the structure of a G-representation by letting $g \in G$ act on $V \otimes W$ as the tensor product of the linear operators $\rho_V(g), \rho_W(g)$.

Concretely, the action of $g \in G$ on an elementary tensor $v \otimes w$ is given by the diagonal action

$$g \cdot (v \otimes w) = (g \cdot v) \otimes (g \cdot w).$$

Fix bases $v_1, \ldots, v_n$ of V and $w_1, \ldots, w_m$ of V, W respectively. If $\rho_V(g)$ is represented by the matrix $A := (a_{ij})$ in the basis $\{v_i\}$ and $\rho_W(g)$ is represented by the matrix $B := (b_{kl})$, one checks that the diagonal action of g on $V \otimes W$ is represented by the matrix $A \otimes B = (a_{ij}b_{kl})$ in the basis $v_i \otimes w_k$. From this, it is easy to show that $\text{tr}(A \otimes B) = \text{tr}(A) \cdot \text{tr}(B)$, hence the character of $V \otimes W$ is given by $\chi_{V \otimes W} = \chi_V \chi_W$. As a special case, the canonical isomorphism

$$\text{Hom}(V, W) \cong V^\vee \otimes W$$

makes $\text{Hom}(V, W)$ into a G-representation. The character of the representation $\text{Hom}(V, W)$ is therefore $\chi_{V^\vee} \chi_W = \overline{\chi_V} \chi_W$. Unwinding the definitions, we find that the action of $g \in G$ on $\varphi \in \text{Hom}(V, W)$ is given by

$$(g \cdot \varphi)(v) = g \cdot (\varphi(g^{-1} \cdot v)) \quad \forall g \in G, v \in V, \varphi \in \text{Hom}(V, W).$$

We note that an element $\varphi \in \text{Hom}(V, W)$ is G-invariant if and only if for all $g \in G$ and $v \in V$ we have

$$g \cdot (\varphi(g^{-1} \cdot v)) = \varphi(v).$$

Replacing v with $g \cdot v$, this formula shows precisely that φ is a G-equivariant homomorphism $V \to W$. We have thus obtained the important formula

$$\text{Hom}_G(V, W) = \text{Hom}(V, W)^G, \tag{4.2}$$

where on the left we take the G-equivariant homomorphisms from V to W, while on the right we take the G-fixed points of G acting on the G-representation $\text{Hom}(V, W)$ defined above.

4.3 Complete Reducibility

Let $\rho : G \to \text{GL}(V)$ be a finite-dimensional complex representation. Fix arbitrarily a positive-definite Hermitian form $\langle \cdot, \cdot \rangle$ on V. By averaging over G, we obtain a G-invariant, positive-definite Hermitian form $\langle \cdot, \cdot \rangle_G : V \times V \to \mathbb{C}$, given by

$$\langle v_1, v_2 \rangle_G := \frac{1}{|G|} \sum_{g \in G} \langle \rho(g)v_1, \rho(g)v_2 \rangle \quad \forall v_1, v_2 \in V.$$

Here G-invariant means that for all $h \in G$ and $v_1, v_2 \in V$ we have

$$\langle \rho(h)v_1, \rho(h)v_2 \rangle_G = \langle v_1, v_2 \rangle_G :$$

this property is trivial to check on the definition, since

$$\langle \rho(h)v_1, \rho(h)v_2 \rangle_G = \frac{1}{|G|} \sum_{g \in G} \langle \rho(g)\rho(h)v_1, \rho(g)\rho(h)v_2 \rangle$$

$$= \frac{1}{|G|} \sum_{g \in G} \langle \rho(gh)v_1, \rho(gh)v_2 \rangle$$

$$= \frac{1}{|G|} \sum_{g' \in G} \langle \rho(g')v_1, \rho(g')v_2 \rangle = \langle v_1, v_2 \rangle_G,$$

where in the next to last equality we have renamed gh to g' (since the set $\{gh : g \in G\}$ coincides with G). Since the Hermitian form $\langle \cdot, \cdot \rangle_G$ is G-invariant, for every G-invariant subspace W of V, the orthogonal complement $W^\perp = \{v \in V : \langle v, w \rangle_G = 0 \ \forall w \in W\}$ is also G-invariant. Moreover, basic linear algebra shows that $V \cong W \oplus W^\perp$. By an immediate induction (on $\dim V$), this implies that every finite-dimensional G-representation is the direct sum of irreducible ones: if V is irreducible there is nothing to prove, and otherwise, if W is a non-trivial G-invariant subspace, we have the decomposition $V \cong W \oplus W^\perp$. By induction, both W and $W^\perp$ are direct sums of irreducible representations, and therefore so is V. This fact is often expressed by saying that the representation V is **completely reducible** (into a direct sum of irreducible representations) or **semisimple**. In other words, we have proved the following theorem:

Theorem 4.1 *Let V be a finite-dimensional complex representation of the finite group G. The representation V is completely reducible, that is, there are finitely many non-zero, G-invariant subspaces $W_1, \ldots, W_n$ of V such that $V \cong W_1 \oplus \cdots \oplus W_n$ as G-representations and each W_i is an irreducible G-representation.*

Example 4.4 (A Representation of S_3, Continued) For the representation V of Example 4.1, the subrepresentations W_1, W_2 are irreducible (this is obvious for W_1, which is 1-dimensional, and is easy to check for W_2: we will do this in Example 4.7 using character theory). A decomposition into irreducible components is therefore $V \cong W_1 \oplus W_2$.

Example 4.5 (Abelian Groups) Let G be abelian and let V be an irreducible representation of G. We claim that $\dim V = 1$. Indeed, by Remark 4.1 each linear operator $\rho(g)$ (for $g \in G$) is diagonalisable. Moreover, for any $g_1, g_2 \in G$, the endomorphisms $\rho(g_1), \rho(g_2)$ commute, because using that $g_1 g_2 = g_2 g_1$ in G we obtain $\rho(g_1)\rho(g_2) = \rho(g_1 g_2) = \rho(g_2 g_1) = \rho(g_2)\rho(g_1)$. Linear algebra then shows that the endomorphisms $\{\rho(g) : g \in G\}$ are all simultaneously diagonalisable, say in the basis $\{v_1, \ldots, v_n\}$. By definition, each $\rho(g)$ for $g \in G$ sends v_1 to a multiple of itself, so the subspace generated by v_1 is G-stable. Since V is an irreducible representation, this implies that V is the 1-dimensional vector space generated by v_1.

Combined with our previous remarks on complete reducibility, the previous example shows that every (finite dimensional, complex) representation of an abelian group G is the direct sum of (finitely many) 1-dimensional representations. Note furthermore that a 1-dimensional representation of G is simply a homomorphism

$$\rho : G \to \mathrm{GL}(\mathbb{C}) = \mathbb{C}^\times.$$

Since G is finite, the image of ρ must be contained in $\mathbb{S}^1$. Thus, ρ is what is usually called a **character**, or an element of the **dual group** $\hat{G} = \mathrm{Hom}(G, \mathbb{S}^1)$. We will explore characters of abelian groups in much greater detail in Sect. 6.1. Note that when G is abelian and V is 1-dimensional, the character of ρ is essentially ρ itself, so the present definition is compatible with the more general definition of characters we have already given.

Example 4.6 (Cyclic Groups) Let $G = \mathbb{Z}/n\mathbb{Z}$ be a cyclic group of order n. Homomorphisms $G \to \mathbb{S}^1$ are uniquely determined by the image ζ of the generator $\overline{1}$ of G; there is one such homomorphism for each $\zeta \in \mathbb{C}^\times$ such that $\zeta^n = 1$. Thus, there are precisely n characters of G, and therefore precisely n (isomorphism classes of) irreducible representations.

4.4 Character Theory

Consider the space $\mathbb{C}^G$ of all complex-valued functions of G and the Hermitian product on $\mathbb{C}^G$ given by the rule

$$\langle \chi_1, \chi_2 \rangle = \sum_{g \in G} \chi_1(g) \overline{\chi_2(g)}. \tag{4.3}$$

Note that this Hermitian product is positive-definite. The following theorem is the technical heart of much of the (complex) representation theory of finite groups:

Theorem 4.2 *Let G be a finite group.*

1. *(Schur's lemma) Let V_1, V_2 be irreducible representations. Any G-equivariant homomorphism $\varphi : V_1 \to V_2$ is either the zero map or an isomorphism. We have*

$$\dim \mathrm{Hom}_G(V_1, V_2) = \begin{cases} 1, & \text{if } V_1, V_2 \text{ are isomorphic} \\ 0, & \text{otherwise}. \end{cases}$$

If $V_1 = V_2$ is irreducible, the space $\mathrm{Hom}_G(V_1, V_1)$ consists of the multiples of the identity.

2. Let V be a representation of G with character χ_V. We have

$$\langle \chi_V, \mathbf{1} \rangle = \dim V^G,$$

where $\mathbf{1}$ (the constant function 1) denotes the character of the trivial representation.
3. Let V_1, V_2 be finite-dimensional complex representations with characters χ_1, χ_2. We have

$$\langle \chi_1, \chi_2 \rangle = \dim \operatorname{Hom}_G(V_1, V_2).$$

If V_1, V_2 are irreducible, we have

$$\langle \chi_1, \chi_2 \rangle = \begin{cases} 1, & \text{if } V_1, V_2 \text{ are isomorphic} \\ 0, & \text{otherwise.} \end{cases}$$

4. Up to isomorphism, there are only finitely many irreducible representations of G.
5. Let $\rho_1 : G \to \operatorname{GL}(V_1), \rho_2 : G \to \operatorname{GL}(V_2)$ be two finite-dimensional complex representations. The characters of ρ_1, ρ_2 are equal if and only if the representations V_1, V_2 are isomorphic.
6. A representation V is irreducible if and only if $\langle \chi_V, \chi_V \rangle = 1$.

Proof

1. If $\varphi : V_1 \to V_2$ is a G-equivariant homomorphism, the definitions imply that the kernel and the image of φ are G-invariant subspaces of V_1, V_2 respectively. Since V_1 is irreducible, $\ker \varphi$ is either trivial or equal to V_1. Similarly, the image of φ is either trivial or all of V_2. This easily implies that if $\varphi : V_1 \to V_2$ is not the zero map, then it is an isomorphism. Now, if the only G-equivariant homomorphism $\varphi : V_1 \to V_2$ is the zero map we are done, so assume that we have an isomorphism $\varphi_0 : V_1 \to V_2$. Composing with φ_0^{-1}, the space $\operatorname{Hom}_G(V_1, V_2)$ is in bijection with $\operatorname{Hom}_G(V_1, V_1)$, so it suffices to treat the case where $V_1 = V_2$ is irreducible. Let $\varphi : V_1 \to V_1$ be G-equivariant and let $\lambda \in \mathbb{C}$ be an eigenvalue of φ. The map $\varphi - \lambda \operatorname{Id} : V_1 \to V_1$ is again G-equivariant and has non-trivial kernel: by the previous arguments, it must be the zero map, and therefore $\varphi = \lambda \operatorname{Id}$. This shows as desired that $\operatorname{Hom}_G(V_1, V_1)$ is 1-dimensional and consists of the multiples of the identity.
2. Consider the linear operator $\pi := \frac{1}{|G|} \sum_{g \in G} \rho(g)$. We claim that π is a projector on the subspace V^G. On the one hand, the image of π is contained in V^G, because

4.4 Character Theory

for all $h \in G$ we have

$$\rho(h)\pi(v) = \frac{1}{|G|}\rho(h)\sum_{g\in G}\rho(g)(v)$$

$$= \frac{1}{|G|}\sum_{g\in G}\rho(h)\rho(g)(v)$$

$$= \frac{1}{|G|}\sum_{g\in G}\rho(hg)(v)$$

$$= \frac{1}{|G|}\sum_{g\in G}\rho(g)(v) = \pi(v),$$

so that $\pi(v)$ is invariant under all $h \in G$. Conversely, if $v \in V^G$, then the definition easily implies $\pi(v) = v$, so the restriction of π to V^G is the identity, and in particular the image of π coincides with V^G. Finally, it is a projector because

$$\pi^2 = \frac{1}{|G|^2}\sum_{g_1,g_2\in G}\rho(g_1)\rho(g_2) = \frac{1}{|G|^2}\sum_{g_1,g_2\in G}\rho(g_1g_2)$$

$$= \frac{1}{|G|^2}\sum_{h\in G}\#\{(g_1,g_2)\in G^2 : g_1g_2 = h\}\cdot\rho(h)$$

$$= \frac{1}{|G|^2}\sum_{h\in G}|G|\cdot\rho(h) = \frac{1}{|G|}\sum_{h\in G}\rho(h) = \pi.$$

Since π is a projector, the only eigenvalues of π are 0 and 1, with the latter having multiplicity equal to the dimension of the image of π, that is, V^G. It follows that

$$\dim V^G = \mathrm{tr}\,\pi = \frac{1}{|G|}\sum_{g\in G}\mathrm{tr}\,\rho(g) = \frac{1}{|G|}\sum_{g\in G}\chi_V(g) = \langle \chi_V, \mathbf{1}\rangle.$$

3. The character of the representation $\mathrm{Hom}(V_1, V_2)$ is $\chi_{\mathrm{Hom}(V_1,V_2)} = \overline{\chi_1}\chi_2$, so

$$\langle \chi_2, \chi_1\rangle = \frac{1}{|G|}\sum_{g\in G}\chi_2(g)\overline{\chi_1}(g) = \frac{1}{|G|}\sum_{g\in G}\chi_{\mathrm{Hom}(V_1,V_2)}(g) = \langle \chi_{\mathrm{Hom}(V_1,V_2)}, \mathbf{1}\rangle.$$

Applying part (2) we obtain

$$\langle \chi_2, \chi_1\rangle = \dim \mathrm{Hom}(V_1, V_2)^G = \dim \mathrm{Hom}_G(V_1, V_2),$$

where the last equality is Eq. (4.2). Finally, we observe that $\langle \chi_1, \chi_2 \rangle$ is the complex conjugate of $\langle \chi_2, \chi_1 \rangle$, which we have just computed to be the integer $\dim \mathrm{Hom}_G(V_1, V_2)$. It follows that we also have

$$\langle \chi_1, \chi_2 \rangle = \overline{\langle \chi_2, \chi_1 \rangle} = \overline{\dim \mathrm{Hom}_G(V_1, V_2)} = \dim \mathrm{Hom}_G(V_1, V_2).$$

The special case where V_1, V_2 are irreducible follows from this and Schur's lemma.

4. Let $V_1, \ldots, V_n$ be non-isomorphic irreducible representations of G and let $\chi_1, \ldots, \chi_n$ be their characters. Note that $\chi_i(\mathrm{id}_G) = \mathrm{tr}\,\mathrm{Id}_{V_i} = \dim V_i > 0$, so χ_i is a non-zero function. By (3), the functions χ_i, χ_j for $i \neq j$ are orthogonal in the finite-dimensional space $\mathbb{C}^G$ (which is endowed with a positive-definite Hermitian product). A collection of such mutually orthogonal, non-zero vectors in a finite-dimensional space is finite; more precisely, we obtain $n \leq \dim \mathbb{C}^G = |G|$.

5. One implication is trivial: if V_1, V_2 are isomorphic, then they have the same character. For the other implication, let $W_1, \ldots, W_n$ be representatives of the (finitely many) isomorphism classes of irreducible representations of G and $\chi_1, \ldots, \chi_n$ be their characters. Consider an arbitrary G-representation V. We already know that V is the direct sum of irreducible representations (Theorem 4.1), and every irreducible representation is isomorphic to one of the W_i, so we can write

$$V \cong W_1^{\oplus k_1} \oplus \cdots \oplus W_n^{\oplus k_n}$$

for certain integers $k_i \geq 0$. It is easy to check that the character χ_V of V is given by $\sum_{i=1}^n k_i \chi_i$. Since the W_i are irreducible and pairwise non-isomorphic, applying part (3) we obtain

$$\langle \chi_V, \chi_j \rangle = \langle \sum_{i=1}^n k_i \chi_i, \chi_j \rangle = \sum_{i=1}^n k_i \langle \chi_i, \chi_j \rangle = \sum_{i=1}^n k_i \delta_{ij} = k_j,$$

where $\delta_{ij} = \begin{cases} 1, & \text{if } i = j \\ 0, & \text{otherwise.} \end{cases}$

Thus, the character of V determines the integers k_j, and therefore the isomorphism class of $V \cong W_1^{\oplus k_1} \oplus \cdots \oplus W_n^{\oplus k_n}$.

6. As above, write $V \cong W_1^{\oplus k_1} \oplus \cdots \oplus W_n^{\oplus k_n}$, where the W_i are representatives of the distinct isomorphism classes of irreducible representations of G. A

4.4 Character Theory

straightforward computation using part (3) yields

$$\langle \chi_V, \chi_V \rangle = \langle \sum_{i=1}^{n} k_i \chi_i, \sum_{j=1}^{n} k_j \chi_j \rangle$$

$$= \sum_{i,j=1}^{n} k_i k_j \langle \chi_i, \chi_j \rangle$$

$$= \sum_{i,j=1}^{n} k_i k_j \delta_{i,j} = \sum_{i=1}^{n} k_i^2.$$

This quantity is equal to 1 if and only if all k_i are zero, except for one that is equal to 1. By the isomorphism $V \cong \bigoplus_{i=1}^{n} W_i^{\oplus k_i}$, this is equivalent to V being irreducible.

□

Remark 4.3 We now know that there are only finitely many isomorphism classes of irreducible representations. The characters of these representations are sometimes called the **irreducible characters** of G.

Remark 4.4 In the proof of part (3) of the theorem we have obtained

$$\dim \mathrm{Hom}_G(V_1, V_2) = \langle \chi_2, \chi_1 \rangle = \dim \mathrm{Hom}_G(V_2, V_1).$$

This symmetry is not clear a priori.

One can be even more precise about the linear subspace of $\mathbb{C}^G$ generated by the characters of the group G. Define the space of **class functions** $\mathrm{Cl}(G)$ as the $\mathbb{C}$-vector subspace of $\mathbb{C}^G$ given by those functions $f : G \to \mathbb{C}$ that are constant on the conjugacy classes of G (that is, $f(hgh^{-1}) = f(g)$ for all $g, h \in G$). It is clear that the space of class functions has dimension equal to the number of conjugacy classes of G. The following result is well known, see for example [1, Theorem 6 in §2.5].

Theorem 4.3 *Let G be a finite group and $V_1, \ldots, V_n$ be representatives for the distinct isomorphism classes of irreducible representations of G. Let χ_i be the character of V_i. The χ_i are class functions and form a basis of $\mathrm{Cl}(G)$, orthonormal with respect to the Hermitian product of Eq. (4.3). In particular, the number of isomorphism classes of irreducible representations of G is equal to the number of conjugacy classes of G.*

Remark 4.5 Note that it is clear that the character χ of any representation ρ is a class function: by definition we have

$$\chi(hgh^{-1}) = \mathrm{tr}\, \rho(hgh^{-1}) = \mathrm{tr}\left(\rho(h)\rho(g)\rho(h)^{-1}\right) = \mathrm{tr}\, \rho(g) = \chi(g),$$

where we have used that the trace of an endomorphism is invariant under conjugation. We have also already established that the characters of non-isomorphic irreducible representations are orthonormal in Theorem 4.2 (3). This already shows that the number of irreducible representations of G is *at most* the number of conjugacy classes in G. The hard part of Theorem 4.3 is to show that the characters of irreducible representations do in fact span the whole space of class functions.

Example 4.7 (Irreducible Representations of S_3) We check Theorem 4.3 in the case $G = S_3$. Note first that S_3 has three conjugacy classes, represented by the elements id, $(1, 2)$ and $(1, 2, 3)$. Thus, we need to find three irreducible representations. One is the trivial one, whose character is $\mathbf{1}$. A second one is given by the representation W_2 of Example 4.1. We check that it is irreducible. Its character is given by $\chi_{W_2} = \chi_V - \mathbf{1}$, where, as explained in Example 4.1, χ_V is the function that sends each permutation to the number of its fixed points. We can then compute

$$\langle \chi_{W_2}, \chi_{W_2}\rangle = \frac{1}{|S_3|} \sum_{g \in S_3} (\chi_V(g) - 1)^2 = \frac{1}{|S_3|} \sum_{g \in S_3} (\#\{i \in \{1, 2, 3\} : g(i) = i\} - 1)^2$$

$$= \frac{1}{6}(1 \cdot 2^2 + 3 \cdot 0^2 + 2 \cdot (-1)^2) = 1,$$

so that by Theorem 4.2 (6) the representation W_2 is irreducible. We are missing at most one irreducible representation (up to isomorphism), which is not too hard to guess. We will instead use Theorem 4.2 (3) to determine it. Let W_3 be an irreducible representation of S_3 not isomorphic to W_1, W_2. Its character χ_3 is a class function, so it is uniquely determined by the values it takes on the three elements id, $(1, 2)$ and $(1, 2, 3)$ of S_3. Write

$$a = \chi_3(\text{id}), \quad b = \chi_3((1, 2)), \quad c = \chi_3((1, 2, 3)).$$

By Theorem 4.2 (3), χ_3 is orthogonal to $\chi_1 := \mathbf{1}$ and $\chi_2 := \chi_{W_2}$. Writing out this condition explicitly, we find $a + 3b + 2c = 0$ and $2a - 2c = 0$. This forces $a = c = -b$. Finally, from Theorem 4.2 (6) we know that

$$1 = \langle \chi_3, \chi_3 \rangle = \frac{1}{6}(|a|^2 + 3|b|^2 + 2|c|^2) = |a|^2.$$

Since a is a positive integer by Remark 4.2, we find that $a = 1$ and that the corresponding representation is 1-dimensional. As such, it coincides with its character, and it is easy to see that the function χ_3 we just constructed is simply the sign homomorphism $S_3 \to \{\pm 1\} \subset \mathbb{C}^\times \cong \text{GL}(\mathbb{C})$.

Remark 4.6 The information about the irreducible representations of a group G is often summarised by a table, called the **character table**, that lists one irreducible representation for each row. The columns of the table are labelled by the conjugacy classes of G, and usually one also records the size of each conjugacy class. The cell

4.4 Character Theory

on the row corresponding to the representation ρ and on the column corresponding to the conjugacy class c contains the value of $\chi_\rho(c)$ (note that this makes sense, because χ_ρ is a class function, so we can evaluate χ_ρ at any element of the conjugacy class c and obtain the same result).

Example 4.8 From the calculations in Example 4.7 we obtain the character table of S_3:

S_3	id 1	(1, 2) 3	(1, 2, 3) 2
1	1	1	1
Sign	1	−1	1
Std	2	0	−1

The representation called Std (for 'standard') in this table is the representation W_2 of Example 4.1.

4.4.1 Further Orthogonality Relations

In our proof of Chebotarev's theorem, in Chap. 7, we will also need additional orthogonality relations, beyond those provided by Theorem 4.2.

Proposition 4.1 (Orthogonality Relations) *Let $\{\chi_1, \ldots, \chi_n\}$ be the distinct irreducible characters of a finite group G. The following hold.*

1. $\sum_{g \in G} \chi_i(g)\overline{\chi_j(g)} = \begin{cases} |G|, & \text{if } i = j \\ 0, & \text{otherwise} \end{cases}$

2. *Fix $g, h \in G$. We have*

$$\sum_{i=1}^{n} \chi_i(g)\overline{\chi_i(h)} = \begin{cases} |Z_G(g)|, & \text{if } g, h \text{ are in the same conjugacy class} \\ 0, & \text{otherwise} \end{cases}$$

where $Z_G(g)$ denotes the centraliser of g in G.

Proof The first statement is a rephrasing of Theorem 4.2 (3). For the second, fix $h \in G$. By the orbit-stabiliser lemma, we have $|G| = |Z_G(h)| \cdot |\text{Cl}_G(h)|$, where $\text{Cl}_G(h)$ is the conjugacy class of h in G. Let f be the class function that takes the value 1 on the conjugacy class of h and 0 elsewhere. From Theorem 4.3 we know that $f = \sum_{i=1}^{n} c_i \chi_i$ for certain coefficients $c_i \in \mathbb{C}$. Since the characters χ_i are mutually orthogonal, the coefficients c_j can be computed as

$$c_j = \sum_{i=1}^n c_i \langle \chi_i, \chi_j \rangle = \langle \sum_{i=1}^n c_i \chi_i, \chi_j \rangle$$

$$= \langle f, \chi_j \rangle = \frac{1}{|G|} \sum_{x \in G} f(x) \overline{\chi_j(x)}$$

$$= \frac{1}{|G|} \sum_{x \in \mathrm{Cl}_G(h)} \overline{\chi_j(x)} = \frac{|\mathrm{Cl}_G(h)|}{|G|} \overline{\chi_j(h)},$$

where in the last step we have used the fact that χ_j is constant on conjugacy classes. We have thus obtained

$$c_j = \frac{|\mathrm{Cl}_G(h)|}{|G|} \overline{\chi_j(h)} = \frac{1}{|Z_G(h)|} \overline{\chi_j(h)},$$

and therefore, for every $g \in G$,

$$|Z_G(h)| \cdot f(g) = |Z_G(h)| \cdot \sum_{i=1}^n c_i \chi_i(g) = \sum_{i=1}^n \chi_i(g) \overline{\chi_i(h)}.$$

By definition of $f(g)$, this implies the claim. □

4.5 Further Operations on Representations

We conclude this chapter by describing three operations on representations that will be important in our study of Artin L-functions: restriction, inflation, and induction. Unlike the constructions we have discussed so far, they involve not just one group G, but rather a group G together with a subgroup H (normal, in the case of inflation). We start with the restriction.

Definition 4.1 (Restriction of a Representation to a Subgroup) Given a G-representation $\rho : G \to \mathrm{GL}(V)$ and a subgroup H of G, the representation $\mathrm{Res}^G_H(V)$, called the **restriction** of V to H, is given by the vector space V and the homomorphism $\rho|_H : H \to \mathrm{GL}(V)$.

In a slightly different context, one can define Res^G_H as a linear operator $\mathrm{Res}^G_H : \mathrm{Cl}(G) \to \mathrm{Cl}(H)$ on class functions. Denoting as usual by χ_V the character of a representation V, we have the obvious formula

$$\mathrm{Res}^G_H(\chi_V) = \chi_{\mathrm{Res}^G_H(V)}.$$

4.5 Further Operations on Representations

Since by Theorem 4.3 the characters of the distinct (isomorphism classes of) irreducible representations of G form a basis of $\mathrm{Cl}(G)$, this formula describes completely the operator $\mathrm{Res}_H^G : \mathrm{Cl}(G) \to \mathrm{Cl}(H)$ in terms of representations.

Next, we consider inflation.

Definition 4.2 (Inflation of a Representation) Let G be a finite group, H be a *normal* subgroup of G, and $\pi : G \to G/H$ be the projection. If $\rho : G/H \to \mathrm{GL}(V)$ is a representation of G/H, the composition

$$G \xrightarrow{\pi} G/H \xrightarrow{\rho} \mathrm{GL}(V)$$

is a representation of G, called the **inflation** of ρ from H to G and denoted by $\mathrm{Inf}_{G/H}^G(\rho)$ or $\mathrm{Inf}_{G/H}^G(V)$.

As before, we can also consider inflation as a linear operator

$$\mathrm{Inf}_{G/H}^G : \mathrm{Cl}(G/H) \to \mathrm{Cl}(G),$$

described either by the formula $\mathrm{Inf}_{G/H}^G(f) = f \circ \pi$, or by extending $\mathbb{C}$-linearly the map

$$\mathrm{Inf}_{G/H}^G(\chi_V) = \chi_{\mathrm{Inf}_{G/H}^G(V)}$$

defined on the characters of the irreducible G/H-representations V. Recall that the characters of irreducible representations of G/H form a basis of $\mathrm{Cl}(G/H)$.

Finally, we have a notion of induced representation, which is the most subtle of the three constructions described in this paragraph.

Definition 4.3 (Induced Representation) Let G be a finite group and H be a subgroup. Let $\rho : H \to \mathrm{GL}(W)$ be a representation of H. The **induced** representation of ρ, denoted by $\mathrm{Ind}_H^G(\rho)$, is the representation of G on the vector space

$$\{f : G \to W \mid f(hg) = \rho(h)(f(g)) \quad \forall h \in H\}$$

with action given by $(g \cdot f)(x) = f(xg)$.

Once again, we can see induction as a $\mathbb{C}$-linear map

$$\mathrm{Ind}_H^G : \mathrm{Cl}(H) \to \mathrm{Cl}(G),$$

but in this case, the interpretation is more complicated, because it is not immediately clear how to define the induction of an arbitrary class function. Before discussing this, we give an alternative, concrete construction of the induced representation.

Definition 4.4 (Induced Representation, Alternative Construction) Let G be a finite group and let H be a subgroup. Let $\rho : H \to \mathrm{GL}(W)$ be a representation of

H and denote by $g_1, \ldots, g_n$ a full set of representatives of the left cosets of H in G (recall that the *left* cosets of a subgroup B of a group A are those of the form aB). Define the vector space $V = \bigoplus_{i=1}^n g_i W$, where each $g_i W$ is simply an isomorphic copy of W. Define an action of G on V in the following way: given a vector w in the copy $g_i W$ of W, which we write for simplicity as $g_i w$, we set

$$g \cdot g_i w = g_j(\rho(h)(w)) \in g_j W, \tag{4.4}$$

where $g_j \in \{g_1, \ldots, g_n\}$ and $h \in H$ are uniquely defined by the equality

$$g g_i = g_j h$$

taking place in G.

Remark 4.7 Omitting the representation ρ for simplicity, Eq. (4.4) can be written in the more intuitive form

$$g \cdot g_i w = g_j(h \cdot w),$$

where $g g_i = g_j h$.

We claim, and we will show in a moment, that the two constructions above give rise to the same representation of G. The construction of Definition 4.4 makes it easy to compute the dimension of the induced representation: we have $\dim \operatorname{Ind}_H^G(V) = [G : H] \cdot \dim V$.

Remark 4.8 The construction of Definition 4.4 also implies that if we restrict $\operatorname{Ind}_H^G(\rho)$ back to the subgroup H of G, the resulting representation of H contains a copy of ρ as a subrepresentation. Indeed, if we take $g_1 = \operatorname{id}_G$ as one of the representatives for the cosets of H in G, for every $h \in H \subseteq G$ Eq. (4.4) shows that $h \cdot (\operatorname{id}_G w) = \operatorname{id}_G \cdot (\rho(h)w)$, so H acts via ρ on the subspace $\operatorname{id}_G W$. Note that H does not necessarily act as ρ on the subspace gW when g represents a non-trivial coset of H in G (in fact, these subspaces need not even be H-stable, if H is not normal in G).

The next proposition shows that Definitions 4.3 and 4.4 are equivalent (at least in our setting of finite groups) and gives a formula for the induction of an arbitrary class function on H.

Proposition 4.2 *The following hold:*

1. *The representation of Definition 4.4 is isomorphic to the representation of Definition 4.3.*
2. *(Frobenius formula) The character of $\operatorname{Ind}_H^G(\rho)$ is given by*

$$g \mapsto \sum_{x \in R} \chi_{\rho,0}(x^{-1} g x),$$

4.5 Further Operations on Representations

where the sum runs over a set R of representatives of the left cosets of H in G and

$$\chi_{\rho,0}(g) = \begin{cases} \operatorname{tr} \rho(g), & \text{if } g \in H \\ 0, & \text{otherwise}. \end{cases}$$

3. There is a well-defined linear map

$$\operatorname{Ind}_H^G : \operatorname{Cl}(H) \to \operatorname{Cl}(G)$$

that sends the character of an irreducible representation ρ of H to the character of $\operatorname{Ind}_H^G(\rho)$, and sends a generic function $\tau \in \operatorname{Cl}(H)$ to

$$g \mapsto \sum_{x \in R} \tau_0(x^{-1} g x),$$

where as above R is a set of representatives of the left cosets of H in G and

$$\tau_0(g) = \begin{cases} \tau(g), & \text{if } g \in H \\ 0, & \text{otherwise}. \end{cases}$$

Proof As in Definition 4.4, we fix once and for all a system of representatives $g_1, \ldots, g_n$ for the left cosets of H in G. This will be the R that appears in the formulas of parts (2) and (3).

1. Let $V_1 = \{f : G \to W \mid f(hg) = \rho(h)(f(g)) \ \forall h \in H\}$ be the vector space of Definition 4.3 and V_2 be the vector space of Definition 4.4. We describe an explicit isomorphism $V_1 \to V_2$: we send $f \in V_1$ to

$$w(f) := \sum_{i=1}^n g_i f(g_i^{-1}) \in \bigoplus_{i=1}^n g_i W = V_2,$$

where $g_i f(g_i^{-1})$ denotes the vector $f(g_i^{-1})$ in the copy $g_i W$ of W. This map is easily seen to be linear and bijective (for the latter, observe that—by the transformation properties of the elements of V_1 under the action of H—a function $f \in V_1$ is uniquely determined by the values $f(g_i^{-1})$). We check that it gives an isomorphism of G-representations. To see this, fix $g \in G$. For every $i = 1, \ldots, n$, let $j(i) \in \{1, \ldots, n\}$ and $h_i \in H$ be the unique elements that satisfy $g g_i = g_{j(i)} h_i$. Note that $i \mapsto j(i)$ is a permutation of $\{1, \ldots, n\}$. By definition of the G-action on V_2, we have

$$g \cdot w(f) = g \cdot \sum_{i=1}^n g_i f(g_i^{-1}) = \sum_{i=1}^n g_{j(i)} \rho(h_i)(f(g_i^{-1})).$$

On the other hand,

$$(g \cdot f)(g_{j(i)}^{-1}) = f(g_{j(i)}^{-1}g) = f(h_i g_i^{-1}) = \rho(h_i)(f(g_i^{-1})),$$

and therefore, using that $i \mapsto j(i)$ is a permutation of $1, \ldots, n$, we obtain

$$w(g \cdot f) = \sum_{i=1}^{n} g_i(g \cdot f)(g_i^{-1}) = \sum_{i=1}^{n} g_{j(i)}(g \cdot f)(g_{j(i)}^{-1})$$

$$= \sum_{i=1}^{n} g_{j(i)} \rho(h_i)(f(g_i^{-1})) = g \cdot w(f),$$

as desired.

2. We use the description of Definition 4.4 to work with $\operatorname{Ind}_H^G(\rho)$. Fix a basis $w_1, \ldots, w_d$ of W. A basis of $V = \operatorname{Ind}_H^G(W)$ is given by $\{g_i w_j\}_{i=1,\ldots,n, j=1,\ldots,d}$. We compute the trace of $\operatorname{Ind}_H^G(\rho)(g)$ using this basis. Since the trace of a matrix is the sum of its diagonal entries, we only need to understand the coordinate along $g_i w_j$ in the basis representation of $g g_i w_j$, for all $i = 1, \ldots, n$ and $j = 1, \ldots, d$. We label the basis vectors of V with the pairs (i, j).

Fix an index $i \in \{1, \ldots, n\}$ and consider the coset $g g_i H = g_k H$. If $k \neq i$, the basis representation of $g(g_i w_j) \in g_k W$ only involves the vectors $\{g_k w_h\}_{h=1,\ldots,d}$, so the coordinate of $g_i w_j$ along the basis vector $g_i w_j$ is clearly 0. Thus, it suffices to consider those indices i for which $g g_i H = g_i H$, that is, $g_i^{-1} g g_i \in H$. For these indices we have

$$g(g_i w_j) = g_i \left(\rho(g_i^{-1} g g_i) w_j \right).$$

Representing $\rho(g_i^{-1} g g_i)$ as a matrix in the basis $\{w_j\}$ and $\operatorname{Ind}_H^G(\rho)(g)$ as a matrix in the basis $\{g_i w_j\}$, we find that the diagonal coefficient in position (i, j) of $\operatorname{Ind}_H^G(\rho)(g)$ is the diagonal coefficient in position j of $\rho(g_i^{-1} g g_i)$. Summing over all i, j we find

$$\operatorname{tr} \operatorname{Ind}_H^G(\rho)(g) = \sum_{\substack{g_i \text{ such that} \\ g_i^{-1} g g_i \in H}} \sum_{j=1}^{d} \rho(g_i^{-1} g g_i)_{jj} = \sum_{\substack{g_i \text{ such that} \\ g_i^{-1} g g_i \in H}} \operatorname{tr} \rho(g_i^{-1} g g_i)$$

$$= \sum_{\substack{g_i \text{ such that} \\ g_i^{-1} g g_i \in H}} \chi_\rho(g_i^{-1} g g_i) = \sum_{g_i} \chi_{\rho,0}(g_i^{-1} g g_i),$$

as desired.

4.5 Further Operations on Representations

3. Follows easily from the previous statement once we check that the function $g \mapsto \sum_{x \in r} \tau_0(x^{-1}gx)$ is a class function. If we conjugate g by $g_1 \in G$, we obtain

$$\sum_{x \in R} \tau_0(x^{-1}g_1^{-1}gg_1x).$$

Since $\{g_1x\}_{x \in R}$ is another set of representatives for the left cosets of H in G, we have $\sum_{x \in R} \tau_0(x^{-1}g_1^{-1}gg_1x) = \sum_{x \in R} \tau_0(x^{-1}gx)$ as desired. □

Remark 4.9 The formula $\sum_{x \in R} \tau_0(x^{-1}gx)$ can also be written without making a choice of representatives for the left cosets of H in G: indeed, it is easy to observe that the sum in question coincides with $\frac{1}{|H|} \sum_{x \in G} \tau_0(x^{-1}gx)$.

Problems

4.1 Describe all the irreducible representations of the quaternion group

$$Q_8 = \langle i, j \mid i^4 = 1, i^2 = j^2, ji = i^{-1}j \rangle$$

up to isomorphism.

4.2 Compute the character table of the dihedral group D_n for all $n \geq 3$.

4.3 Let $n \geq 2$ be an integer. The symmetric group S_n acts on $\mathbb{C}^n$ by permutation of the coordinates. The subspace $W = \{{}^t(x_1, \ldots, x_n) \in \mathbb{C}^n \mid x_1 + \cdots + x_n = 0\}$ is invariant under the G-action, hence it gives a representation of the symmetric group called the *standard representation*.

1. Show that W is an irreducible representation of S_n.
2. Denote by $\mathrm{Fix}(\sigma) \subseteq \{1, \ldots, n\}$ the set of fixed points of the permutation $\sigma \in S_n$. Show that

$$\sum_{\sigma \in S_n} (\#\mathrm{Fix}(\sigma) - 1)^2 = n!$$

4.4 (Regular Representation) Let G be a finite group and let $V = \mathbb{C}^G$ be the space of complex-valued functions on G. We make V into a representation of G by letting $g \in G$ act on functions as

$$(g \cdot f)(x) = f(xg).$$

The space V is called the *regular representation* of G.

1. Show that $V \cong \mathrm{Ind}^G_{\{\mathrm{id}_G\}}(\mathbf{1})$.
2. Show that
$$\chi_V(g) = \begin{cases} |G|, & \text{if } g = \mathrm{id} \\ 0, & \text{otherwise} \end{cases}$$

3. Let $V_1, \ldots, V_n$ be the irreducible representations of G. By computing $\langle \chi_V, \chi_{V_i} \rangle$ (or otherwise), show that $V \cong \bigoplus_{i=1}^n V_i^{\oplus \dim V_i}$.
4. Deduce that
$$\sum_{i=1}^n (\dim V_i)^2 = |G|.$$

4.5 Let G be a finite group.

1. Show that there exists a representation $\rho : G \to \mathrm{GL}(V)$ that is injective as a group homomorphism (such representations are called *faithful*).
2. Without using Theorem 4.3, show that G is abelian if and only if every irreducible representation of G is 1-dimensional.

4.6 Describe the 1-dimensional (automatically irreducible) representations of S_n for $n \geq 5$. Compute the decomposition into irreducible components of $\mathrm{Ind}^{S_n}_{A_n}(\mathbf{1})$.

4.7 Let V be an irreducible complex representation of the finite group G. Describe all subrepresentations of $V \oplus V$.

Reference

1. Serre, J.P.: Linear representations of finite groups. Graduate Texts in Mathematics, vol. 42, French edn. Springer-Verlag, New York-Heidelberg (1977)

Chapter 5
The L-function of a Complex Galois Representation

Abstract In this chapter we define Artin's L-functions and prove their functoriality properties. We further show that every L-function we encountered in the previous chapters is in fact an abelian L-function in the sense of Artin. Finally, assuming the results of Chapter 1, we prove that all Artin L-functions admit meromorphic continuation to the whole complex plane.

We are now ready to define a large class of L-functions which includes all those we have already encountered. It was Artin's fundamental insight that L-functions should not just be associated with a single number field, but rather with a *Galois extension* of number fields, and that the data that goes into defining an L-function is representation-theoretic in nature.

Definition 5.1 (Artin L-function) Let L/K be a Galois extension of number fields with group G. Let $\rho : G \to \mathrm{GL}(V)$ be a finite-dimensional complex representation of G. For every non-zero prime $\mathfrak{p}$ of O_K, fix a prime $\mathfrak{P}$ of O_L lying over $\mathfrak{p}$ and let $I_\mathfrak{P} < G$ be the corresponding inertia subgroup. We define the **Artin L-function** of ρ to be

$$L(s, \rho) := \prod_{\mathfrak{p} \text{ nonzero prime of } O_K} \det\left(\mathrm{Id} - \rho\left(\left(\frac{L/K}{\mathfrak{P}}\right)\right) N(\mathfrak{p})^{-s} \mid V^{I_\mathfrak{P}}\right)^{-1}, \tag{5.1}$$

where $V^{I_\mathfrak{P}}$ is the subspace of V on which $I_\mathfrak{P}$ acts trivially via ρ. Notice that, when $\mathfrak{P}$ is ramified over $\mathfrak{p}$, the Artin symbol $\left(\frac{L/K}{\mathfrak{P}}\right)$ is not a well-defined element of G (see Remark 3.4). In this case, we simply take an arbitrary choice of Frobenius element at $\mathfrak{P}$ to represent the Artin symbol: Remark 5.1 below shows that the definition is independent of this choice. The factor $\det\left(\mathrm{Id} - \rho\left(\left(\frac{L/K}{\mathfrak{P}}\right)\right) N(\mathfrak{p})^{-s} \mid V^{I_\mathfrak{P}}\right)$ appearing in the above product is often called the **local factor at** $\mathfrak{p}$ of the Artin L-function.

Remark 5.1 With the notation of the previous definition, we show that

1. $V^{I_\mathfrak{P}}$ is stable under the action of $\rho\left(\left(\frac{L/K}{\mathfrak{P}}\right)\right)$;
2. $\rho\left(\left(\frac{L/K}{\mathfrak{P}}\right)\right)$ is a well-defined endomorphism of $V^{I_\mathfrak{P}}$ even when $I_\mathfrak{P}$ is non-trivial (recall that in this case we have made an arbitrary choice in the definition of $\left(\frac{L/K}{\mathfrak{P}}\right)$, see Definition 3.2).

Write for simplicity $g \in G$ for a fixed choice of Frobenius element at $\mathfrak{P}$. Any other choice of Frobenius element at $\mathfrak{P}$ is of the form $g \cdot h$ for some $h \in I_\mathfrak{P}$, see also Remark 3.4. Let furthermore $x \in V$ be a vector fixed by $\rho(I_\mathfrak{P})$. Since $I_\mathfrak{P}$ is a normal subgroup of the decomposition group of $\mathfrak{P}$, for every $h' \in I_\mathfrak{P}$ we have $h'g = gh''$ for some $h'' \in I_\mathfrak{P}$, and therefore

$$\rho(h') \cdot (\rho(gh) \cdot v) = \rho(h'gh) \cdot v = \rho(gh''h) \cdot v = \rho(g) \cdot (\rho(h''h) \cdot v) = \rho(g) \cdot v,$$

where we used that v is fixed by the group $\rho(I_\mathfrak{P})$. The above equality shows that $\rho(gh)v$ is again in $V^{I_\mathfrak{P}}$, for any $v \in I_\mathfrak{P}$ and any choice gh of Frobenius element at $\mathfrak{P}$. In particular, it makes sense to consider $\mathrm{Id} - \rho\left(\left(\frac{L/K}{\mathfrak{P}}\right)\right) N(\mathfrak{p})^{-s}$ as an endomorphism of $V^{I_\mathfrak{P}}$. The same calculation also shows that $\mathrm{Id} - \rho\left(\left(\frac{L/K}{\mathfrak{P}}\right)\right) N(\mathfrak{p})^{-s}$ is independent of the choice of the Frobenius element $\left(\frac{L/K}{\mathfrak{P}}\right)$ at $\mathfrak{P}$.

It is not completely clear that Definition 5.1 is well-posed, in the sense that it does not depend on the arbitrary choices made. We will prove this in the next section.

The rest of the chapter is divided into five relatively short sections that cover the following topics:

1. Definition 5.1 does not depend on the choice of the prime $\mathfrak{P}$ of O_L lying over $\mathfrak{p}$, and the dimension of the subspace $V^{I_\mathfrak{P}}$ itself does not depend on the choice of $\mathfrak{P}$;
2. the Euler product defining $L(\rho, s)$ converges for all s with $\Re s > 1$;
3. every L-function we have encountered so far is in fact an Artin L-function;
4. Artin's conjecture on analytic continuation;
5. the canonical factorisation of the Dedekind zeta function of a number field which is a Galois extension of another.

5.1 Independence of the Choice of $\mathfrak{P}$

Proposition 5.1 *Let L/K be a Galois extension of number fields with group G and let $\mathfrak{p}$ be a prime of O_K. Let $\mathfrak{P}, \mathfrak{P}'$ be two primes of O_L lying over $\mathfrak{p}$, and let $I_\mathfrak{P}, I_{\mathfrak{P}'}$ be the corresponding inertia groups. Finally, let $\rho : G \to \mathrm{GL}(V)$ be a finite-dimensional complex representation of G. The following hold:*

5.1 Independence of the Choice of $\mathfrak{P}$

1. *the subspaces $V^{I_\mathfrak{P}}$ and $V^{I_{\mathfrak{P}'}}$ have the same dimension;*
2. *the complex numbers*

$$\det\left(\mathrm{Id} - \rho\left(\left(\frac{L/K}{\mathfrak{P}}\right)\right) N(\mathfrak{p})^{-s} \mid V^{I_\mathfrak{P}}\right) \text{ and } \det\left(\mathrm{Id} - \rho\left(\left(\frac{L/K}{\mathfrak{P}'}\right)\right) N(\mathfrak{p})^{-s} \mid V^{I_{\mathfrak{P}'}}\right)$$

are equal.

Proof By Theorem 3.6 (1) we know that there exists $\sigma \in G$ such that $\sigma(\mathfrak{P}) = \mathfrak{P}'$. One checks immediately that $I_{\mathfrak{P}'} = \sigma I_\mathfrak{P} \sigma^{-1}$.

1. By definition, $v \in V^{I_{\mathfrak{P}'}}$ means

$$\rho(i) \cdot v = v \quad \forall i \in I_{\mathfrak{P}'} = \sigma I_\mathfrak{P} \sigma^{-1},$$

which is equivalent to

$$\rho(\sigma i \sigma^{-1}) \cdot v = v \quad \forall i \in I_\mathfrak{P},$$

and therefore also to

$$\rho(i)\rho(\sigma)^{-1} \cdot v = \rho(\sigma)^{-1} \cdot v \quad \forall i \in I_\mathfrak{P}.$$

Thus, v is in $V^{I_{\mathfrak{P}'}}$ if and only if $\rho(\sigma)^{-1} \cdot v$ is in $V^{I_\mathfrak{P}}$. It follows that $V^{I_{\mathfrak{P}'}} = \rho(\sigma) V^{I_\mathfrak{P}}$, and in particular, since $\rho(\sigma)$ is an invertible linear transformation, these two spaces have the same dimension.

2. It is useful to interpret the factor $\det\left(\mathrm{Id} - \rho\left(\left(\frac{L/K}{\mathfrak{P}}\right)\right) N(\mathfrak{p})^{-s} \mid V^{I_\mathfrak{P}}\right)$ as the evaluation at $t = N(\mathfrak{p})^{-s}$ of the (inverse) characteristic polynomial

$$f_\mathfrak{P}(t) := \det\left(\mathrm{Id} - t\rho\left(\left(\frac{L/K}{\mathfrak{P}}\right)\right) \mid V^{I_\mathfrak{P}}\right)$$

of $\left(\frac{L/K}{\mathfrak{P}}\right)$ acting on $V^{I_\mathfrak{P}}$, and similarly for the factor corresponding to $\mathfrak{P}'$. The claim now follows from the fact that $\rho(\sigma)$ acts as a change of basis between $\rho\left(\left(\frac{L/K}{\mathfrak{P}}\right)\right)$ and $\rho\left(\left(\frac{L/K}{\mathfrak{P}'}\right)\right)$.

$\square$

Remark 5.2 (Local Factors at the Unramified Places) Suppose that $\mathfrak{p}$ is unramified in L. In this case, $V^{I_\mathfrak{P}}$ is simply V, and the interpretation of the local factor at $\mathfrak{p}$ as a characteristic polynomial shows that it is independent of the choice of the Frobenius element in the conjugacy class $\left(\frac{L/K}{\mathfrak{p}}\right)$. We may therefore write the local factor without any reference to the choice of $\mathfrak{P}$ as

$$\det(\mathrm{Id} - t\rho(\mathrm{Frob}_{\mathfrak{p}}));$$

see Definition 3.3 for the symbol $\mathrm{Frob}_{\mathfrak{p}}$, which denotes any element in the conjugacy class $\left(\frac{L/K}{\mathfrak{p}}\right)$.

Finally, we introduce a common notation for the local factors of Artin L-functions: for a prime $\mathfrak{p}$ of O_K, choose a prime $\mathfrak{P}$ of O_L lying over it and define

$$L_{\mathfrak{p}}(\rho, t) := \det\left(\mathrm{Id} - t\rho\left(\left(\frac{L/K}{\mathfrak{P}}\right)\right) \mid V^{I_{\mathfrak{P}}}\right).$$

We then have

$$L(s, \rho) = \prod_{\mathfrak{p}} \frac{1}{L_{\mathfrak{p}}(\rho, N(\mathfrak{p})^{-s})}.$$

5.2 Convergence of the Euler Product

We now know that the single factors appearing in Definition 5.1 are well-defined. However, it is not *a priori* clear that their *product* gives a holomorphic function over some open subset of $\mathbb{C}$. The proof of the next proposition reduces the convergence of the Euler product in Definition 5.1 to the convergence of the Euler product for the Riemann ζ function.

Proposition 5.2 *The Euler product in Eq.* (5.1) *converges absolutely for all s with $\Re s > 1$.*

Proof Notice that, since $G := \mathrm{Gal}(L/K)$ is a finite group, for every $g \in G$ (hence for every Artin symbol $\left(\frac{L/K}{\mathfrak{P}}\right) \in G$) the element $\rho(g)$ has finite order, hence all its eigenvalues are roots of unity. In particular, each factor appearing in the product is well-defined and non-zero for $\Re s > 1$, and as a consequence, we can safely ignore a finite number of factors in the product. For the rest of the proof we only consider primes $\mathfrak{p}$ of O_K that are unramified in O_L. This only excludes a finite set of primes, and for such primes we have $V^{I_{\mathfrak{P}}} = V$.

For each prime $\mathfrak{P}$ of O_L lying over the prime $\mathfrak{p}$ of O_K, we have

$$\det\left(\mathrm{Id} - t\rho\left(\left(\frac{L/K}{\mathfrak{P}}\right)\right) \mid V\right) = \prod_{j=1}^{\dim V} (1 - t\zeta_{\mathfrak{P}, j}),$$

where the $\zeta_{\mathfrak{P},j}$, for $j = 1, \ldots, \dim V$, are roots of unity, corresponding to the eigenvalues of $\rho\left(\left(\frac{L/K}{\mathfrak{P}}\right)\right)$ acting on V. Thus, in particular, writing $\operatorname{char}(\mathfrak{P})$ for the characteristic of the residue field $O_L/\mathfrak{P}$, we have

$$\prod_{\mathfrak{p}} \left|\det\left(\operatorname{Id} - N(\mathfrak{p})^{-s}\rho\left(\left(\frac{L/K}{\mathfrak{P}}\right)\right) \mid V\right)\right|^{-1} = \prod_{\mathfrak{p}} \prod_{j=1}^{\dim V} \left|1 - N(\mathfrak{p})^{-s}\zeta_{\mathfrak{P},j}\right|^{-1}$$

$$\leq \prod_{\mathfrak{p}} \prod_{j=1}^{\dim V} (1 - |\operatorname{char}(\mathfrak{P})^{-s}|)^{-1} \leq \prod_{j=1}^{\dim V} \prod_{\mathfrak{p}} (1 - \operatorname{char}(\mathfrak{P})^{-\Re(s)})^{-1}$$

$$= \prod_{j=1}^{\dim V} \prod_{p} \prod_{\mathfrak{p} \mid p} (1 - p^{-\Re(s)})^{-1} \leq \zeta(\Re(s))^{\dim V \cdot [K:\mathbb{Q}]},$$

which converges for all s with $\Re s > 1$ as claimed (see Problem 1.4). In the last step we used that there are $\dim V$ factors in the product over j, and at most $[K : \mathbb{Q}]$ in the product over $\mathfrak{p}$, since each rational prime factors into at most $[K : \mathbb{Q}]$ primes in O_K by Theorem 3.2. □

5.3 The Riemann and Dedekind ζ Functions, and Dirichlet's L-functions, Are Artin L-functions

We now show that all the L-functions introduced so far are special cases of Definition 5.1.

Proposition 5.3

1. *The Riemann ζ function is an Artin L-function.*
2. *Let K be a number field. The Dedekind ζ function of K is an Artin L-function.*
3. *Let χ be a **primitive** Dirichlet character modulo m. The Dirichlet L-function $L(s, \chi)$ is an Artin L-function.*
4. *Let χ be an arbitrary Dirichlet character modulo m. There exists an Artin L-function $L_{\text{Artin}}(s, \rho)$ such that $L_{\text{Dirichlet}}(s, \chi) = f(s) L_{\text{Artin}}(s, \rho)$, where the factor $f(s)$ is a holomorphic function of s which is non-vanishing on $\{\Re s > 0\}$.*

Proof

1. Clearly, Riemann's ζ function is the Dedekind ζ function of $\mathbb{Q}$, so it suffices to prove 2.
2. In Definition 5.1 we take $L = K$ and ρ to be the unique 1-dimensional trivial representation of $G = \operatorname{Gal}(L/K)$. With reference to Eq. (5.1), the inertia groups $I_{\mathfrak{P}}$ are all trivial, so $V^{I_{\mathfrak{P}}} = V$ for all $\mathfrak{p}$, while $\rho\left(\left(\frac{L/K}{\mathfrak{P}}\right)\right) = 1$. Thus, Eq. (5.1) reads simply

$$L(\rho, s) = \prod_{\mathfrak{p}} \det(1 - N(\mathfrak{p})^{-s})^{-1} = \prod_{\mathfrak{p}}(1 - N(\mathfrak{p})^{-s})^{-1} = \zeta_K(s),$$

where the last equality follows from Problem 1.4 in Chap. 1.

3. In Definition 5.1 we take $K = \mathbb{Q}$, $L = \mathbb{Q}(\zeta_m)$ and define a Galois representation of $G = \text{Gal}(L/K) \cong (\mathbb{Z}/m\mathbb{Z})^\times$ by setting

$$\rho(\sigma_a) = \chi(a)\,\text{Id} \in \text{GL}(V),$$

where $V = \mathbb{C}$ and $\sigma_a \in G$ is the automorphism $\zeta_m \mapsto \zeta_m^a$, corresponding to $a \in (\mathbb{Z}/m\mathbb{Z})^\times$. On the one hand, since χ is a completely multiplicative function, by Theorem 1.3 we have

$$L(s, \chi) = \prod_p (1 - \chi(p) p^{-s})^{-1}.$$

On the other hand, the definition of $L(s, \rho)$ gives

$$L(s, \rho) = \prod_p \det\left(1 - \rho\left(\left(\frac{\mathbb{Q}(\zeta_m)/\mathbb{Q}}{\mathfrak{P}}\right)\right) N(p)^{-s} \mid V^{I_\mathfrak{P}}\right)^{-1}.$$

We now prove equality by matching up the local factors. Let p be a rational prime and let $\mathfrak{P}$ be a prime of $\mathbb{Q}(\zeta_m)$ lying over p.

(a) Suppose that $p \nmid m$. Then p is unramified in $\mathbb{Q}(\zeta_m)$, so $I_\mathfrak{P}$ is trivial, while $\left(\frac{\mathbb{Q}(\zeta_m)/\mathbb{Q}}{\mathfrak{P}}\right) \in G$ is given by σ_p, see Example 3.2. It follows that the local factor of $L(s, \rho)$ at p is $\det\left(1 - \rho(\sigma_p) p^{-s} \mid \mathbb{C}\right)^{-1} = (1 - \chi(p) p^{-s})^{-1}$, which precisely matches the local factor of $L(s, \chi)$ at p.

(b) Suppose now that $p \mid m$. The local factor at p of $L(s, \chi)$ is trivially 1. To conclude the proof, it suffices to show that $V^{I_\mathfrak{P}}$ is trivial, that is, that $\rho(I_\mathfrak{P})$ is not the trivial group. If this were the case, χ would be trivial on the (nontrivial) subgroup J_p of $(\mathbb{Z}/m\mathbb{Z})^\times$ corresponding to $I_\mathfrak{P}$ under the canonical isomorphism $G \cong (\mathbb{Z}/m\mathbb{Z})^\times$. Writing $m = p^s m'$ with $(p, m') = 1$, we claim that J_p is precisely the kernel of the canonical projection $(\mathbb{Z}/m\mathbb{Z})^\times \to (\mathbb{Z}/m'\mathbb{Z})^\times$. Assuming the claim, we obtain that if $\rho(I_\mathfrak{P})$ is trivial, then χ factors via $(\mathbb{Z}/m\mathbb{Z})^\times / J_p \cong (\mathbb{Z}/m'\mathbb{Z})^\times$, contradicting the fact that χ is primitive. The claim is an exercise in algebraic number theory, and is left to the reader (see Problem 5.1).

4. Construct $L_{\text{Artin}}(s, \rho)$ as in part 3. For all primes $p \nmid m$, the local factors at p of $L_{\text{Dirichlet}}(s, \chi)$ and $L_{\text{Artin}}(s, \rho)$ match, so their ratio

$$f(s) = \frac{L_{\text{Dirichlet}}(s, \chi)}{L_{\text{Artin}}(s, \rho)}$$

is a finite product of factors of the form $1-\chi(p)p^{-s}$ for (certain) primes dividing m. These functions are clearly holomorphic on all of $\mathbb{C}$, and their zeroes arise for $1-\chi(p)p^{-s}=0$. Taking absolute values we get $|p^{-s}|=1$, that is, $\Re s=0$. It follows that $f(s)$ is holomorphic and non-vanishing in $\{\Re s>0\}$.

□

Remark 5.3 Let χ be a Dirichlet character modulo m, induced from the primitive Dirichlet character $\tilde{\chi}$ modulo the conductor d of χ. The proof of Proposition 5.3 shows that $L_{\text{Dirichlet}}(s,\tilde{\chi})$ is the Artin L-function attached to the representation

$$\text{Gal}(\mathbb{Q}(\zeta_d)/\mathbb{Q}) \cong (\mathbb{Z}/d\mathbb{Z})^\times \xrightarrow{\tilde{\chi}} \mathbb{C}^\times.$$

By Theorem 5.2 (2) below, this implies that $L_{\text{Dirichlet}}(s,\tilde{\chi})$ is also the Artin L-function of the representation

$$\text{Gal}(\mathbb{Q}(\zeta_m)/\mathbb{Q}) \twoheadrightarrow \text{Gal}(\mathbb{Q}(\zeta_d)/\mathbb{Q}) \cong (\mathbb{Z}/d\mathbb{Z})^\times \xrightarrow{\tilde{\chi}} \mathbb{C}^\times.$$

This shows that the language of Artin L-function is less sensitive to the notion of conductor, or rather, that the construction of the Artin L-function attached to a character implicitly selects the minimal conductor.

This is also a good time to informally introduce our last family of (abelian) L-functions, namely, **Hecke L-functions**. We give a provisional definition, which captures a large part of the family of Hecke L-functions (technically, those corresponding to characters with finite image), and postpone the complete definition to Chap. 14.

Definition 5.2 (Hecke L-functions, Provisional Definition for Characters with Finite Image) Let K be a number field, let L/K be a finite Galois extension with group G, and let $\chi: G \to \mathbb{S}^1$ be a character (equivalently, a representation of dimension 1) of G. The **Hecke L-function** attached to this data is the Artin L-function $L(s,\chi)$.

Remark 5.4 Hecke did not introduce his L-functions in this way at all. His definition was

$$L(s,\chi) = \sum_{I \triangleleft \mathcal{O}_K} \frac{\chi(I)}{N(I)^s},$$

where χ is a map from the ideals of $\mathcal{O}_K$ to $\mathbb{C}$ satisfying a complicated set of properties that generalise those of Dirichlet characters. The fact that (many) Hecke L-functions can be expressed as Artin L-functions as in Definition 5.2 is related to the fact that the abelian extensions of a number field are well-understood in terms of class field theory. We will come back to this point in Chap. 14, where we will also describe the exact conditions imposed by Hecke on the function χ to correctly generalise Dirichlet characters. See also Sect. 1.5 for a discussion of the intrinsically automorphic nature of Hecke L-functions and their analytic properties.

At this point, the reader will not be surprised by the following result:

Theorem 5.1 (Analytic Continuation for the Hecke L-functions) *Let L/K be a finite abelian extension with group G and let $\chi : G \to \mathbb{S}^1$ be a nontrivial character. The Hecke L-function $L(s, \chi)$ admits analytic continuation to the full complex plane.*

Artin L-functions also satisfy a (fairly complicated) functional equation relating s to $1 - s$ and ρ to its dual representation, but we will not need its precise form, so we omit the exact statement. The interested reader can refer to [2, Theorem 12.6 in Chapter VII].

5.4 The Formalism of Artin L-functions

The main result of this section collects some important properties of Artin L-functions, which together are known as the *functoriality of Artin L-functions*. We will use these facts extensively in the rest of the course.

Theorem 5.2 (Functoriality of Artin L-functions) *Let $L/F/K$ be a tower of extensions of number fields, with L/K Galois. Let $G = \mathrm{Gal}(L/K)$, $N = \mathrm{Gal}(L/F)$, and (when F/K is Galois) $H = \mathrm{Gal}(F/K)$. Let ρ_1, ρ_2 be representations of G, let σ be a representation of H, and let η be a representation of N. The following hold:*

1. $L(s, \rho_1 \oplus \rho_2) = L(s, \rho_1)L(s, \rho_2)$.
2. *If F/K is Galois, then $L(s, \mathrm{Inf}_H^G(\sigma)) = L(s, \sigma)$, where the inflation of σ from H to G is defined in Definition 4.2.*
3. $L(s, \mathrm{Ind}_N^G(\eta)) = L(s, \eta)$, *where $\mathrm{Ind}_N^G(\tau)$ is the induced representation of Definition 4.3.*

Proof

1. Recall the following elementary fact from linear algebra: if M_1, M_2 are matrices in $\mathrm{GL}_{d_1}(\mathbb{C})$, $\mathrm{GL}_{d_2}(\mathbb{C})$ respectively, and if $M_1 \oplus M_2$ denotes their block-sum in $\mathrm{GL}_{d_1+d_2}(\mathbb{C})$, then the characteristic polynomial of $M_1 \oplus M_2$ is the product of the characteristic polynomials of M_1, M_2. This allows one to identify the local factors of $L(s, \rho_1 \oplus \rho_2)$ with the product of the corresponding local factors of $L(s, \rho_1), L(s, \rho_2)$.
2. It suffices to prove that $L(s, \mathrm{Inf}_H^G(\sigma))$ and $L(s, \sigma)$ have the same local factors at each prime. Fix a place $\mathfrak{P}$ of F lying over $\mathfrak{p}$ and a place $\mathfrak{Q}$ of L lying over $\mathfrak{P}$.

    We denote by D, I (respectively D', I') the decomposition and inertia group of $\mathfrak{P}$ over $\mathfrak{p}$ (respectively, of $\mathfrak{Q}$ over $\mathfrak{p}$). Let $\pi : G \to H$ be the canonical projection. It is a standard fact in algebraic number theory that $I = \pi(I')$, see

5.4 The Formalism of Artin L-functions

Problem 5.2. We now compare (representatives of) the Artin symbols

$$\left(\frac{F/K}{\mathfrak{p}}\right) \quad \text{and} \quad \left(\frac{L/K}{\mathfrak{p}}\right).$$

Fix an element $\varphi \in D' \subseteq G$ that represents a Frobenius at $\mathfrak{Q}$. This means that the congruence

$$\varphi(x) \equiv x^{N\mathfrak{p}} \mod \mathfrak{Q} \iff \varphi(x) - x^{N\mathfrak{p}} \in \mathfrak{Q}$$

holds for all $x \in O_L$, hence, in particular, for every $x \in O_F$ we have

$$\varphi(x) - x^{N\mathfrak{p}} \in \mathfrak{Q} \cap O_F$$

since F/K is Galois (and therefore $\varphi(F) = F$, $\varphi(O_F) = O_F$). The previous equation can be rewritten as

$$\varphi|_F(x) - x^{N\mathfrak{p}} \in \mathfrak{P} \quad \forall x \in O_F \iff \varphi|_F(x) \equiv x^{N\mathfrak{p}} \mod \mathfrak{P} \quad \forall x \in O_F,$$

which shows that $\varphi|_F = \pi(\varphi)$ is a representative for the Frobenius at $\mathfrak{P}$.

We are finally ready to prove the statement. Since $\pi(\varphi)$ gives a Frobenius at $\mathfrak{P}$, the local factor at $\mathfrak{p}$ of the L-function $L(s, \sigma)$ is given by

$$\det\left(1 - N(\mathfrak{p})^{-s} \sigma(\pi(\varphi)) \mid V_\sigma^{\sigma(I)}\right)^{-1},$$

where V_σ is the underlying vector space of the representation σ.

On the other hand, the local factor of $L(s, \operatorname{Inf}_H^G(\sigma))$ at the same place $\mathfrak{p}$ is given by

$$\det\left(1 - N(\mathfrak{p})^{-s} \operatorname{Inf}_H^G(\sigma)(\varphi) \mid V_\sigma^{\operatorname{Inf}_H^G(\sigma)(I')}\right)^{-1}.$$

Since $\operatorname{Inf}_H^G(\sigma)(\varphi) = \sigma(\pi(\varphi))$ and $\operatorname{Inf}_H^G(\sigma)(I') = \sigma(\pi(I')) = \sigma(I)$, the claim follows.

3. Again we try to match local factors, but more complications arise in this case. The proof we give is adapted from [2, Proposition 10.4]. Let $\mathfrak{p}$ be a place of K, let $\mathfrak{P}_1, \ldots, \mathfrak{P}_r$ be the primes of F lying over $\mathfrak{p}$, and for each $i = 1, \ldots, r$ let $\mathfrak{Q}_i$ be a prime of L lying over $\mathfrak{P}_i$ (see Fig. 5.1). Further denote by

$$D_i = D(\mathfrak{Q}_i \mid \mathfrak{p}), \quad I_i = I(\mathfrak{Q}_i \mid \mathfrak{p})$$

Fig. 5.1 Fields and primes in the proof of Theorem 5.2 (3)

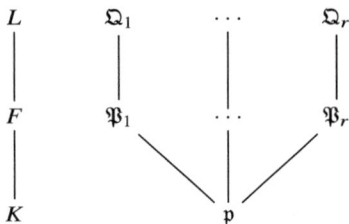

the decomposition and inertia groups of the $\mathfrak{Q}_i$ over $\mathfrak{p}$. From the definitions one easily obtains that

$$D'_i := D(\mathfrak{Q}_i \mid \mathfrak{P}_i) = N \cap D_i, \quad I'_i := I(\mathfrak{Q}_i \mid \mathfrak{P}_i) = N \cap I_i.$$

The inertia degree f_i of $\mathfrak{P}_i$ over $\mathfrak{p}$ is

$$f(\mathfrak{P}_i \mid \mathfrak{p}) = \#\operatorname{Gal}(\kappa(\mathfrak{P}_i) \mid \kappa(\mathfrak{p})) = \#\frac{D_i/I_i}{D'_i/I'_i} = \#\frac{D_i}{D'_i I_i}.$$

Note that by definition we have $N(\mathfrak{P}_i) = N(\mathfrak{p})^{f_i}$. We now set some further notation for Frobenius elements. For simplicity, we choose elements $\tau_i \in G$ such that $\tau_i^{-1}(\mathfrak{Q}_1) = \mathfrak{Q}_i$ for each i (with $\tau_1 = \mathrm{id}$), and choose $\varphi_1 \in D_1$ that represents the Frobenius of $\mathfrak{Q}_1$ over $\mathfrak{p}$. We then have the relations

$$D_i = \tau_i^{-1} D_1 \tau_i, \quad I_i = \tau_i^{-1} I_1 \tau_i,$$

and the element

$$\varphi_i := \tau_i^{-1} \varphi_1 \tau_i$$

represents a Frobenius of $\mathfrak{Q}_i$ over $\mathfrak{p}$. Moreover, $\varphi_i^{f_i}$ represents a Frobenius of $\mathfrak{Q}_i$ over $\mathfrak{P}_i$.

Let W be the underlying vector space of the representation η, and let

$$V = \operatorname{Ind}_N^G(W) = \{f : G \to W \mid f(\tau g) = \eta(\tau) f(g) \quad \forall \tau \in N\}$$

the vector space underlying $\operatorname{Ind}_N^G(\eta)$. For ease of notation, denote by $\psi : G \to \mathrm{GL}(V)$ the map giving the induced representation of G. In order to establish the theorem, it is enough to prove that

$$\det\left(1 - N(\mathfrak{p})^{-s} \psi(\varphi_1) \mid V^{I_1}\right) = \prod_{i=1}^{r} \det\left(1 - N(\mathfrak{P}_i)^{-s} \eta(\varphi_i^{f_i}) \mid W^{I'_i}\right).$$

5.4 The Formalism of Artin L-functions

In fact, since $N(\mathfrak{P}_i) = N(\mathfrak{p})^{f_i}$, it suffices to prove the polynomial identity

$$\det\left(1 - t\psi(\varphi_1) \mid V^{I_1}\right) = \prod_{i=1}^{r} \det\left(1 - t^{f_i}\eta(\varphi_i^{f_i}) \mid W^{I'_i}\right).$$

We now observe that, under the action of N, the subspace id $W = W$ of the induced representation V is a copy of the representation η (see Remark 4.8). In the vector space V it makes sense to conjugate using $\psi(\tau_i)$, and we obtain (omitting ψ for simplicity)

$$\det\left(1 - t^{f_i}\eta(\varphi_i^{f_i}) \mid W^{I'_i}\right) = \det\left(1 - t^{f_i}\tau_i\eta(\varphi_i^{f_i})\tau_i^{-1} \mid \tau_i\left(W^{I'_i}\right)\right)$$

$$= \det\left(1 - t^{f_i}\eta(\tau_i\varphi_i^{f_i}\tau_i^{-1}) \mid \tau_i\left(W^{I'_i}\right)\right)$$

$$= \det\left(1 - t^{f_i}\eta(\varphi_1^{f_i}) \mid (\tau_i W)^{\tau_i I'_i\tau_i^{-1}}\right)$$

$$= \det\left(1 - t^{f_i}\eta(\varphi_1^{f_i}) \mid (\tau_i W)^{\tau_i(N\cap I_i)\tau_i^{-1}}\right)$$

$$= \det\left(1 - t^{f_i}\eta(\varphi_1^{f_i}) \mid (\tau_i W)^{\tau_i N\tau_i^{-1}\cap I_1}\right).$$

Similarly,

$$f_i = [D_i : D'_i I_i]$$
$$= [\tau_i^{-1}D_1\tau_i : (N \cap \tau_i^{-1}D_1\tau_i)\tau_i^{-1}I_1\tau_i]$$
$$= [D_1 : (\tau_i N\tau_i^{-1} \cap D_1)I_1].$$

For every i we now choose a system of representatives $\sigma_{i,j}$ for the left cosets of $\tau_i N\tau_i^{-1} \cap D_1$ in D_1. Since the τ_i are representatives for the cosets of D_1 in G, we obtain (Problem 5.4) that $\{\sigma_{i,j}\tau_i\}$ represent the left cosets of N in G. Hence,

$$V = \bigoplus_{i,j} \sigma_{i,j}\tau_i W,$$

and each $V_i = \bigoplus_j \sigma_{i,j}\tau_i W$ is a D_1-submodule of V, with $V \cong \bigoplus_{i=1}^r V_i$ as D_1-representations.

Remark 5.5 To check that V_i is stable under the action of D_1 one can proceed as follows. Since the $\{\sigma_{i,j}\tau_i\}$ are representatives for the left cosets of N in G, for every $d_1 \in D_1$ there exists $n \in N$ and indices i', j' such that $d_1\sigma_{i,j}\tau_i = \sigma_{i',j'}\tau_{i'}n$. We rewrite this as $\tau_i^{-1}\sigma_{i,j}^{-1}d_1^{-1}\sigma_{i',j'} = n^{-1}\tau_{i'}^{-1}$. Applying both sides of this equality to $\mathfrak{Q}_1$ and recalling that by definition $\sigma_{i,j}^{-1}$, d_1^{-1} and $\sigma_{i',j'}$ all belong to D_1 we obtain

$$\mathfrak{Q}_i = n^{-1}\mathfrak{Q}_{i'}.$$

Intersecting with O_F (on which n^{-1} acts trivially) we then obtain $\mathfrak{P}_i = \mathfrak{P}_{i'}$ and therefore $i = i'$. Thus, $d_1 \sigma_{i,j} \tau_i = \sigma_{i,j'} \tau_i n$ for some $n \in N$. By definition of the action of G on $V_i = \bigoplus_j \sigma_{i,j} \tau_i W$, this implies as desired $d_1 \cdot V_i \subseteq V_i$.

Thus, since φ_1 is in D_1, we obtain

$$\det(1 - t\psi(\varphi_1) \mid V^{I_1}) = \prod_{i=1}^r \det(1 - t\psi(\varphi_1) \mid V_i^{I_1}),$$

and it suffices to show that

$$\det(1 - t\psi(\varphi_1) \mid V_i^{I_1}) = \det\left(1 - t^{f_i} \eta(\varphi_1^{f_i}) \mid (\tau_i W)^{\tau_i N \tau_i^{-1} \cap I_1}\right).$$

Notice that V_i is the induced representation of W from $D_1 \cap \tau_i N \tau_i^{-1}$ to D_1 (indeed, V_i is obtained by summing over representatives for the cosets of $D_1 \cap \tau_i N \tau_i^{-1}$ in D_1). Thus, renaming

$$G := D_1,\ I := I_1,\ N := D_1 \cap \tau_i N \tau_i^{-1},\ f := f_i,\ V := V_i,\ W := \tau_i W,\ \varphi := \varphi_1,$$

the desired equality can be rewritten as

$$\det(1 - t\psi(\varphi) \mid V^I) = \det\left(1 - t^f \eta(\varphi^f) \mid W^{N \cap I}\right),$$

so that we are essentially reduced to the case $r = 1$ and $D_1 = G$. The next and final reduction is to the case $I = \{1\}$. We claim that

$$V^I = \operatorname{Ind}_{N/I \cap H}^{G/I}(W^{N \cap I}).$$

We show this from the definition (using Definition 4.3 for the induced representation). An element of V^I is by definition a function $f : G \to W$ that satisfies

(a) $f(g) = f(gi)$ for every $g \in G, i \in I$ (this is the condition $i \cdot f = f$ that ensures $f \in V^I$). Notice that, since I is normal in G, right-invariance and left-invariance are equivalent, and therefore we also have $f(ig) = f(g)$ for all $g \in G, i \in I$
(b) $h \cdot f(g) = f(hg)$ for every $g \in G, h \in N$ (this is the condition that ensures that f is in $\operatorname{Ind}_N^G(W)$).

Now, the first condition is equivalent to f factoring via G/I. Moreover, any such function takes values in $W^{N \cap I}$, because

$$i \cdot f(g) = f(ig) = f(g).$$

for all $i \in N \cap I$. The claim follows. So, replacing G by G/I, I by $\{1\}$, and $W^{N \cap I}$ by W, we are reduced[1] to proving

$$\det(1 - t\psi(\varphi) \mid \operatorname{Ind}_1^G(W)) = \det\left(1 - t^f \eta(\varphi^f) \mid W\right).$$

In this case G is cyclic, generated by φ, and $V = \bigoplus_{i=0}^{f-1} \varphi^i W$. If A is the matrix of $\eta(\varphi)$ (with respect to any basis $w_1, \ldots, w_d$ of W), then the matrix of $\psi(\varphi)$ with respect to the basis $\{\varphi^i w_j\}$ is

$$M := \begin{pmatrix} & & & & \boxed{A} \\ \boxed{\operatorname{Id}_d} & & & & \\ & \boxed{\operatorname{Id}_d} & & & \\ & & \ddots & & \\ & & & \boxed{\operatorname{Id}_d} & \end{pmatrix},$$

a block matrix with f blocks on each row/column. An easy exercise in linear algebra (Problem 5.3) now shows that

$$\det(1 - tM) = \det(1 - t^f \det A^f), \tag{5.2}$$

which finishes the proof.

$\square$

In the next two sections, we will apply Theorem 5.2 to obtain quick proofs of two very non-trivial statements, namely, that Artin L-functions admit *meromorphic* continuation to the whole complex plane, and that—given a number field F which is a Galois extension of a number field K—one can obtain a factorisation of the ζ function of F, with one of the factors being the ζ function of K.

5.5 Artin's Conjecture on Analytic Continuation

The following is one of the most important conjectures on Artin's L-functions:

Conjecture 5.1 (Artin) Let L/K be a Galois extension of number fields with group G. If ρ is a non-trivial irreducible[2] complex representation of G, then $L(s, \rho)$ has analytic continuation to the whole complex plane.

---

[1] This essentially amounts to replacing L by L^I, which is normal over F since we are now assuming $G = D$.

[2] We recall that a representation $\rho : G \to \operatorname{GL}(V)$ is called *irreducible* if $V \neq \{0\}$ and the only vector subspaces of V that are stable under the action of $\rho(G)$ are $\{0\}$ and V itself.

While Conjecture 5.1 is wide open, we show that $L(s, \rho)$ admits *meromorphic continuation* to the full complex plane.

Theorem 5.3 (Meromorphic Continuation for Artin L-functions) *For any Galois extension L/K with group G and every complex representation ρ of G, the L-function $L(s, \rho)$ has meromorphic extension to the complex plane.*

For the proof, we will need the following result in representation theory:

Theorem 5.4 (Brauer's Induction Theorem [3, Theorem 20 in §10.5]) *Let G be a finite group and let $\rho : G \to \mathrm{GL}_n(\mathbb{C})$ be a finite-dimensional complex representation. There exists finitely many subgroups $H_1, \ldots, H_r$ of G, characters $\lambda_i : H_i \to \mathbb{S}^1$ for $i = 1, \ldots, r$, and integers $n_1, \ldots, n_r$ such that*

$$\mathrm{tr}(\rho(g)) = \sum_{i=1}^{r} n_i \,\mathrm{tr}\left(\mathrm{Ind}_{H_i}^{G}(\lambda_i)\right)(g)$$

for all $g \in G$.

Proof of Theorem 5.3 Let $\chi = \mathrm{tr}(\rho)$ be the character of the representation ρ. Write $\chi = \sum n_i \,\mathrm{tr}\,\mathrm{Ind}_{H_i}^{G}(\lambda_i)$ as in Brauer's theorem. Rearranging this equation, we get an equality of the form

$$\chi + \sum n_i \,\mathrm{tr}\,\mathrm{Ind}_{H_i}^{G}(\lambda_i) = \sum n'_j \,\mathrm{tr}\,\mathrm{Ind}_{H'_j}^{G}(\lambda'_j),$$

where the integers n_i, n'_j are now all strictly positive. Since the character of a complex representation of a finite group determines the representation itself, we obtain

$$\rho \oplus \bigoplus \left(\mathrm{Ind}_{H_i}^{G}(\lambda_i)\right)^{\oplus n_i} \cong \bigoplus \left(\mathrm{Ind}_{H'_j}^{G}(\lambda'_j)\right)^{\oplus n'_j}.$$

In particular, we have

$$L\left(s, \rho \oplus \bigoplus \left(\mathrm{Ind}_{H_i}^{G}(\lambda_i)\right)^{\oplus n_i}\right) = L\left(s, \bigoplus \left(\mathrm{Ind}_{H'_j}^{G}(\lambda'_j)\right)^{\oplus n'_j}\right)$$

The general formalism of Artin L-functions (Theorem 5.2) implies first

$$L(s, \rho) \cdot \prod_i L\left(s, \mathrm{Ind}_{H_i}^{G}(\lambda_i)\right)^{n_i} = \prod_j L\left(s, \mathrm{Ind}_{H'_j}^{G}(\lambda'_j)\right)^{n'_j}$$

and then

$$L(s,\rho) \cdot \prod_i L(s,\lambda_i)^{n_i} = \prod_j L\left(s,\lambda'_j\right)^{n'_j}.$$

Since the λ_i, λ'_j are characters with values in $\mathbb{S}^1$, they factor via some abelian (Galois) group, hence they are Hecke L-functions. Thus, each $L(s,\lambda_i)$ and $L\left(s,\lambda'_j\right)$ admits meromorphic continuation by Theorem 5.1. Since the above formula expresses $L(s,\rho)$ as a ratio of (products of) such Hecke L-functions, $L(s,\rho)$ also has meromorphic continuation to the complex plane. □

Remark 5.6 This theorem is perhaps part of the motivation for Artin's conviction that abelian L-functions should be sufficient to understand arbitrary extensions of number fields. In the report on the Princeton University Bicentennial Conference on the Problems of Mathematics of 1946, Artin is quoted to have stated: *My own belief is that we know it already, though no one will believe me—that whatever can be said about non-Abelian class field theory follows from what we know now, since it depends on the behavior of the broad field over the intermediate fields— and there are sufficiently many Abelian cases* (see [1, p. 312]). In other words, class field theory (the description of *abelian* extensions of a number field) should contain in itself all the necessary ingredients to describe arbitrary (Galois) extensions of number fields. No one has yet been able to fully realise Artin's dream, and extending class field theory to arbitrary non-abelian extensions remains a central problem in number theory. However, since Artin had enormous insight into the theory of L-functions, one probably shouldn't be too quick to dismiss his intuition!

5.6 Factorisation of the Dedekind ζ-function

The general formalism of Artin L-functions easily implies an interesting relation between the ζ functions of different number fields, when one is a Galois extension of the other:

Theorem 5.5 (Factorisation of the Dedekind ζ Function in Terms of Artin L-functions) *Let L/K be a Galois extension of number fields with group G. The function $\zeta_L(s)$ factors as*

$$\zeta_L(s) = \zeta_K(s) \prod_{\rho \neq 1} L(s,\rho)^{\dim \rho},$$

where the product runs over the non-trivial irreducible complex representations of G.

Proof This is a special case of Theorem 5.2. Specifically, in the setting of that theorem, take $F = L$ (so that $N = \{1\}$) and σ to be the trivial representation of the trivial group. Then $\mathrm{Ind}_N^G(\sigma)$ is the regular representation of G, which decomposes as $\bigoplus_{\rho \text{ irrep of } G} \rho^{\oplus \dim \rho}$ (see Problem 4.4 or [3, Corollary 1 in §2.4]). Hence, applying properties (3) and (1) in Theorem 5.2 we get

$$L(s, 1) = L(s, \mathrm{Ind}_N^G(1)) = L\left(s, \bigoplus_{\rho \text{ irrep of } G} \rho^{\oplus \dim \rho}\right) = \prod_{\rho \text{ irrep of } G} L(s, \rho)^{\dim \rho}.$$

Finally, the trivial representation 1 of N gives the Dedekind ζ function of L, while the trivial representation 1 of G gives ζ_K. Thus, the previous equation can be rewritten as

$$\zeta_L(s) = \zeta_K(s) \prod_{\substack{\rho \text{ irrep of } G \\ \rho \neq 1}} L(s, \rho)^{\dim \rho},$$

as desired. □

Corollary 5.1 *For every integer $n \geq 2$ we have the factorisation*

$$\zeta_{\mathbb{Q}(\zeta_n)} = \zeta(s) \prod_{\substack{\chi \text{ non-trivial Dirichlet} \\ \text{character modulo } n}} L(s, \tilde{\chi}),$$

where $\tilde{\chi}$ is the primitive character corresponding to χ.

Proof Follows from Theorem 5.5 together with our identification of Dirichlet's L-functions as special Artin L-functions (see Proposition 5.3 and Remark 5.3). Also note that every complex representation of the abelian group $G = \mathrm{Gal}\,(\mathbb{Q}(\zeta_n)/\mathbb{Q})$ is 1-dimensional, hence a character. □

In the interest of keeping this book as accessible as possible, we will give below in Proposition 6.4 an independent proof of (a slightly weaker version of) Corollary 5.1 that avoids all the machinery of representation theory and Artin L-functions. This weaker version will be sufficient for the proof of Dirichlet's theorem on arithmetic progressions.

The general factorisation provided by Theorem 5.5 also implies the following (conditional) non-vanishing statement for the values of Artin L-functions at $s = 1$.

Proposition 5.4 *Assume Artin's conjecture (Conjecture 5.1). Let L/K be a Galois extension of number fields with group G and let $\rho : G \to \mathrm{GL}(V)$ be a non-trivial irreducible representation. The function $L(s, \rho)$ is non-vanishing at $s = 1$.*

5.6 Factorisation of the Dedekind ζ-function

Proof Start from the relation

$$\zeta_L(s) = \zeta_K(s) \prod_{\rho \neq 1} L(s, \rho)^{\dim \rho}$$

given by Theorem 5.5. Theorem 1.4 shows that $\zeta_K(s)$ and $\zeta_L(s)$ both have a simple pole at $s = 1$. This implies that $\prod_{\rho \neq 1} L(s, \rho)^{\dim \rho}$ is non-zero and finite at $s = 1$. Assuming Artin's conjecture, each factor in the product is holomorphic, so $L(1, \rho)$ is a well-defined complex number. Since the product $\prod_{\rho \neq 1} L(1, \rho)^{\dim \rho}$ is non-zero, each $L(1, \rho)$ is non-zero. □

Problems

5.1 Let p be a prime number, s be a positive integer, and m' be a positive integer prime to p. Set $m = p^s m'$. Let $G = \text{Gal}(\mathbb{Q}(\zeta_m)/\mathbb{Q})$, let $\mathfrak{p}$ be a prime of $\mathbb{Q}(\zeta_m)$ lying over p, and let $I_\mathfrak{p}$ be the corresponding inertia subgroup.

1. Show that $I_\mathfrak{p}$ depends only on p and not on the choice of the prime $\mathfrak{p}$ lying over it. We will therefore denote the group $I_\mathfrak{p}$ simply by I_p.
2. Prove (or recall) that $\mathbb{Q}(\zeta_{m'})/\mathbb{Q}$ is unramified at p.
3. Deduce that I_p is contained in the kernel of the canonical map

$$\text{Gal}(\mathbb{Q}(\zeta_m)/\mathbb{Q}) \to \text{Gal}(\mathbb{Q}(\zeta_{m'})/\mathbb{Q}).$$

4. Conclude that I_p is in fact *equal* to the kernel of the natural projection $\text{Gal}(\mathbb{Q}(\zeta_m)/\mathbb{Q}) \to \text{Gal}(\mathbb{Q}(\zeta_{m'})/\mathbb{Q})$ (one way to do this is to argue by cardinality, showing that $\#I_p = \varphi(p^s)$).

5.2 Let $L/F/K$ be extensions of number fields, with L/K and F/K both Galois. Let $\mathfrak{Q}$ be a place of L, and let $\mathfrak{P}, \mathfrak{p}$ be the places of F and K lying under $\mathfrak{Q}$. Let $G = \text{Gal}(L/K)$, $H = \text{Gal}(F/K)$, and $\pi : G \to H$ be the projection map. Show that $\pi(I(\mathfrak{Q} \mid \mathfrak{p})) = I(\mathfrak{P} \mid \mathfrak{p})$.

Hint. It is easy to show that π sends $I(\mathfrak{Q} \mid \mathfrak{p})$ into $I(\mathfrak{P} \mid \mathfrak{p})$. For the surjectivity, take a pre-image in G, then try to modify it using elements in $\ker(G \to H)$.

5.3 Prove the formula in Eq. (5.2).

5.4 With notation as in the proof of Theorem 5.2(3), prove that $\{\sigma_{i,j} \tau_i\}$ is a set of representatives for the left cosets of N in G.

Hint. To show that they are the correct number, observe that

$$\sum_i \left| \frac{D_1}{\tau_i N \tau_i^{-1} \cap D_1} \right| = \sum_i \left| \frac{\tau_i^{-1} D_1 \tau_i}{N \cap \tau_i^{-1} D_1 \tau_i} \right| = \sum_i \left| \frac{D_i}{N \cap D_i} \right|$$

$$= \sum_i \left| \frac{D_i}{D_i'} \right| = \sum_i \frac{e(\mathfrak{Q}_i \mid \mathfrak{p}) f(\mathfrak{Q}_i \mid \mathfrak{p})}{e(\mathfrak{Q}_i \mid \mathfrak{P}_i) f(\mathfrak{Q}_i \mid \mathfrak{P}_i)}$$

$$= \sum_i e(\mathfrak{P}_i \mid \mathfrak{p}) f(\mathfrak{P}_i \mid \mathfrak{p}) = [F : K] = [G : N].$$

To show that they represent distinct cosets, assume

$$\sigma_{i,j} \tau_i N = \sigma_{i',j'} \tau_{i'} N.$$

Prove that $\tau_{i'}^{-1} \sigma_{i',j'}^{-1} \sigma_{i,j} \tau_i$ sends $\mathfrak{Q}_i$ to $\mathfrak{Q}_{i'}$ and $\mathfrak{P}_i$ to $\mathfrak{P}_i$. Conclude that $i = i'$ and then that $j = j'$.

5.5 Let L/K be a Galois extension of number fields with group G and let ρ be a finite-dimensional Galois representation of G. Set $\chi(g) = \operatorname{tr} \rho(g)$. Prove the following formula for (a determination of) the logarithm of $L(s, \rho)$, valid for $\Re s > 1$:

$$\log L(s, \rho) = \sum_{\mathfrak{p}} \sum_{m \geq 1} \frac{\chi(\mathfrak{p}^m)}{m(N\mathfrak{p})^{ms}},$$

where $\chi(\mathfrak{p}^m)$ is defined as follows: let $\mathfrak{P}$ be a prime of L lying over $\mathfrak{p}$ and let $e = e(\mathfrak{P} \mid \mathfrak{p})$. Fix a representative $\sigma \in G$ for the Artin symbol $\left(\frac{L/K}{\mathfrak{p}} \right)$. We set

$$\chi(\mathfrak{p}^m) := \frac{1}{e} \sum_{\tau \in I(\mathfrak{P}\mid\mathfrak{p})} \chi(\sigma^m \tau).$$

Hint. It can be useful to write the determinant of a linear map as the product of its eigenvalues. This leads to the identity

$$\log \det(1 - tA) = \log \prod_i (1 - \lambda_i t) = \sum_i \log(1 - \lambda_i t)$$

$$= -\sum_i \sum_{m \geq 1} \frac{(\lambda_i t)^m}{m} = -\sum_{m \geq 1} \frac{t^m}{m} \left(\sum_i \lambda_i^m \right)$$

$$= -\sum_{m \geq 1} \operatorname{tr}(A^m) \frac{t^m}{m}.$$

Up to some fiddling at the ramified places, this gives the required formula.

5.6 Let K be the splitting field over $\mathbb{Q}$ of $x^3 - 2$, let G be the Galois group of K over $\mathbb{Q}$, and let ρ be the standard (2-dimensional) representation of G. Determine the behaviour of $L(s, \rho)$ around $s = 1$ (does it have a zero, a pole, or neither?) and, if $L(1, \rho)$ exists and is nonzero, determine its value in terms of arithmetic quantities.

5.7 For a prime number $p \neq 3, 7$, define

$$a_p := \#\{x \in \mathbb{F}_p^\times : x^3 \equiv 7 \pmod{p}\}.$$

Prove that a_p is the p-th Dirichlet coefficient of the ζ function of the field $K = \mathbb{Q}(\sqrt[3]{7})$ (that is: writing $\zeta_K(s)$ as the Dirichlet series $\sum_{n\geq 1} \frac{b_n}{n^s}$, one has $b_p = a_p$ for all primes $p \neq 3, 7$). Let L be the Galois closure of $K/\mathbb{Q}$ and let G be the Galois group of $L/\mathbb{Q}$: describe a representation ρ of G such that $L(s, \rho) = \zeta_K(s)$.

5.8 Let $K = \mathbb{Q}(i, \sqrt{5})$.

1. Express the residue of $\zeta_K(s)$ at $s = 1$ in terms of L-functions of Dirichlet characters.
2. Prove that a fundamental unit of $\mathcal{O}_{\mathbb{Q}(\sqrt{5})}$ is also a fundamental unit of $\mathcal{O}_K$.
   Hint. Consider the norm from K to $\mathbb{Q}(\sqrt{5})$ of a fundamental unit of K.
3. Using $h(\mathbb{Q}(i)) = 1$ and $h(\mathbb{Q}(\sqrt{-5})) = 2$, compute $h(K)$.

References

1. Duren, P., Askey, R.A., Merzbach, U.C. (eds.): A century of mathematics in America. Part II. History of Mathematics, vol. 2. American Mathematical Society, Providence (1989)
2. Neukirch, J.: Algebraic number theory. Grundlehren der Mathematischen Wissenschaften [Fundamental Principles of Mathematical Sciences], vol. 322. Springer-Verlag, Berlin (1999). https://doi.org/10.1007/978-3-662-03983-0. Translated from the 1992 German original and with a note by Norbert Schappacher, With a foreword by G. Harder
3. Serre, J.P.: Linear representations of finite groups. Graduate Texts in Mathematics, vol. 42, French edn. Springer-Verlag, New York-Heidelberg (1977)

Chapter 6
Dirichlet's Theorem on Arithmetic Progressions

Abstract In this chapter we prove Dirichlet's theorem on primes in arithmetic progressions. We also briefly discuss the philosophy of *special values* of *L*-functions.

In this chapter we *assume* Theorems 1.4 and 1.5 from Chap. 1 and show that these analytic results have deep arithmetic consequences. In particular, we will show that they imply rather easily Dirichlet's famous theorem on primes in arithmetic progressions, namely, Theorem 3.8. We start by discussing a notion of duality for (finite) abelian groups.

6.1 Pontryagin Duality: Finite Case

The central notion in this section is that of *dual* group.

Definition 6.1 (Dual Group, Finite Case) Let G be a finite abelian group. The **dual group** (or **group of characters**) of G is the set

$$\hat{G} := \mathrm{Hom}\left(G, \mathbb{C}^\times\right),$$

equipped with the operation of pointwise product (that is, $(\chi_1 \chi_2)(g) = \chi_1(g)\chi_2(g)$). The elements of $\hat{G}$ are called the **characters** of G.

Remark 6.1 (Connection to Representation Theory) Recall that every irreducible complex representation of an abelian group G is 1-dimensional. The elements of $\hat{G}$ are in bijection with the 1-dimensional complex representations of G: a 1-dimensional representation coincides with its character, which justifies the name.

As G is finite, the image of any character $\chi : G \to \mathbb{C}^\times$ has order dividing $|G|$, which implies that $\chi(G) \subseteq \mu_{|G|}$, where $\mu_{|G|}$ is the set of roots of unity of order dividing $|G|$. Hence, we can identify $\hat{G}$ with $\mathrm{Hom}(G, \mu_{|G|})$. With this observation in hand, and using the structure theorem for finite abelian groups, the isomorphism between G and $\hat{G}$ becomes an easy exercise, see Problem 6.1.

Remark 6.2 Let $\chi \in \hat{G}$ be a character. We have already observed that $\chi(g)$ is a root of unity for every $g \in G$. The inverse and the complex conjugate of any root of unity coincide, hence we obtain

$$\overline{\chi(g)} = \chi(g)^{-1} = \chi(g^{-1}).$$

Remark 6.3 The group $\hat{G}$ should be compared with the dual of a vector space. As in the linear-algebraic setting, $\hat{G}$ and G are isomorphic, but only non-canonically. On the other hand, we will prove below that $\hat{\hat{G}} \cong G$ canonically, see Proposition 6.1

Remark 6.4 There is a pairing $\langle \cdot, \cdot \rangle : G \times \hat{G} \to \mathbb{C}$ given by $\langle g, \chi \rangle = \chi(g)$. This pairing is perfect: the only element $g \in G$ such that $\langle g, \chi \rangle = 1$ for all $\chi \in \hat{G}$ is the identity of G, and the only character χ such that $\langle g, \chi \rangle = 1$ for all $g \in G$ is the trivial character (the identity element of $\hat{G}$, that is, the homomorphism that sends every element of G to 1).

Proposition 6.1 (Canonical Isomorphism of G with $\hat{\hat{G}}$) *The map*

$$\Psi : G \to \hat{\hat{G}}$$
$$g \mapsto \psi_g,$$

where the homomorphism $\psi_g : \hat{G} \to \mathbb{C}^\times$ (which is an element of $\hat{\hat{G}}$) is defined by

$$\psi_g : \hat{G} \to \mathbb{C}^\times$$
$$\chi \mapsto \chi(g),$$

gives an isomorphism $G \cong \hat{\hat{G}}$.

Proof The homomorphism Ψ is injective by Remark 6.4. By Problem 6.1, one has $|G| = |\hat{G}| = \left|\hat{\hat{G}}\right|$, so Ψ is also surjective, hence an isomorphism. □

Remark 6.5 Proposition 6.1, which is almost trivial when G is finite, can be generalised to a suitable class of infinite abelian groups G. This more general statement often goes under the name of Pontryagin duality, see Theorem 9.1.

To finish our introduction to the dual group, we remark on its functorial properties.

Proposition 6.2 (Functoriality of $G \mapsto \hat{G}$) *The following hold:*

1. *The association $G \mapsto \hat{G}$ can be extended to a contravariant functor from the category of finite abelian groups to itself by letting it act on arrows as follows: if*

6.1 Pontryagin Duality: Finite Case

$f : G \to H$ is a group homomorphism, we define

$$\hat{f} : \hat{H} \to \hat{G}$$
$$\chi \mapsto \chi \circ f.$$

2. *This functor is exact: for every short exact sequence*

$$0 \to H \xrightarrow{\iota} G \xrightarrow{\pi} G/H \to 0,$$

the dual sequence

$$0 \to \widehat{G/H} \xrightarrow{\hat{\pi}} \hat{G} \xrightarrow{\hat{\iota}} \hat{H} \to 0$$

is also exact. Note that $\hat{\iota}$ is simply the restriction map: given an element $\chi \in \hat{G}$, that is, a homomorphism $\chi : G \to \mathbb{C}^\times$, the character $\hat{\iota}(\chi) \in \hat{H}$ is simply $\chi|_H$.

After these formal preliminaries, we are ready to state and prove the orthogonality relations, which we will then use to prove a Fourier inversion theorem for functions on abelian groups. We note that the next two results are special cases of Proposition 4.1, but we prefer to also give more direct and elementary proofs.

Proposition 6.3 (Orthogonality Relations I) *Fix $g_0 \in G$ and $\chi_0 \in \hat{G}$. We have*

$$\sum_{\chi \in \hat{G}} \chi(g_0) = \begin{cases} 0, & \text{if } g_0 \neq \mathrm{id}_G \\ |G|, & \text{otherwise} \end{cases}$$

and

$$\sum_{g \in G} \chi_0(g) = \begin{cases} 0, & \text{if } \chi_0 \neq \mathrm{id}_{\hat{G}} \\ |G|, & \text{otherwise.} \end{cases}$$

Proof We begin with the second statement. If χ_0 is the identity of $\hat{G}$, the statement is trivial. Otherwise, let $a \in G$ be an element such that $\chi_0(a) \neq 1$. Setting $S := \sum_{g \in G} \chi_0(g)$, we obtain

$$\chi_0(a)S = \sum_{g \in G} \chi_0(a)\chi_0(g) = \sum_{g \in G} \chi_0(ag) = \sum_{g \in G} \chi_0(g) = S,$$

hence $(\chi_0(a) - 1)S = 0$. Since $\chi_0(a) \neq 1$, this implies $S = 0$.

The first statement follows upon applying the first to $\hat{G}$ and using the identification $\hat{\hat{G}} \cong G$ provided by Proposition 6.1. □

Corollary 6.1 (Orthogonality Relations II)

1. Let χ_1, χ_2 be elements of $\hat{G}$. We have

$$\sum_{g \in G} \chi_1(g)\overline{\chi_2(g)} = \begin{cases} 0, & \text{if } \chi_1 \neq \chi_2 \\ |G|, & \text{otherwise.} \end{cases}$$

2. Let g, h be elements of G. We have

$$\sum_{\chi \in \hat{G}} \chi(g)\overline{\chi(h)} = \begin{cases} 0, & \text{if } g \neq h \\ |G|, & \text{otherwise.} \end{cases}$$

Proof For the first statement, observe that we have

$$\chi_1(g)\overline{\chi_2(g)} = \chi_1(g)\chi_2(g)^{-1} = (\chi_1\chi_2^{-1})(g)$$

by Remark 6.2 and apply Proposition 6.3 to the character $\chi_1\chi_2^{-1}$.

Similarly, for the second statement observe that $\chi(g)\overline{\chi(h)} = \chi(gh^{-1})$ and apply Proposition 6.3 to the element gh^{-1}. □

For later use, we introduce the following handy notation:

Definition 6.2 Let G be a group. The function $\delta : G^2 \to \{0, 1\}$ is defined by

$$\delta(g, h) = \begin{cases} 1, & \text{if } g = h \\ 0, & \text{otherwise.} \end{cases}$$

We will usually write $\delta_{g,h}$ instead of $\delta(g, h)$.

Definition 6.3 (Abstract Fourier Transform, Finite Case) Let f be any function $G \to \mathbb{C}$. The **Fourier transform** of f, denoted by $\hat{f}$, is the function

$$\hat{f} : \hat{G} \to \mathbb{C}$$
$$\chi \mapsto \tfrac{1}{|G|} \sum_{g \in G} f(g)\overline{\chi}(g).$$

Remark 6.6 Exactly as in the real case, there are several natural normalisations for the Fourier transform. The abstract theory we will discuss in Sect. 9 helps clarify the nature of these normalisations.

Theorem 6.1 (Fourier Inversion, Finite Case) Let $f : G \to \mathbb{C}$ be any function and let $\hat{f} : \hat{G} \to \mathbb{C}$ be its Fourier transform. We have

$$f(g) = \sum_{\chi \in \hat{G}} \hat{f}(\chi)\chi(g).$$

6.2 Densities

Proof Replacing the definition of $\hat{f}$ we obtain

$$\sum_{\chi \in \hat{G}} \hat{f}(\chi)\chi(g) = \sum_{\chi \in \hat{G}} \left(\frac{1}{|G|} \sum_{h \in G} f(h)\overline{\chi}(h)\right) \chi(g)$$

$$= \frac{1}{|G|} \sum_{h \in G} f(h) \left(\sum_{\chi \in \hat{G}} \overline{\chi}(h)\chi(g)\right)$$

We can now use Corollary 6.1 to rewrite the inner sum as $|G|\delta_{g,h}$, obtaining

$$\sum_{\chi \in \hat{G}} \hat{f}(\chi)\chi(g) = \frac{1}{|G|} \sum_{h \in G} f(h)|G|\delta_{g,h} = \sum_{h \in G} f(h)\delta_{g,h} = f(g).$$

□

Example 6.1 (Fourier Transform of the Characteristic Function of a Singleton)
Let G be a finite abelian group, let $a \in G$ be a fixed element, and let $f := \mathbf{1}_a$ be the function $\mathbf{1}_a(g) = \delta_{a,g}$. Its Fourier transform is given by

$$\hat{f}(\chi) = \frac{1}{|G|} \sum_{g \in G} \mathbf{1}_a(g)\overline{\chi(g)} = \frac{1}{|G|}\overline{\chi(a)} = \frac{1}{|G|}\chi(a)^{-1}.$$

Applying Theorem 6.1 we obtain the following representation for the function $\mathbf{1}_a$:

$$\mathbf{1}_a(g) = \frac{1}{|G|} \sum_{\chi \in \hat{G}} \chi(a)^{-1}\chi(g),$$

which recovers part of Corollary 6.1.

6.2 Densities

We now define two notions of *density* for sets of prime numbers (or, more generally, prime ideals in a number field): *natural density* and *Dirichlet density*. Even though the natural density is (as the name suggests) more 'natural', we will mostly focus on the notion of Dirichlet density, which is easier to treat from an analytic point of view.

Definition 6.4 Let K be a number field and let S be a set of (non-zero) prime ideals of O_K. We define the **natural density** of S as

$$\lim_{T\to\infty} \frac{\#\{\mathfrak{p} \in S : N(\mathfrak{p}) \leq T\}}{\#\{\mathfrak{p} \text{ non-zero prime of } O_K : N(\mathfrak{p}) \leq T\}},$$

provided that the limit exists.

Definition 6.5 (Dirichlet Density) Let K be a number field and let S be a subset of the set of nonzero prime ideals of O_K. We define the **Dirichlet density** of S as

$$\mathrm{Dens}_K(S) = \lim_{s\to 1^+} \frac{\sum_{\mathfrak{p} \in S} N(\mathfrak{p})^{-s}}{\sum_{\mathfrak{p} \text{ nonzero prime of } O_K} N(\mathfrak{p})^{-s}}, \qquad (6.1)$$

provided that the limit exists. We will omit the subscript K if it is clear from context.

Remark 6.7 There is at least another variant of the definition of Dirichlet density in the literature. For the sake of simplicity, we describe it only in the case of the number field K being $\mathbb{Q}$: if S is a set of prime numbers, the **logarithmic density** of S is

$$\lim_{x\to\infty} \frac{\sum_{p \in S, p \leq x} \frac{1}{p}}{\sum_{p \leq x} \frac{1}{p}},$$

if the limit exists.

Problem 6.5 shows that the notion of Dirichlet density is a genuine extension of the notion of natural density. On the other hand, it is a non-trivial exercise in analytic number theory to show that the Dirichlet and logarithmic densities of any set of primes coincide, see Problem 6.6 for extended hints.

Remark 6.8 The Dirichlet density satisfies the following basic properties:

1. $\mathrm{Dens}(S) \in [0, 1]$ for every set S of primes admitting Dirichlet density.
2. Let S be the set of all prime numbers: then it follows immediately from the definition that $\mathrm{Dens}(S) = 1$.
3. Let S be a finite set: again, it follows immediately from the definition that $\mathrm{Dens}(S) = 0$. In particular, if $\mathrm{Dens}(S)$ exists and is positive, then S is infinite.
4. Let S_1, S_2 be sets of primes, each admitting Dirichlet density, and suppose that $S_1 \cup S_2$ admits density. Then,

$$\mathrm{Dens}(S_1 \cup S_2) \leq \mathrm{Dens}(S_1) + \mathrm{Dens}(S_2).$$

A sufficient condition for equality to hold is $S_1 \cap S_2 = \emptyset$.

Furthermore, we will show below that $\sum_{\mathfrak{p} \text{ nonzero prime of } O_K} N(\mathfrak{p})^{-s}$ is asymptotic to $\log\left(\frac{1}{s-1}\right)$ as $s \to 1^+$ (see Proposition 6.5 and Problem 6.7), so that the denominator in Eq. (6.1) can be replaced by $\log\left(\frac{1}{s-1}\right)$.

Even though the density doesn't necessarily exist for an arbitrary set of primes, the following variants certainly do:

Definition 6.6 (Upper and Lower Density) Let K be a number field and let S be a subset of the set of prime ideals of O_K. We define the **upper and lower density** of S as

$$\operatorname{Dens}_K^+(S) = \limsup_{s \to 1^+} \frac{\sum_{\mathfrak{p} \in S} N(\mathfrak{p})^{-s}}{\sum_{\mathfrak{p} \text{ nonzero prime of } O_K} N(\mathfrak{p})^{-s}}$$

and

$$\operatorname{Dens}_K^-(S) = \liminf_{s \to 1^+} \frac{\sum_{\mathfrak{p} \in S} N(\mathfrak{p})^{-s}}{\sum_{\mathfrak{p} \text{ nonzero prime of } O_K} N(\mathfrak{p})^{-s}},$$

respectively. We will omit the subscript K if it is clear from the context.

We also recall another general fact: prime ideals whose residue field has prime order have full density, see Problem 6.8 for details.

6.3 Factorisation of the Cyclotomic Dedekind ζ Function

We are almost ready to prove Dirichlet's theorem on primes in arithmetic progression. Before doing so, as promised, we give an essentially self-contained proof of (a version of) Corollary 5.1, which will be used in our proof of Dirichlet's theorem.

Proposition 6.4 (Slightly Weaker Version of Corollary 5.1) *For every integer $n \geq 2$ we have a factorisation*

$$\zeta_{\mathbb{Q}(\zeta_n)} = f(s) \zeta(s) \prod_{\substack{\chi \text{ Dirichlet character} \\ \text{modulo } n}} L(s, \chi),$$

where $f(s)$ is holomorphic and non-vanishing in $\{\Re s > 0\}$.

Proof Both sides are given by suitable Euler products. If p is a fixed prime, the local contributions at p to both sides of the desired equality are of the form $(1 - \zeta_n^j p^{-s})^{\pm 1}$. Each such function is meromorphic on all of $\mathbb{C}$, and all of its zeroes and poles lie on the line $\Re(s) = 0$, where $|\zeta_n^j p^{-s}| = 1$. Thus, ignoring the finitely many ramified primes (which contribute to the function $f(s)$), it suffices to prove that

$$\prod_{\mathfrak{p}|p}(1 - N(\mathfrak{p})^{-s})^{-1} = \prod_{\chi}(1 - \chi(p)p^{-s})^{-1}, \tag{6.2}$$

where $\mathfrak{p}$ ranges over the prime divisors of p in $\mathbb{Z}[\zeta_n]$ (the ring of integers of $\mathbb{Q}(\zeta_n)$) and χ ranges over *all* Dirichlet characters modulo n, including the principal one. Suppose that the multiplicative order of p modulo n is equal to f: then, the Frobenius at p is an element of $G := \mathrm{Gal}(\mathbb{Q}(\zeta_n)/\mathbb{Q})$ of order f, and by definition this means that the finite field $\mathbb{F}_{\mathfrak{p}}$ is isomorphic to $\mathbb{F}_{p^f}$ (since $\mathrm{Gal}(\overline{\mathbb{F}_p}/\mathbb{F}_p)$ is isomorphic to the subgroup of G generated by the Frobenius at p). It follows from Eq. (3.2) that there are precisely $r = \varphi(n)/f$ primes $\mathfrak{p}$ of $\mathbb{Q}(\zeta_n)$ lying over p, and for each of them one has $N(\mathfrak{p}) = p^f$. Thus, the left-hand side of (6.2) can be rewritten as $(1 - p^{-fs})^{-\varphi(n)/f}$.

We now turn to the right-hand side of (6.2). Let H be the cyclic subgroup of $(\mathbb{Z}/n\mathbb{Z})^{\times}$ generated by the class of p. By definition, $|H| = f$. In (6.2), χ ranges over the dual group $\hat{G}$, but clearly only the image of χ in $\hat{H}$ is important. Since $\hat{G} \to \hat{H}$ is surjective with kernel given by $\widehat{G/H}$ (Proposition 6.2), every character of H appears $|\widehat{G/H}| = |G/H| = \varphi(n)/f$ times in this product. Moreover, since $\hat{H}$ is also cyclic, generated by the character that sends p to a primitive f-th root of unity, we obtain that the right-hand side of (6.2) can be written as

$$\prod_{j=1}^{f}\left(1 - \zeta_f^j p^{-s}\right)^{-\varphi(n)/f}.$$

The claim now follows from the elementary identity $\prod_{j=1}^{f}\left(1 - \zeta_f^j T\right) = 1 - T^f$. □

6.4 Infinitely Many Primes in Arithmetic Progressions

The tools we have developed (or assumed) are enough to give a quick proof of Dirichlet's theorem (Theorem 3.8), which we can now restate in a stronger form:

Theorem 6.2 (Dirichlet's Theorem, Quantitative Form) *Let a, m be positive integers with $(a, m) = 1$. The set*

$$\{p \text{ prime} : p \equiv a \pmod{m}\}$$

has Dirichlet density $\frac{1}{\varphi(m)}$.

Note that this form clearly implies Theorem 3.8 by Remark 6.8. Theorem 6.2 is also true with 'Dirichlet density' replaced by 'natural density', but the proof is

6.4 Infinitely Many Primes in Arithmetic Progressions

harder (it involves essentially the same difficulties that one faces when proving the Prime Number Theorem). By Problem 6.5, the version for natural density implies the version given above.

The crux of the proof lies in the following fact:

Theorem 6.3 (Dirichlet) *Let $m \geq 1$ be an integer and χ be a non-principal character of $(\mathbb{Z}/m\mathbb{Z})^\times$. The 'special value' $L(1, \chi)$ is finite and nonzero.*

Proof By Theorem 1.5, $L(1, \chi)$ is well-defined. Consider the formula of Corollary 5.1,

$$\zeta_{\mathbb{Q}(\zeta_m)}(s) = \zeta(s) \prod_{\substack{\chi \text{ non-principal Dirichlet} \\ \text{character modulo } m}} L(s, \tilde{\chi}).$$

Since both $\zeta_{\mathbb{Q}(\zeta_m)}$ and $\zeta(s)$ have a simple pole at $s = 1$ (Theorem 1.4), this shows that $\prod_{\chi \text{ non-principal Dirichlet}} L(s, \tilde{\chi})$ is bounded and nonzero around $s = 1$. Since each $L(s, \tilde{\chi})$ is holomorphic around $s = 1$ (Theorem 1.5), this implies that each $L(1, \chi)$ is nonzero. Note that the same argument also goes through if one uses Proposition 6.4 instead of Corollary 5.1. □

Lemma 6.1 *The sum $\sum_{p, k \geq 2} p^{-ks}$ remains bounded as $s \to 1^+$.*

Proof This is an easy computation:

$$\sum_{p, k \geq 2} \frac{1}{p^{ks}} = \sum_p \frac{1}{p^s(p^s - 1)} \leq \sum_p \frac{1}{p(p-1)} \leq \sum_{n \geq 2} \frac{1}{n(n-1)} = 1.$$

□

Proposition 6.5 *Let m be a positive integer and let χ be a character of $(\mathbb{Z}/m\mathbb{Z})^\times$. We consider the function $\sum_p \chi(p) \frac{1}{p^s}$.*

1. *If χ is the trivial character, $\sum_p \frac{1}{p^s}$ is asymptotic to $\log\left(\frac{1}{s-1}\right)$ as $s \to 1^+$.*
2. *If χ is non-trivial, $\sum_p \chi(p) \frac{1}{p^s}$ stays bounded as $s \to 1$.*

Proof

1. We have

$$\log \zeta(s) = \log \prod_p (1 - p^{-s})^{-1} = -\sum_p \log(1 - p^{-s})$$

$$= \sum_p \sum_k \frac{1}{kp^{ks}} = \sum_p p^{-s} + \sum_{p, k \geq 2} \frac{1}{kp^{ks}}.$$

Since $\zeta(s) \sim \frac{1}{s-1}$ as $s \to 1^+$ (Theorem 1.1), the first claim follows from the fact that $\sum_{p,k\geq 2} \frac{1}{kp^{ks}}$ stays bounded as $s \to 1$. Since $\sum_{p,k\geq 2} \frac{1}{kp^{ks}} \leq \sum_{p,k\geq 2} \frac{1}{p^{ks}}$, this is true by Lemma 6.1.

2. We proceed in a similar fashion. We have

$$\log L(s, \chi) = \log \prod_p (1 - \chi(p)p^{-s})^{-1}$$

$$= -\sum_p \log(1 - \chi(p)p^{-s})$$

$$= \sum_p \sum_{k \geq 1} \frac{\chi(p)^k}{kp^{ks}},$$

and this time we know (from Theorems 1.5 and 6.3) that $L(s, \chi)$ is holomorphic **and nonzero** in a neighbourhood of $s = 1$, so that $\log L(s, \chi)$ is also holomorphic at $s = 1$.

In the above expression, the sum for $k \geq 2$ stays bounded as $s \to 1^+$ (take absolute values and apply Lemma 6.1), so we obtain as desired that $\sum_p \frac{\chi(p)}{p^s}$ is bounded as $s \to 1^+$.

$\square$

Proof of Theorem 6.2 Fix $s > 1$ and consider the sum

$$\sum_{\substack{p \leq x \\ p \equiv a \pmod{m}}} \frac{1}{p^s} = \sum_{p \leq x} \mathbf{1}_a(p) \frac{1}{p^s},$$

where $\mathbf{1}_a(x)$ is the characteristic function of the subset $\{n \in \mathbb{Z} : n \equiv a \pmod{m}\}$ of $\mathbb{Z}$. Clearly, $\mathbf{1}_a(x)$ factors via $\mathbb{Z} \to \mathbb{Z}/m\mathbb{Z}$. By Example 6.1, we can then rewrite the above sum as

$$\frac{1}{\varphi(m)} \sum_{p \leq x} \sum_{\chi \in (\mathbb{Z}/m\mathbb{Z})^\times} \chi(a)^{-1} \chi(p) \frac{1}{p^s} = \frac{1}{\varphi(m)} \sum_{\chi \in (\mathbb{Z}/m\mathbb{Z})^\times} \chi(a)^{-1} \sum_{p \leq x} \chi(p) \frac{1}{p^s}.$$

Passing to the limit for $x \to \infty$ and applying Proposition 6.5 we obtain

$$\sum_{p \equiv a \pmod{m}} \frac{1}{p^s} = \frac{1}{\varphi(m)} \log\left(\frac{1}{s-1}\right)(1 + o(1)) + \sum_{\substack{\chi \in (\mathbb{Z}/m\mathbb{Z})^\times \\ \chi \neq 1}} O(1) \quad \text{as } s \to 1^+,$$

which, dividing by $-\log(s-1)$ and passing to the limit $s \to 1$, yields that the Dirichlet density of $\{p \text{ prime} \mid p \equiv a \pmod{m}\}$ is equal to $\frac{1}{\varphi(m)}$, as claimed. $\square$

6.5 The Philosophy of Special Values

Dirichlet's original proof of Theorem 6.2 follows basically the approach outlined above, with the main difference being in the proof of Theorem 6.3. In this section, we sketch briefly the main idea, which helps demonstrate one of the key features of L-functions: values of L-functions (especially at points where they are not naturally defined) encode arithmetic information.

Sketch of Proof of Theorem 6.3 Consider the factorisation in Theorem 5.5,

$$\zeta_{\mathbb{Q}(\zeta_m)}(s) = \zeta(s) \prod_{\chi \neq 1} L(s, \tilde{\chi}). \tag{6.3}$$

Since $L(s, \tilde{\chi}) = f(s) L(s, \chi)$ for some function $f(s)$ which is holomorphic and non-vanishing near $s = 1$ (see Proposition 5.3 (4)), the quantity $L(1, \tilde{\chi})$ is nonzero if and only if $L(s, \chi)$ is. Suppose that for some χ with $\overline{\chi} \neq \chi$ we had $L(1, \chi) = 0$. Then it is immediate to see that also $\overline{L(1, \chi)} = L(1, \overline{\chi}) = 0$, hence the right-hand side of (6.3) vanishes at $s = 1$ (since the *simple* pole of $\zeta(s)$ cancels out with the zero of $L(s, \chi)$, and then $L(1, \overline{\chi}) = 0$ implies the vanishing). This is clearly a contradiction, because

$$\limsup_{s \to 1^+} \zeta_{\mathbb{Q}(\zeta_m)}(s) \geq \limsup_{s \to 1^+} \sum_{I \triangleleft \mathbb{Z}[\zeta_m]} N(I)^{-s} \geq \limsup_{s \to 1^+} 1 = 1.$$

Hence, it only remains to show that $L(1, \chi) \neq 0$ when $\chi = \overline{\chi}$, that is, when χ takes values in $\{\pm 1\}$. Now, such a character (seen as a character $(\mathbb{Z}/m\mathbb{Z})^\times \to \{\pm 1\}$) defines—via taking the fixed field of the kernel—a quadratic extension $\mathbb{Q}(\sqrt{a})$ of $\mathbb{Q}$. Using Theorem 5.5 again (in a very simple case), one obtains

$$\zeta_{\mathbb{Q}(\sqrt{a})}(s) = \zeta(s) L(s, \tilde{\chi}).$$

The claim now follows from Theorem 6.4 below, where the crucial point is that the *arithmetic* interpretation of the limit immediately implies its non-vanishing. □

The following result was proved by Dirichlet in the case K is a quadratic number field (the general case was then established by Dedekind):

Theorem 6.4 (Analytic Class Number Formula) *Let K be a number field, with standard invariants d_K (discriminant), h_K (class number), (r_1, r_2) (signature), and R_K (regulator). Let w_K be the number of roots of unity in K. We have*

$$\operatorname{Res}_{s=1} \zeta_K(s) = \lim_{s \to 1} (s - 1) \zeta_K(s) = \frac{2^{r_1} (2\pi)^{r_2} h_K R_K}{\sqrt{|d_K|} w_K}.$$

Remark 6.9 (Philosophical Principle: Analytic Information Versus Arithmetic Information) Analytic objects can encode arithmetic information! Theorem 6.4

is extremely remarkable, in that it relates something which is purely analytic (the residue of a holomorphic function) to something arithmetic (information about unique factorisation, roots of unity, etc.).

There is also another important point of view on Theorem 6.4:

Remark 6.10 (Local-global Principles) One can also reformulate Theorem 6.4 as

$$\mathrm{Res}_{s=1} \prod_{\mathfrak{p}} (1 - N\mathfrak{p}^{-s})^{-1} = \frac{2^{r_1}(2\pi)^{r_2} h_K R_K}{\sqrt{|d_K|} w_K}.$$

This shows that we can get 'global' information on the arithmetic of K from the knowledge of (an infinite amount of) 'local' data, namely, the sizes of the residue fields of $\mathcal{O}_K$.

6.5.1 Further Special Values

The functional equation (Theorem 3.13) and the analytic class number formula (Theorem 6.4) imply that $\zeta_K(s)$ has a zero of order $r = \mathrm{rk}\,\mathcal{O}_K^\times = r_1 + r_2 - 1$ at $s = 0$, and one has

$$\lim_{s \to 0} s^{-r} \zeta_K(s) = -\frac{h_K R_K}{w_K}.$$

Conjecturally, the values (or more precisely, the first non-zero terms in the local series development) of ζ_K at all integers should have an arithmetic interpretation. Discussing this in detail would take us quite far afield, so we limit ourselves to recalling a striking formula due to Euler and its connection with the arithmetic of cyclotomic extensions.

Theorem 6.5 (Euler) *Let n be a positive integer. We have*

$$\zeta(2n) = (-1)^{n+1} \frac{B_{2n}(2\pi)^{2n}}{2(2n)!},$$

where B_{2n} are the **Bernoulli numbers**, *defined by the power series development*

$$\frac{t}{e^t - 1} = \sum_{k=0}^{\infty} B_k \frac{t^k}{k!}.$$

Even though this result does not seem to have much arithmetic content (after all, the numbers B_k seem to have nothing to do with arithmetic!), Kummer, Herbrand and Ribet have found a remarkable interpretation for these numbers:

6.5 The Philosophy of Special Values

Theorem 6.6 (Kummer; Herbrand [1], Ribet [5]) *A prime p divides the class number of $\mathbb{Q}(\zeta_p)$ if and only if p divides the numerator of some Bernoulli number B_n for some even n with $0 < n < p - 1$.*

Special values of L-functions are linked to extremely deep conjectures in number theory. To get a sense of just *how deep* these problems lie, the reader can have a look at Question 4.2 in [3], and then start learning about K-theory, higher regulators, and the Beilinson and Bloch-Kato conjectures (see for example the surveys [4] and [2]).

Problems

6.1 Prove that, for any finite abelian group G, the groups G and $\hat{G}$ are isomorphic.

6.2 Prove the claims made in Remark 6.4.

6.3 Prove Proposition 6.2.

6.4 (Squaring the Fourier Transform) Let G be a finite abelian group and let $f : G \to \mathbb{C}$ be a function. The Fourier transform $\hat{f}$ is a function $\hat{G} \to \mathbb{C}$, so we can take *its* Fourier transform, obtaining $\hat{\hat{f}} : \hat{\hat{G}} \to \mathbb{C}$. Using Proposition 6.1 we may identify $\hat{\hat{f}}$ to a function $\hat{\hat{f}} : G \to \mathbb{C}$. Prove that $\hat{\hat{f}}(g) = \frac{1}{|G|} f(g^{-1})$ for all $g \in G$.

6.5 Let K be a number field and let S be a set of non-zero prime ideals. Prove that, if S admits natural density, then it also admits Dirichlet density, and the two coincide.

6.6 Show that, for $K = \mathbb{Q}$, a set of primes admits Dirichlet density if and only if it admits logarithmic density, and the two coincide.

Hint. Here is a possible strategy.

1. Define
$$a_n = \begin{cases} 1/n, & \text{if } n \in S \\ 0, & \text{otherwise} \end{cases}$$
and $A(u) = \sum_{n \leq u} a_n$. Show that S admits logarithmic density δ_S if and only if $A(u) = (\delta_S + o(1)) \log \log u$.

2. Assume that S does admit logarithmic density. By a suitable summation by parts, show that for $s > 1$ we have
$$\sum_{p \in S} \frac{1}{p^s} = (s - 1) \int_1^\infty A(u) u^{-s} \, du.$$

3. Prove that

$$(s-1)\int_1^\infty u^{-s}\log\log u\, du \sim \log\left(\frac{1}{s-1}\right) \text{ as } s\to 1^+.$$

Deduce that, if S admits logarithmic density, then it admits Dirichlet density and the two coincide.

4. Show the following general estimate on primes: one has $\sum_{p\le t}\left(\frac{1}{p}-\frac{1}{p^{1+1/\log t}}\right)=O(1)$, and in particular the sum is $o(\log\log t)$.

5. Show that $\sum_{p>t}p^{-1-1/\log t}=O(1)=o(\log\log t)$. (You may need some weak version of the prime number theorem.)

6. Suppose now that S admits Dirichlet density δ. Introduce $N(s)=\frac{\sum_{p\in S}p^{-s}}{\sum_p p^{-s}}$ and observe that $N(s)\sim(\delta+o(1))\log(1/(s-1))$ for $s\to 1^+$. Prove that $\sum_{p\in S, p\le t}\frac{1}{p}=N(1+1/\log t)+o(\log\log t)$ and deduce that S admits logarithmic density equal to δ.

6.7 Prove that the denominator $\sum_{\mathfrak{p}\text{ nonzero prime of }O_K}N(\mathfrak{p})^{-s}$ appearing in (6.1) is asymptotic to $\log(\frac{1}{s-1})$ as $s\to 1^+$.

Hint. You can start by looking at Proposition 6.5.

6.8 Let K be a number field and S be a set of primes of K. Denote by $S^{(1)}$ the subset of S consisting of those prime ideals $\mathfrak{p}$ for which the cardinality of the residue field $O_K/\mathfrak{p}$ is a prime number (such prime ideals are called *of degree 1*). Prove that $\text{Dens}^+(S^{(1)})=\text{Dens}^+(S)$ and $\text{Dens}^-(S^{(1)})=\text{Dens}^-(S)$. In particular, if S admits density, then so does $S^{(1)}$, and the densities coincide.

6.9 Let χ_5 be the unique primitive Dirichlet character modulo 5 taking values in $\{\pm 1\}$. Compute $L(1,\chi_5)$.

References

1. Herbrand, J.: Sur les classes des corps circulaires. J. Math. Pures Appl. (9) **11**, 417–441 (1932)
2. Kings, G.: The Bloch-Kato Conjecture on Special Values of L-Functions. A Survey of Known Results, pp. 179–198 (2003). Les XXIIèmes Journées Arithmetiques, Lille (2001)
3. Lichtenbaum, S.: Values of zeta-functions, étale cohomology, and algebraic K-theory. In: Algebraic K-Theory, II: "Classical" Algebraic K-Theory and Connections with Arithmetic (Proc. Conf., Battelle Memorial Inst., Seattle, Wash., 1972). Lecture Notes in Mathematics, vol. 342, pp. 489–501. Springer, Berlin (1973)
4. Nekovář, J.: Beĭlinson's conjectures. In: Motives (Seattle, WA, 1991), Proceedings of Symposia in Pure Mathematics, vol. 55, Part 1, pp. 537–570. American Mathematics Society, Providence (1994). https://doi.org/10.1090/pspum/055.1/1265544
5. Ribet, K.A.: A modular construction of unramified p-extensions of $\mathbb{Q}(\mu_p)$. Invent. Math. **34**(3), 151–162 (1976). https://doi.org/10.1007/BF01403065

Chapter 7
The Chebotarev Density Theorem

Abstract In this chapter we give two proofs of the fundamental density theorem of Chebotarev. The first directly generalises our proof of Dirichlet's theorem. The second is more algebraic in nature and provides a nice illustration of the power of Frobenius elements in algebraic number theory.

In this chapter we give two different proofs of Chebotarev's theorem, which we can now state in its quantitative form.

Theorem 7.1 (Chebotarev, Quantitative Form) *Let K/F be a Galois extension with group G and let $C \subseteq G$ be a union of conjugacy classes. The set*

$$S = \{\mathfrak{p} \text{ prime of } O_F \mid \left(\frac{K/F}{\mathfrak{p}}\right) \subseteq C\}$$

admits (Dirichlet) density, given by $\operatorname{Dens}(S) = \frac{\#C}{\#G}$.

Remark 7.1 Note that $\left(\frac{K/F}{\mathfrak{p}}\right)$ is a conjugacy class (see Definition 3.3). Since C is a union of conjugacy classes, one may equivalently rephrase the condition in the theorem as $\left(\frac{K/F}{\mathfrak{p}}\right) \in C$, where $\left(\frac{K/F}{\mathfrak{p}}\right)$ now means *any element of the conjugacy class*.

Remark 7.2 Even though we will not use it much in this course, Chebotarev's theorem really is one of the fundamental tools in modern number theory, and its importance is hard to overstate. To provide some context, we mention some rather different applications of the theorem:

1. Let F_1, F_2 be two finite extensions of the number field K. Then F_1, F_2 have the same Galois closure over K if and only if the primes of K that split completely in F_1, F_2 are the same, or even the same up to a subset of density 0.
2. Let p be a prime number. An element $a \in \mathbb{Q}^\times$ is a p-th power if and only if it is a p-th power modulo ℓ for almost all primes ℓ. Even more: if $f(x) \in \mathbb{Z}[x]$ is irreducible and has a root modulo almost every prime p, then $\deg f = 1$.

3. A sufficiently strong version of Chebotarev with error term (unfortunately, one *so strong* that we can only prove it under the assumption of the Generalised Riemann Hypothesis for Dedekind ζ functions) implies Artin's conjecture on primitive roots, see [2] for the precise statement:

   Conjecture 7.1 (Artin) Let a be a non-zero integer which is not a square and is different from -1. There exist infinitely many primes p such that a is a primitive root modulo p.

4. Let $f(x) \in \mathbb{Z}[x]$ be any polynomial. There exist infinitely many primes p such that $f(x)$ mod p splits completely (this can also be proven elementarily, without relying on the Chebotarev theorem).

5. Let K be a number field. The 'probability' that a prime of O_K is principal (that is, the density of principal primes) is $1/h_K$. This partially justifies the oft-repeated claim that 'h_K measures the failure of unique factorisation'—recall that unique factorisation is equivalent to the ring of integers being a PID, which in turn is equivalent to every prime ideal being principal, so h_K really is a measure of 'how much' unique factorisation fails.

6. The density of primes p that divide at least one number of the form $2^n + 1$ (for $n \geq 0$ an integer) exists and is equal to $\frac{17}{24}$ (a result of Hasse [1] later extended to a much more general context [4, 6]).

7. (Elementary consequence of the theorem) Let $f(x) \in \mathbb{Z}[x]$, of degree n, have Galois group G. Fix a 'cycle type', that is, a conjugacy class of permutations in S_n (which we think as a tuple $m_1, \ldots, m_n$ with $1m_1 + 2m_2 + \cdots + nm_n = n$—here m_i is the number of cycles of length i). Let d be the number of elements of G with the given cycle type. Then, the set

$$S = \left\{ p \text{ prime} : \begin{array}{l} f(x) \text{ mod } p \text{ factors with } m_1 \text{ factors of degree } 1, \\ m_2 \text{ factors of degree } 2, \ldots, m_n \text{ factors of degree } n \end{array} \right\}$$

   admits density, and this density is $\frac{d}{\#G}$.

8. Let a, b, c be integers with $(a, b, c) = 1$ and $d := b^2 - 4ac < 0$. Suppose that d is a *fundamental discriminant*, that is, d is not divisible by the square of any odd prime, and it satisfies either $d \equiv 1 \pmod 4$ or $d = 4d'$ with $d' \equiv 2, 3 \pmod 4$. As x, y range over the integers, the positive-definite quadratic form $ax^2 + bxy + cy^2$ represents infinitely many primes. In fact, the density of the set

$$\{p \text{ prime} : \exists x, y \in \mathbb{Z} \text{ such that } p = ax^2 + bxy + cy^2\}$$

   is positive and can be expressed in terms of invariants of the number field $\mathbb{Q}(\sqrt{-d})$. (This is related to example 5 in this list; the density in question is $\frac{1}{2h_{\mathbb{Q}(\sqrt{-d})}}$.)

9. Let $f(x) \in \mathbb{Z}[x]$ be an irreducible polynomial. The average number of zeroes of f modulo primes is 1, that is to say,

$$\lim_{T \to \infty} \frac{\sum_{p \leq T} \#\{a \in \mathbb{F}_p : f(a) = 0 \pmod{p}\}}{\#\{p \text{ prime} : p \leq T\}} = 1.$$

Having mentioned some of the many applications of Chebotarev's theorem, we now turn to its proof. We give both an algebraic proof and an analytic one, starting with the latter.

7.1 Analytic Proof

The proof we give in this section is inspired by a famous paper by Lagarias and Odlyzko [5]. However, the aim of that paper is to give effective estimates, which involves some rather heavy analytic number theory. We give a streamlined argument so as to arrive at the result *for the Dirichlet density* in the most efficient way possible; note that Lagarias and Odlyzko work with the natural density. We think that the resulting proof is fairly straightforward, and very interesting in the way it directly generalises the proof of Theorem 6.2. According to the MathOverflow discussion [7], a similar approach appears in unpublished notes of Serre, but to the best of the author's knowledge, Serre's proof has not appeared in print.

We start with the following lemma:

Lemma 7.1 *Let $L(s, \rho)$ be the Artin L-function of some representation $\rho : \mathrm{Gal}(L/K) \to \mathrm{GL}(V)$. We have the asymptotic relation*

$$\log L(s, \rho) = \sum_{\mathfrak{p} \text{ unramified in } L} \mathrm{tr}\, \rho\left(\left(\frac{L/K}{\mathfrak{p}}\right)\right) N(\mathfrak{p})^{-s} + O(1) \text{ as } s \to 1^+.$$

Proof Start from the formula given in Problem 5.5, namely

$$\log L(s, \rho) = \sum_{\mathfrak{p}} \sum_{m \geq 1} \frac{\mathrm{tr}(\rho(\mathfrak{p}^m))}{m(N\mathfrak{p})^{ms}}.$$

We may ignore the finitely many ramified primes (which give a bounded contribution), and bound

$$\left| \sum_{\mathfrak{p}} \sum_{m \geq 2} \frac{\chi(\mathfrak{p}^m)}{m(N\mathfrak{p})^{ms}} \right| \leq \sum_{\mathfrak{p}} \sum_{m \geq 2} \left| \frac{\chi(\mathfrak{p}^m)}{m(N\mathfrak{p})^{ms}} \right| \leq \sum_{\mathfrak{p}} \sum_{m \geq 2} \frac{\dim V}{(N\mathfrak{p})^{m\Re s}}$$

$$\leq [K : \mathbb{Q}] \sum_{p} \sum_{m \geq 2} \frac{\dim V}{p^{m\Re s}} = O(1)$$

by Lemma 6.1. The result now follows, because we have

$$\log L(s, \rho) = \sum_{\mathfrak{p}} \sum_{m \geq 1} \frac{\mathrm{tr}(\rho(\mathfrak{p}^m))}{m(N\mathfrak{p})^{ms}}$$

$$= \sum_{\mathfrak{p} \text{ unramified}} \sum_{m \geq 1} \frac{\mathrm{tr}(\rho(\mathfrak{p}^m))}{m(N\mathfrak{p})^{ms}} + O(1)$$

$$= \sum_{\mathfrak{p} \text{ unramified}} \frac{\mathrm{tr}(\rho(\mathfrak{p}))}{(N\mathfrak{p})^s} + O(1)$$

$$= \sum_{\mathfrak{p} \text{ unramified}} \mathrm{tr}\, \rho\left(\left(\frac{L/K}{\mathfrak{p}}\right)\right) N(\mathfrak{p})^{-s} + O(1),$$

as claimed. □

Let C be a conjugacy class in G. Define the class function

$$f_C = \frac{|G|}{|C|} \mathbf{1}_C. \qquad (7.1)$$

Fix $g \in C$. The orthogonality relations (Proposition 4.1) easily imply that

$$f_C = \sum_{\varphi} \overline{\varphi(g)} \varphi, \qquad (7.2)$$

where the sum is over all the characters of the finite-dimensional irreducible representations of G (up to isomorphism). Let furthermore $H = \langle g \rangle$ and let τ be the class function on H defined by

$$\tau = |H| \cdot \mathbf{1}_{\{g\}},$$

where $\mathbf{1}_{\{g\}}$ is the characteristic function of the set $\{g\}$. Note that this is a class function since H is abelian (hence all functions are class functions). By the usual abelian orthogonality relations (Corollary 6.1) we have

$$\tau = \sum_{\chi \in \hat{H}} \overline{\chi(g)} \chi.$$

Let $Z_G(g)$ be the centraliser of g in G. A simple calculation (see Lemma 7.2 below) shows that

$$\mathrm{Ind}_H^G(\tau) = |Z_G(g)| \cdot \mathbf{1}_C = \frac{|G|}{|C|} \mathbf{1}_C = f_C, \qquad (7.3)$$

7.1 Analytic Proof

and therefore

$$\sum_\varphi \overline{\varphi(g)}\varphi = f_C = \mathrm{Ind}_H^G(\tau) = \sum_{\chi \in \hat{H}} \overline{\chi(g)}\, \mathrm{Ind}_H^G(\chi). \tag{7.4}$$

Lemma 7.2 *Equation* (7.3) *holds.*

Proof By Proposition 4.2 we have

$$\mathrm{Ind}_H^G(\tau)(x) = \sum_{s \in S} \tau_0(s^{-1}xs),$$

where

$$\tau_0(g) = \begin{cases} \tau(g), & \text{if } g \in H \\ 0, & \text{otherwise} \end{cases}$$

and S is a set of representatives for the left cosets $\{sH\}$. In the case of our function τ, we obtain

$$\mathrm{Ind}_H^G(\tau)(x) = |H| \cdot \#\{s \in S : s^{-1}xs = g\} = |H| \cdot |\{s \in S : x = sgs^{-1}\}|.$$

We claim that this quantity is 0 if g, x are in different conjugacy classes in G, and is $|Z_G(g)|$ otherwise. To see this, simply observe that every element of $H = \langle g \rangle$ commutes with g, so (since $G = SH$) we obtain

$$|H| \cdot \#\{s \in S : x = sgs^{-1}\} = |\{(s, h) \in S \times H : x = shgh^{-1}s^{-1}\}|$$
$$= |\{y \in G : x = ygy^{-1}\}|.$$

The statement is now clear: if x, g belong to different conjugacy classes no such y exists, while if x, g are in the same conjugacy class, the set above is a coset for $Z_G(g)$. □

We are now ready to prove Theorem 7.1.

Proof of Theorem 7.1 We can assume that C itself is a conjugacy class. We wish to estimate

$$\sum_{\mathfrak{p} \subset O_K} \mathbf{1}_C\left(\left(\frac{L/K}{\mathfrak{p}}\right)\right)(N\mathfrak{p})^{-s}$$

as $s \to 1$. For simplicity, if φ is a character of $G = \mathrm{Gal}(L/K)$ and $\mathfrak{p}$ is a prime of K unramified in L, we denote by $\varphi(\mathfrak{p})$ the value of φ at $\left(\frac{L/K}{\mathfrak{p}}\right)$. Note that this is well-defined, because different choices of Frobenius all lie in the same conjugacy class of G, and φ is constant on conjugacy classes.

Using (7.1) and (7.2), we write the above as

$$\sum_{\mathfrak{p} \subset O_K} 1_C\left(\left(\frac{L/K}{\mathfrak{p}}\right)\right)(N\mathfrak{p})^{-s} = \frac{|C|}{|G|} \sum_{\mathfrak{p} \subset O_K} \sum_{\varphi} \overline{\varphi(g)}\varphi(\mathfrak{p})(N\mathfrak{p})^{-s},$$

where as before φ ranges over the irreducible characters of G. We now use (7.4) to arrive at

$$\sum_{\mathfrak{p} \subset O_K} 1_C\left(\left(\frac{L/K}{\mathfrak{p}}\right)\right)(N\mathfrak{p})^{-s} = \frac{|C|}{|G|} \sum_{\mathfrak{p} \subset O_K} \sum_{\chi \in \hat{H}} \overline{\chi(g)} \operatorname{Ind}_H^G(\chi)(\mathfrak{p})(N\mathfrak{p})^{-s}$$

$$= \sum_{\chi \in \hat{H}} \overline{\chi(g)} \frac{|C|}{|G|} \sum_{\mathfrak{p} \subset O_K} \operatorname{Ind}_H^G(\chi)(\mathfrak{p})(N\mathfrak{p})^{-s}.$$

Lemma 7.1 allows us to recognise the inner sum as the series development of $L(s, \operatorname{Ind}_H^G(\chi))$ around $s = 1$, up to a bounded error term:

$$\sum_{\mathfrak{p} \subset O_K} 1_C\left(\left(\frac{L/K}{\mathfrak{p}}\right)\right)(N\mathfrak{p})^{-s} = \sum_{\chi \in \hat{H}} \overline{\chi(g)} \frac{|C|}{|G|} \log L(s, \operatorname{Ind}_H^G(\chi)) + O(1).$$

Using the formalism of Artin L-functions (in particular, Theorem 5.2(3)), we can replace $L(s, \operatorname{Ind}_H^G(\chi))$ with $L(s, \chi)$ to arrive at

$$\sum_{\mathfrak{p} \subset O_K} 1_C\left(\left(\frac{L/K}{\mathfrak{p}}\right)\right)(N\mathfrak{p})^{-s} = \sum_{\chi \in \hat{H}} \overline{\chi(g)} \frac{|C|}{|G|} \log L(s, \chi) + O(1) \text{ as } s \to 1,$$

(7.5)

where now all the involved Artin L-functions are Hecke L-functions.

Letting $E = L^H$ be the subfield fixed by $H = \langle g \rangle$, Theorem 5.5 shows that

$$\zeta_L(S) = \zeta_E(s) \prod_{\chi \neq 1, \chi \in \hat{H}} L(s, \chi).$$

(Recall that H is a cyclic group, so all its irreducible complex representations are 1-dimensional.) Arguing as in Theorem 6.3, and using Theorem 5.1 to ensure analyticity of $L(s, \chi)$ around $s = 1$ for χ a non-principal character, we obtain that $L(s, \chi)$ is finite and non-vanishing at $s = 1$ when χ is a non-principal character. Thus, $\log L(s, \chi)$ is bounded as $s \to 1$ for $\chi \neq 1$. Using this information in

7.1 Analytic Proof

Eq. (7.5) we finally obtain

$$\sum_{\mathfrak{p} \subset O_K} \mathbf{1}_C\left(\left(\frac{L/K}{\mathfrak{p}}\right)\right)(N\mathfrak{p})^{-s} = \sum_{\chi=1} \overline{\chi(g)}\frac{|C|}{|G|} \log L(s,\chi) + O(1)$$

$$= \frac{|C|}{|G|} \log L(s, 1_H) + O(1)$$

as $s \to 1$. Notice that in this last formula 1_H is the trivial representation of the Galois group H of L over E, so $L(s, 1_H)$ is the ζ-function of E. Using Theorem 1.4 both for E and for K, we obtain that (as $s \to 1$) we have the asymptotic

$$\log L(s, 1_H) = \log \zeta_E(s) \sim \log\left(\frac{1}{s-1}\right) \sim \log \zeta_K(s).$$

Interpreting $\zeta_K(s)$ as $L(s, 1_G)$ for the trivial representation of $G = \text{Gal}(L/K)$, Lemma 7.1 yields $\log \zeta_K(s) \sim \sum_{\mathfrak{p} \subset O_K}(N\mathfrak{p})^{-s}$, again for $s \to 1$. Combining these relations, we have $\sum_{\mathfrak{p} \subset O_K}(N\mathfrak{p})^{-s} \sim L(s, 1_H)$, and we are done:

$$\lim_{s \to 1^+} \frac{\sum_{\mathfrak{p} \subset O_K} \mathbf{1}_C\left(\left(\frac{L/K}{\mathfrak{p}}\right)\right)(N\mathfrak{p})^{-s}}{\sum_{\mathfrak{p} \subset O_K}(N\mathfrak{p})^{-s}} = \lim_{s \to 1^+} \frac{\frac{|C|}{|G|} \log L(s, 1_H) + O(1)}{\log L(s, 1_H)} = \frac{|C|}{|G|}.$$

$\square$

Remark 7.3 The reader won't fail to notice the parallelism between the argument just given and the proof of Dirichlet's theorem in Sect. 6.4. Indeed, in both cases we use some version of discrete Fourier analysis to write the characteristic function of a conjugacy class (a singleton, in the case of Dirichlet's theorem) as a linear combination of characters of the relevant group. In the proof of Dirichlet's theorem, these are characters of abelian representations. In the proof of Chebotarev's theorem this is not necessarily the case, but we know from Theorem 5.4 that one can always write any character as a linear combination of characters induced from 1-dimensional representations of subgroups. In the proof of Chebotarev's theorem we write a decomposition of this form explicitly, inducing from characters of the subgroup H. The formalism of Artin L-functions then allows us to only deal with the L-functions of these 1-dimensional characters, and the rest of the proof is virtually identical.

Remark 7.4 (The 'Ideal' Proof of Chebotarev's Theorem) If one knew that Artin's conjecture on analytic continuation (Conjecture 5.1) holds, the proof of Chebotarev's theorem would be even simpler, and it would mirror even more closely the proof of Dirichlet's theorem. Specifically, by Theorem 4.3 we know that the class function $\mathbf{1}_C$ is a linear combination of characters of irreducible representations of G. Let χ_0 be the trivial character and $\chi_1, \ldots, \chi_r$ be the other irreducible characters of G. Writing $\mathbf{1}_C = c_0\chi_0 + \sum_{i=1}^r c_i\chi_i$, the orthogonality relations (Theorem 4.2 (3) or Proposition 4.1) yield

$$c_0 = \langle \mathbf{1}_C, \chi_0 \rangle = \frac{1}{|G|} \sum_{g \in G} \mathbf{1}_C(g) = \frac{|C|}{|G|}.$$

For every irreducible character χ, let ρ_χ be the corresponding irreducible representation. By the (easy) Lemma 7.1 we obtain

$$\log L(s, \rho_{\chi_i}) = \sum_{\mathfrak{p} \text{ unramified in } L} \chi_i\left(\left(\frac{L/K}{\mathfrak{p}}\right)\right) N(\mathfrak{p})^{-s} + O(1) \text{ as } s \to 1^+. \quad (7.6)$$

For $i \neq 0$, Artin's conjecture implies that $L(1, \rho_{\chi_i})$ is finite and non-zero (see Proposition 5.4), so for all $i \neq 0$ we have

$$\log L(s, \rho_{\chi_i}) = O(1) \text{ as } s \to 1^+.$$

Multiplying Eq. (7.6) by c_i and summing over i we obtain (as $s \to 1^+$)

$$\sum_{i=0}^{r} c_i \log L(s, \rho_{\chi_i}) = \sum_{\mathfrak{p} \text{ unramified in } L} \sum_{i=0}^{r} c_i \chi_i\left(\left(\frac{L/K}{\mathfrak{p}}\right)\right) N(\mathfrak{p})^{-s} + O(1)$$

$$= \frac{|C|}{|G|} \sum_{\mathfrak{p} \text{ unramified in } L} N(\mathfrak{p})^{-s} + \sum_{i=1}^{r} c_i \chi_i\left(\left(\frac{L/K}{\mathfrak{p}}\right)\right) N(\mathfrak{p})^{-s} + O(1)$$

$$= \frac{|C|}{|G|} \sum_{\mathfrak{p} \text{ unramified in } L} N(\mathfrak{p})^{-s} + O(1),$$

and on the other hand

$$\sum_{\mathfrak{p} \text{ unramified in } L} \sum_{i=0}^{r} c_i \chi_i\left(\left(\frac{L/K}{\mathfrak{p}}\right)\right) N(\mathfrak{p})^{-s}$$

$$= \sum_{\mathfrak{p} \text{ unramified in } L} \mathbf{1}_C\left(\left(\frac{L/K}{\mathfrak{p}}\right)\right) N(\mathfrak{p})^{-s}.$$

Comparing these two expressions, dividing by $\sum_\mathfrak{p} N(\mathfrak{p})^{-s}$ and passing to the limit we obtain Chebotarev's theorem for the Dirichlet density. Finally, we mention that if one further assumes the Generalised Riemann Hypothesis (GRH) for the functions $L(s, \rho_\chi)$, it is not hard to also obtain the version for the natural density, see for example [3, p. 143].

7.2 Algebraic Proof

We now give a second, more algebraic proof of Chebotarev's theorem. This proof is based on Schoof's version given in [8, Chapter 15], with several adjustments, especially in the proof of Proposition 7.3.

The first step is to establish an analogue of Dirichlet's theorem for arbitrary number fields. Recall that we have already remarked that Dirichlet's theorem is precisely Chebotarev's theorem for the extensions $\mathbb{Q}(\zeta_m)/\mathbb{Q}$ (see the proof we gave for Theorem 3.8). Our first proposition is then a direct generalisation of Dirichlet's theorem to cyclotomic extensions of arbitrary number fields.

Proposition 7.1 *Let F be a number field and let $m \geq 1$ be an integer. Write H for the group* $\mathrm{Gal}(F(\zeta_m)/F)$. *For every $h \in H$, the set*

$$S_h := \left\{ \mathfrak{p} \text{ prime of } \mathcal{O}_F \mid \left(\frac{F(\zeta_m)/F}{\mathfrak{p}} \right) = h \right\}$$

admits Dirichlet density, equal to $1/|H|$.

Remark 7.5 Since H is abelian, $\left(\frac{F(\zeta_m)/F}{\mathfrak{p}} \right)$ is well-defined as an element of H (it is a conjugacy class consisting of a single element).

Proof This closely parallels the proof of Theorem 6.2. By restriction, H can be identified with a subgroup $\tilde{H}$ of $\mathrm{Gal}(\mathbb{Q}(\zeta_m)/\mathbb{Q}) \cong (\mathbb{Z}/m\mathbb{Z})^\times$. For every character $\chi : \tilde{H} \to \mathbb{C}^\times$ (that we interpret as a character on a subset of $(\mathbb{Z}/m\mathbb{Z})^\times$) we introduce the series

$$L(s, \chi) = \prod_{\mathfrak{p} \subset \mathcal{O}_F} \left(1 - \frac{\chi(N(\mathfrak{p}))}{N(\mathfrak{p})^s} \right)^{-1},$$

with the usual convention that $\chi(n) = 0$ if $(n, m) > 1$. This is not a Dirichlet L-function, but it *is* a Hecke L-function, corresponding to the abelian extension $F(\zeta_m)/F$ and the representation

$$\mathrm{Gal}(F(\zeta_m)/F) \xrightarrow{\sim} \tilde{H} \xrightarrow{\chi} \mathbb{C}^\times.$$

This is proved exactly as in Proposition 5.3(2). We limit ourselves to showing equality of the local factors at the unramified primes, which is enough for our purposes.

Given a prime $\mathfrak{p}$ of F unramified in $F(\zeta_m)$ and a prime $\mathfrak{P}$ of $F(\zeta_m)$ lying over it, we show that the Artin symbol $\left(\frac{F(\zeta_m)/F}{\mathfrak{P}} \right)$ maps to $N(\mathfrak{p})$ under the isomorphism $H \to \tilde{H} \subseteq \mathrm{Gal}(\mathbb{Q}(\zeta_m)/\mathbb{Q}) \cong (\mathbb{Z}/m\mathbb{Z})^\times$. To see this, notice that $\sigma := \left(\frac{F(\zeta_m)/F}{\mathfrak{P}} \right)$ is determined by its restriction to $\mathbb{Q}(\zeta_m)$, and that it satisfies

$$\sigma(\zeta_m) \equiv \zeta_m^{N(\mathfrak{p})} \pmod{\mathfrak{P} \cap \mathbb{Z}[\zeta_m]}.$$

By uniqueness of Frobenius elements (in the unramified case), since the automorphism $\zeta_m \mapsto \zeta_m^{N(\mathfrak{p})}$ satisfies the condition above, we see that $\left(\frac{F(\zeta_m)/F}{\mathfrak{P}}\right)$ does indeed map to $\zeta_m \mapsto \zeta_m^{N(\mathfrak{p})}$, which—under the canonical isomorphism $\text{Gal}(\mathbb{Q}(\zeta_m)/\mathbb{Q}) \cong (\mathbb{Z}/m\mathbb{Z})^\times$—corresponds to $N(\mathfrak{p})$, as claimed. Thus, for every non-trivial character χ, the function $L(s, \chi)$ agrees (possibly up to finitely many factors of the form $(1 - \zeta_m^k/N(\mathfrak{p})^s)$, which are holomorphic and non-vanishing near $s = 1$) with a Hecke L-function. Hence, by Theorem 5.1, it extends to a holomorphic function near $s = 1$. On the other hand, $L(s, 1)$ is the ζ function of F.

As in Corollary 5.1 one then shows $\zeta_{F(\zeta_m)}(s) = g(s) \prod_\chi L(s, \chi)$, where the product runs over the characters χ of H and $g(s)$ is holomorphic and non-vanishing around $s = 1$ (in fact, $g(s)$ is the constant 1, but we won't need this). We then deduce that $L(1, \chi) \neq 0$ for every nontrivial character, and the rest of the proof of Theorem 6.2 goes through. □

Next we prove the key case of Chebotarev's theorem, namely, that of cyclic extensions. We remark that the argument we give applies without change to any *abelian* extension (but of course this is still not enough, since the full Chebotarev theorem applies to arbitrary, possibly non-abelian, Galois extensions).

Proposition 7.2 *Let F be a number field and let K/F be a cyclic extension with group G. For every $\sigma \in G$, the set*

$$S_\sigma := \left\{ \mathfrak{p} \text{ prime of } \mathcal{O}_F \mid \left(\frac{K/F}{\mathfrak{p}}\right) = \sigma \right\}$$

has density equal to $\frac{1}{\#G}$.

Since the proof is long, we divide it into several lemmas and propositions. In what follows, we will tacitly exclude from any set of primes those that ramify in the relevant extensions: since we are going to argue about densities, and the number of ramified primes is always finite, this does not affect any of our arguments.

We begin with the following general setup. Let $n = [K : F] = |G|$ and put $N = n^k$, where k is any positive integer (we will eventually take the limit $k \to \infty$). By Dirichlet's theorem (Theorem 6.2), there exist infinitely many primes $q \equiv 1 \pmod{N}$. In particular, there exists such a prime q that is unramified in $K/\mathbb{Q}$. We will work with this auxiliary prime q and consider the extension $K(\zeta_q)$, which will be easier to treat, since it is a cyclotomic extension: see Proposition 7.1.

7.2 Algebraic Proof

Lemma 7.3 *We have $K \cap \mathbb{Q}(\zeta_q) = F \cap \mathbb{Q}(\zeta_q) = \mathbb{Q}$, hence the restriction homomorphisms*

$$\mathrm{res}_K : \mathrm{Gal}(K(\zeta_q)/K) \to \mathrm{Gal}(\mathbb{Q}(\zeta_q)/\mathbb{Q})$$

$$\mathrm{res}_F : \mathrm{Gal}(F(\zeta_q)/F) \to \mathrm{Gal}(\mathbb{Q}(\zeta_q)/\mathbb{Q})$$

are isomorphisms.

Proof The only prime ramified in $\mathbb{Q}(\zeta_q)$ is q, which is unramified in K. In particular, no prime ramifies both in K and in $\mathbb{Q}(\zeta_q)$, so $K \cap \mathbb{Q}(\zeta_q)$ is everywhere unramified. By Minkowski's theorem (Theorem 3.5), this implies $K \cap \mathbb{Q}(\zeta_q) = \mathbb{Q}$. As a consequence, also $F \cap \mathbb{Q}(\zeta_q) \subseteq K \cap \mathbb{Q}(\zeta_q)$ coincides with $\mathbb{Q}$. The last statement follows from basic Galois theory. □

Let $H = \mathrm{Gal}(F(\zeta_q)/F)$ and consider the following diagram of field extensions:

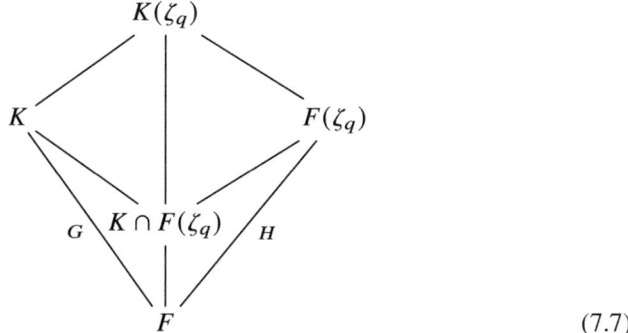

(7.7)

The restriction homomorphism

$$\mathrm{Gal}(K(\zeta_q)/K) \to \mathrm{Gal}(F(\zeta_q)/(K \cap F(\zeta_q)))$$

is injective (Galois theory), so

$$\begin{aligned}[F(\zeta_q) : K \cap F(\zeta_q)] &\geq [K(\zeta_q) : K] = \#\mathrm{Gal}(K(\zeta_q)/K) \\ &= \#\mathrm{Gal}(\mathbb{Q}(\zeta_m)/\mathbb{Q}) = \#\mathrm{Gal}(F(\zeta_q)/F) \\ &= [F(\zeta_q) : F],\end{aligned}$$

which implies $K \cap F(\zeta_q) = F$. (An alternative proof of the same equality can be obtained as follows: any prime of F of characteristic q is unramified in $K \cap F(\zeta_q)$, because this is a sub-extension of K. On the other hand, it is totally ramified in $K \cap F(\zeta_q)$, because this is a sub-extension of $F(\zeta_q)$. The intersection must therefore be F, since it is both unramified and totally ramified at the same prime.) We deduce that

$$\mathrm{Gal}(K(\zeta_q)/F) \cong \mathrm{Gal}(K/F) \times \mathrm{Gal}(F(\zeta_q)/F) = G \times H. \tag{7.8}$$

Notice that this implies in particular $[K(\zeta_q) : F(\zeta_q)] = |G|$, and in fact, more precisely, $\mathrm{Gal}(K(\zeta_q)/F(\zeta_q))$ is identified with $G \times \{\mathrm{id}\}$ under the isomorphism (7.8).

Remark 7.6 Before giving a formal proof of Proposition 7.2, we try to describe the basic idea. The point is that the set S_σ of primes of F with Frobenius σ in the extension K/F is the union of the sets $S_{\sigma,\tau}$, where for each $\tau \in H$ we write

$$S_{\sigma,\tau} := \{P \text{ prime of } O_F \mid \left(\frac{K(\zeta_q)/F}{P}\right) = (\sigma, \tau) \in G \times H\}. \tag{7.9}$$

The trick is that for most[1] (but not all) $\tau \in H$ one can compute the density of $S_{\sigma,\tau}$ using Proposition 7.1. Hence, by summing over these 'good' τ, we can at least estimate the density of $S_\sigma = \cup_\tau S_{\sigma,\tau}$. By choosing H appropriately (that is, by choosing q), the fraction of the elements of H for which we can compute the density of $S_{\sigma,\tau}$ tends to 1, and this will lead to the desired estimate for the density of S_σ.

Proposition 7.3 *Let $\tau \in H$ have order multiple of n. Then*

$$d_F^-(S_{\sigma,\tau}) = \frac{1}{[K(\zeta_q) : F]}.$$

Proof Let J be the cyclic subgroup of $G \times H$ generated by (σ, τ) and let L be the subfield of $K(\zeta_q)$ fixed by J. Let furthermore

$$T = \{P \text{ prime of } L \mid \left(\frac{K(\zeta_q)/L}{P}\right) = (\sigma, \tau)\}$$

We first prove that

$$d_L(T) = \frac{1}{[K(\zeta_q) : L]}. \tag{7.10}$$

By Proposition 7.1, it suffices to check that $K(\zeta_q)/L$ is a cyclotomic extension. We now show this.

Notice that $J \cap (G \times \{\mathrm{id}\}) = \{(\mathrm{id}, \mathrm{id})\}$: one has $(\sigma, \tau)^h = (\rho, \mathrm{id})$ for some $\rho \in G$ if and only if $\mathrm{ord}(\tau) \mid h$, but then $\rho = \sigma^h = \mathrm{id}$ since $n \mid \mathrm{ord}(\tau) \mid h$. Galois theory

---

[1] Specifically, for those τ of order multiple of n: see Proposition 7.3.

7.2 Algebraic Proof

then gives

$$K(\zeta_q) = K(\zeta_q)^{J\cap(G\times\{\mathrm{id}\})} = K(\zeta_q)^J K(\zeta_q)^{G\times\{\mathrm{id}\}} = LF(\zeta_q) = L(\zeta_q),$$

that is, $K(\zeta_q)$ is generated over L by a root of unity, as desired.

Finally, let

$$T' = \{P \text{ prime of } L \mid f(P \mid P \cap F) = 1 \text{ and } \left(\frac{K(\zeta_q)/L}{P}\right) = (\sigma, \tau)\}.$$

By Problem 6.8 we have $d_L(T) = d_L(T') = d_L^-(T')$. To finish the proof, we show that

$$d_L^-(T') = [L:F]d_F^-(S_{\sigma,\tau}). \tag{7.11}$$

It suffices to prove:

1. for every prime P in T', the prime $P \cap O_F$ of F is in $S_{\sigma,\tau}$;
2. for every prime $\mathfrak{p}$ in $S_{\sigma,\tau}$ there are precisely $[L:F]$ primes $P_1, \ldots, P_{[L:F]}$ in L lying over $\mathfrak{p}$, and each of those is in T' (in particular, $N(P_i) = N(\mathfrak{p})$, since by definition the primes in T' satisfy $f(P_i \mid \mathfrak{p}) = 1$).

Indeed, if we have these two properties, we can compute the lower density of $S_{\sigma,\tau}$ in F as

$$\mathrm{Dens}_F^-(S_{\sigma,\tau}) = \liminf_{s\to 1^+} \frac{\sum_{\mathfrak{p}\in S_{\sigma,\tau}} N(\mathfrak{p})}{\sum_{\mathfrak{p}\text{ nonzero prime of } O_F} N(\mathfrak{p})^{-s}}$$

$$= \frac{1}{[L:F]} \liminf_{s\to 1^+} \frac{\sum_{\mathfrak{p}\in S_{\sigma,\tau}}\sum_{P\mid\mathfrak{p}, P\in T'} N(\mathfrak{p})}{\log\left(\frac{1}{s-1}\right)}$$

$$= \frac{1}{[L:F]} \liminf_{s\to 1^+} \frac{\sum_{\mathfrak{p}\in S_{\sigma,\tau}}\sum_{P\mid\mathfrak{p}, P\in T'} N(P)}{\log\left(\frac{1}{s-1}\right)}$$

$$= \frac{1}{[L:F]} \liminf_{s\to 1^+} \frac{\sum_{P\in T'} N(P)}{\log\left(\frac{1}{s-1}\right)}$$

$$= \frac{1}{[L:F]} \mathrm{Dens}_L^-(T'),$$

as desired. Note that in the next-to-last equality we have used the fact that every prime of T' lies over a prime in $S_{\sigma,\tau}$, that is, (1). We now establish properties (1) and (2) above.

1. Let P be a prime in T'. Since the norm of $P \cap O_F$ is equal to the norm of P, the Artin symbols $\left(\frac{K(\zeta)/L}{P}\right) = (\sigma, \tau)$ and $\left(\frac{K(\zeta)/F}{P \cap O_F}\right)$ coincide. Hence, $P \cap O_F$ is in $S_{\sigma,\tau}$.
2. Conversely, let $\mathfrak{p}$ be in $S_{\sigma,\tau}$. Recall that $\mathfrak{p}$ is unramified in $K(\zeta)$ (by convention: we exclude the ramified primes) and let $\mathfrak{Q}$ be a prime of $K(\zeta)$ lying over $\mathfrak{p}$. The decomposition group $D(\mathfrak{Q} \mid \mathfrak{p})$ is by definition generated by $\left(\frac{K(\zeta)/F}{\mathfrak{p}}\right) = (\sigma, \tau)$. Hence, $K(\zeta)^{\langle(\sigma,\tau)\rangle} = L$ is by definition the decomposition field of $\mathfrak{p}$ (that is, $\mathfrak{p}$ is totally split in L), so that there are $r = [L : F]$ primes $P_1, \ldots, P_r$ of L lying over $\mathfrak{p}$. Each of these has degree 1 over $\mathfrak{p}$, and, as in (1), their Frobenius elements are given by (σ, τ).

Finally, combining Eqs. (7.10) and (7.11) we obtain

$$d_F^-(S_{\sigma,\tau}) = \frac{1}{[L:F]} d_L^-(T) = \frac{1}{[L:F]} \frac{1}{[K(\zeta):L]} = \frac{1}{[K(\zeta):F]},$$

as claimed. □

We are now ready to prove Proposition 7.2.

Proof With $S_{\sigma,\tau}$ as in Eq. (7.9) we have

$$S_\sigma = \bigsqcup_{\tau \in H} S_{\sigma,\tau}.$$

(Recall that we are excluding from all our sets the finitely many primes that ramify in $K(\zeta_q)$.) In particular,

$$d_F^-(S_\sigma) \geq \sum_{\tau \in H} d_F^-(S_{\sigma,\tau})$$

by Problem 7.3 and the fact that the sets $S_{\sigma,\tau}$ are clearly pairwise disjoint. We can further estimate the lower density of S_σ as

$$d_F^-(S_\sigma) \geq \sum_{\substack{\tau \in H \\ n \mid \mathrm{ord}(\tau)}} d_F^-(S_{\sigma,\tau}).$$

Using Proposition 7.3 we obtain

$$d_F^-(S_\sigma) \geq \sum_{\substack{\tau \in H \\ n \mid \mathrm{ord}\,\tau}} \frac{1}{[K(\zeta_q):F]} = \frac{\#\{\tau \in H : n \mid \mathrm{ord}\,\tau\}}{[K(\zeta_q):F]},$$

7.2 Algebraic Proof

hence, by Problem 7.2,

$$d_F^-(S_\sigma) \geq \frac{\#H}{[K(\zeta_q) : F]} \prod_{p_i \mid n} \left(1 - \frac{1}{p_i^{\alpha_i+1-\beta_i}}\right),$$

where $\alpha_i = v_p(\#H) = v_p(q-1)$ and $\beta_i = v_p(n)$. Since $n^k m \mid q - 1$, for every prime p that divides n we have $\alpha_i \geq k v_p(n)$, so $\alpha_i - \beta_i + 1 \geq (k-1)v_p(n) + 1 \geq k$. Thus, we have shown

$$d_F^-(S_\sigma) \geq \frac{\#H}{[K(\zeta_q) : F]} \cdot \prod_{p \mid n} \left(1 - \frac{1}{p^k}\right)$$

$$= \frac{\varphi(q)}{[K : F]\varphi(q)} \cdot \prod_{p \mid n} \left(1 - \frac{1}{p^k}\right)$$

$$= \frac{1}{[K : F]} \cdot \prod_{p \mid n} \left(1 - \frac{1}{p^k}\right)$$

for every $k \geq 1$. By passing to the limit $k \to \infty$, we get $d_F^-(S_\sigma) \geq \frac{1}{[K:F]}$.

In order to gain information about the *upper* density, it suffices to notice that the set of all primes of F (with finitely many exceptions, namely the primes that ramify in K) is the disjoint union of the sets $S_{\sigma'}$ for $\sigma' \in G$. This immediately implies

$$d_F^+(S_\sigma) \leq 1 - \sum_{\sigma' \neq \sigma} d_F^-(S_{\sigma'}) \leq 1 - \frac{[K : F] - 1}{[K : F]} = \frac{1}{[K : F]},$$

which concludes the proof. $\square$

Finally, we prove Chebotarev's theorem:

Proof of Theorem 7.1 We may and do assume that C itself is a conjugacy class. We can also replace S by its subset S' of places for which $N(\mathfrak{p})$ is a prime number (by Problem 6.8, this does not alter its density).

Choose an element $g \in C$ and let $E = K^{\langle g \rangle}$ be the field fixed by the subgroup $H = \langle g \rangle$. Consider the set

$$T_g = \{\mathfrak{q} \text{ prime of } E \mid \left(\frac{K/E}{\mathfrak{q}}\right) = g, N(\mathfrak{q}) \text{ is prime}\}.$$

Suppose that $\mathfrak{q}$ is in T_g: we claim that $\mathfrak{p} := \mathfrak{q} \cap O_F$ is in S'. Indeed, $N(\mathfrak{p})$ divides $N(\mathfrak{q})$, so $N(\mathfrak{p}) = N(\mathfrak{q})$ is a prime number. If $\mathfrak{Q}$ is a prime of K lying over $\mathfrak{q}$, this implies that $\left(\frac{K/F}{\mathfrak{Q}}\right) = \left(\frac{K/E}{\mathfrak{Q}}\right) = g \in C$, hence $\mathfrak{p} \in S'$. Moreover, we claim that $\mathfrak{Q}$ is the *unique* prime of K lying over $\mathfrak{q}$. To see this, notice that $D(\mathfrak{Q} \mid \mathfrak{q})$ is by definition generated by $\left(\frac{K/E}{\mathfrak{Q}}\right) = g$, so $D(\mathfrak{Q} \mid \mathfrak{q}) = H$: the whole Galois group of K over E sends $\mathfrak{Q}$ to itself, and therefore $\mathfrak{Q}$ is the only prime of K over $\mathfrak{q}$.

Conversely, given $\mathfrak{p} \in S'$, by definition there exists a prime $\mathfrak{Q}$ of K lying over $\mathfrak{p}$ with $\left(\frac{K/F}{\mathfrak{Q}}\right) \in C$. Replacing $\mathfrak{Q}$ by a conjugate if necessary, we can assume that $\left(\frac{K/F}{\mathfrak{Q}}\right) = g$. If we define $\mathfrak{q} = \mathfrak{Q} \cap E$, then $\mathfrak{q}$ lies over $\mathfrak{p}$ (obvious), and we claim that it is in T_g. To see this, notice that again we have

$$E = K^H = K^{\langle g \rangle} = K^{D(\mathfrak{Q}|\mathfrak{p})},$$

so E is the decomposition field of $\mathfrak{p}$. This means that $f(\mathfrak{q} \mid \mathfrak{p}) = 1$ and $f(\mathfrak{Q} \mid \mathfrak{q}) = |H| = \operatorname{ord}(g)$. In particular, $N(\mathfrak{q}) = N(\mathfrak{p})^{f(\mathfrak{q}|\mathfrak{p})}$ is prime, and as above it follows that $\left(\frac{K/F}{\mathfrak{Q}}\right) = \left(\frac{K/E}{\mathfrak{Q}}\right) = g \in C$. Hence $\mathfrak{q}$ is in T_g as claimed.

Summarising, there is a bijection between the primes in T_g lying over $\mathfrak{p} \in S'$ and the primes $\mathfrak{Q}$ of K that divide $\mathfrak{p}$ and satisfy $\left(\frac{K/F}{\mathfrak{Q}}\right) = g$.

Now let $Z_G(g)$ be the centraliser of g in G. By the orbit-stabiliser lemma, $|C| = \frac{|G|}{|Z_G(g)|}$. We count the primes in T_g lying over each prime $\mathfrak{p} \in S'$. By the previous paragraph, it suffices to count the primes $\mathfrak{Q}$ of K lying over $\mathfrak{p}$ with $\left(\frac{K/F}{\mathfrak{Q}}\right) = g$. We have already shown that there is at least one such prime, call it $\mathfrak{Q}_1$. Any other prime $\mathfrak{Q}'$ of K lying over $\mathfrak{p}$ is conjugate to $\mathfrak{Q}_1$ by an element $\sigma \in G$, say $\mathfrak{Q}' = \sigma \mathfrak{Q}_1$. Then, the Artin symbol $\left(\frac{K/F}{\mathfrak{Q}'}\right)$ is given by

$$\left(\frac{K/F}{\mathfrak{Q}'}\right) = \sigma \left(\frac{K/F}{\mathfrak{Q}_1}\right) \sigma^{-1} = \sigma g \sigma^{-1}.$$

Hence, $\left(\frac{K/F}{\mathfrak{Q}'}\right) = g$ if and only if $\sigma \in Z_G(g)$. By the orbit-stabiliser lemma again, the number of *distinct* primes $\mathfrak{Q}'$ with $\left(\frac{K/F}{\mathfrak{Q}'}\right) = g$ is

$$\frac{|Z_G(g)|}{|\operatorname{Stab}(\mathfrak{Q}_1) \cap Z_G(g)|} = \frac{|Z_G(g)|}{|D(\mathfrak{Q}_1 \mid \mathfrak{p}) \cap Z_G(g)|} = \frac{|Z_G(g)|}{|D(\mathfrak{Q}_1 \mid \mathfrak{p})|},$$

where we have used both the definition of $D(\mathfrak{Q}_1 \mid \mathfrak{p})$ and the fact that $D(\mathfrak{Q}_1 \mid \mathfrak{p}) = \langle g \rangle \subseteq Z_G(g)$. In conclusion, the number of primes of K lying over $\mathfrak{p}$ and having Artin symbol g is $\frac{|Z_G(g)|}{|D(\mathfrak{Q}_1|\mathfrak{p})|} = \frac{|G|}{|C| \cdot f}$, where $f = |D(\mathfrak{Q}_1 \mid \mathfrak{p})| = \operatorname{ord}(g) = |H|$. By what we already argued above, this is also the number of primes in T_g lying over $\mathfrak{p}$.

On the other hand, by Proposition 7.2, Chebotarev's theorem holds for the extension $E \subset K$, hence we have

$$\operatorname{Dens}_E(T_g) = \frac{1}{|H|} = \frac{1}{f}. \tag{7.12}$$

7.2 Algebraic Proof

Recall that we replaced S by its subset S' of primes whose residue field is a prime field. From the above discussion, noticing that by definition the norm of a prime $\mathfrak{p} \in S'$ is the same as the norm of any prime $\mathfrak{q} \in T_g$ lying over $\mathfrak{p}$, we obtain

$$\frac{|G|}{f|C|} \sum_{\mathfrak{p} \in S'} N(\mathfrak{p})^{-s} = \sum_{\mathfrak{q} \in T_g} N(\mathfrak{q})^{-s}.$$

Dividing by $\log\left(\frac{1}{s-1}\right)$ and passing to the limit for $s \to 1^+$, we obtain

$$\frac{|G|}{f|C|} \lim_{s \to 1^+} \frac{\sum_{\mathfrak{p} \in S'} N(\mathfrak{p})^{-s}}{\log\left(\frac{1}{s-1}\right)} = \lim_{s \to 1^+} \frac{\sum_{\mathfrak{q} \in T_g} N(\mathfrak{q})^{-s}}{\log\left(\frac{1}{s-1}\right)} = \mathrm{Dens}_E(T_g) = \frac{1}{f}.$$

This shows as desired that $\mathrm{Dens}_F(S')$ exists and equals $\frac{|C|}{|G|}$. $\square$

Remark 7.7 In its broad outline, the algebraic proof is not very different from the analytic one, at least in the sense that one reduces from a generic extension to a cyclic one (in the analytic proof, this is achieved by taking the cyclic subgroup of G generated by g; in the algebraic case, it is clear from the argument above that this is one of the main steps). On the other hand, once we are reduced to the cyclic case, the analytic proof—relying on the formalism of Artin L-functions—is much easier: one should compare the analytic argument for the cyclic case, which is essentially the same as for Dirichlet's theorem, to the (hard) proof of Proposition 7.2. The main reason why the argument using L-functions is simpler is that Hecke L-functions allow us to handle directly any abelian extension, while in the algebraic proof we further have to reduce to the cyclotomic case by introducing an artificial auxiliary extension.

Problems

7.1 Prove as many of the statements listed in Remark 7.2 as you can.

Indication You should avoid 3 (which is really hard) and 6 (which is quite hard). Only try 8 if you know what the Hilbert class field of a quadratic number field is. Everything else you should be able to prove; it is especially important to make sure you understand how statement 7 follows from Chebotarev's theorem.

7.2 Let G be a cyclic group of order $n = \prod p_i^{\alpha_i}$ and let $m = \prod p_i^{\beta_i}$ be a divisor of n (so that $0 \leq \beta_i \leq \alpha_i$ for every i). Prove that

$$\#\{g \in G : \text{ord}(g) \equiv 0 \pmod{m}\} = n \prod_{p_i \mid m} \left(1 - \frac{1}{p_i^{\alpha_i+1-\beta_i}}\right).$$

7.3 Let S_1, S_2 be disjoint sets of primes in a number field F. Prove that $d_F^-(S_1 \cup S_2) \geq d_F^-(S_1) + d_F^-(S_2)$.

7.4 Let K_1, K_2 be two number fields such that $\zeta_{K_1}(s) = \zeta_{K_2}(s)$. Prove that $[K_1 : \mathbb{Q}] = [K_2 : \mathbb{Q}]$. If at least one among K_1, K_2 is Galois over $\mathbb{Q}$, prove that $K_1 = K_2$.

7.5 Let K be a number field and write $\zeta_K(s) = \sum_{n \geq 1} \frac{a_n}{n^s}$. Suppose that, for all primes $p \equiv 1 \pmod 5$, the coefficient a_p is equal to 4. Prove that $K = \mathbb{Q}(\zeta_5)$.
Hint.

1. Prove that $[K : \mathbb{Q}] \geq 4$.
2. Prove that there exists a prime $p \equiv 1 \pmod 5$ that is totally split in K.
3. Deduce that $[K : \mathbb{Q}] = 4$.
4. Conclude by comparing the primes that split completely in K and $\mathbb{Q}(\zeta_5)$.

References

1. Hasse, H.: über die Dichte der Primzahlen p, für die eine vorgegebene ganzrationale Zahl $a \neq 0$ von durch eine vorgegebene Primzahl $\ell \neq 2$ teilbarer bzw. unteilbarer Ordnung mod p ist. Math. Ann. **162**, 74–76 (1965/1966). https://doi.org/10.1007/BF01361933
2. Hooley, C.: On Artin's conjecture. J. Reine Angew. Math. **225**, 209–220 (1967). https://doi.org/10.1515/crll.1967.225.209
3. Iwaniec, H., Kowalski, E.: Analytic number theory. American Mathematical Society Colloquium Publications, vol. 53. American Mathematical Society, Providence (2004). https://doi.org/10.1090/coll/053
4. Jones, R., Rouse, J.: Galois theory of iterated endomorphisms. Proc. Lond. Math. Soc. (3) **100**(3), 763–794 (2010). https://doi.org/10.1112/plms/pdp051. Appendix A by Jeffrey D. Achter
5. Lagarias, J.C., Odlyzko, A.M.: Effective versions of the Chebotarev density theorem. In: Algebraic Number Fields: L-Functions and Galois Properties (Proc. Sympos., Univ. Durham, Durham, 1975), pp. 409–464. Academic Press, London (1977)
6. Lombardo, D., Perucca, A.: Reductions of points on algebraic groups. J. Inst. Math. Jussieu **20**(5), 1637–1669 (2021). https://doi.org/10.1017/S1474748019000598
7. MathOverflow Contributors: Effective Chebotarev Without Artin's Conjecture. https://mathoverflow.net/questions/131543/effective-chebotarev-without-artins-conjecture, Accessed 11 Oct, 2024
8. Schoof, R.: Catalan's conjecture. Universitext. Springer-Verlag London, Ltd., London (2008)

Part II
Prerequisites for Tate's Thesis

Chapter 8
The Haar Measure

Abstract This chapter develops the theory of the Haar measure. We prove its existence for an arbitrary locally compact topological group and show uniqueness up to scalars in the abelian case.

In this chapter we develop the theory of the Haar measure, the essentially canonical choice of measure on a locally compact topological group. We review all the necessary definitions and show the existence of this special measure. We also prove uniqueness (up to constants) in the abelian case. Our method is a modification of that of Weil [4, Chapter 2], as described by Cohn [1].

8.1 Preliminaries

Definition 8.1 Let X be a Hausdorff topological space. The Borel σ-algebra on X, denoted $\mathcal{B}(X)$, is the σ-algebra generated by the open subsets of X. The elements of $\mathcal{B}(X)$ are called **Borel subsets** of X.

Definition 8.2 Let X be a Hausdorff topological space and let $\mathcal{A}$ be a σ-algebra on X such that $\mathcal{B}(X) \subseteq \mathcal{A}$. A measure μ on $\mathcal{A}$ is called **regular** (or **Radon**) if the following conditions all hold:

1. $\mu(K) < \infty$ for all compact subsets K of X;
2. $\mu(A) = \inf\{\mu(U) : U \text{ is open and } A \subseteq U\}$ for all $A \in \mathcal{A}$;
3. $\mu(U) = \sup\{\mu(K) : K \text{ is compact and } K \subseteq U\}$ for all open subsets U of X.

A measure satisfying condition (2) is called **outer regular**, and a measure satisfying condition (3) is called **inner regular**.

Definition 8.3 Let G be a group, let $a \in G$, and let A and B be subsets of G. We define the following sets.

1. $aB = \{ab : b \in B\}$.
2. $Ba = \{ba : b \in B\}$.

3. $AB = \{ab : a \in A, b \in B\}$.
4. $A^{-1} = \{a^{-1} : a \in A\}$.

A set A such that $A = A^{-1}$ is called **symmetric**.

Definition 8.4 A **topological group** is a group G endowed[1] with a topology τ such that the product $m : G \times G \to G$ and the inverse $i : G \to G$ are continuous with respect to $(\tau \times \tau, \tau)$ and (τ, τ), respectively. We will denote by e the identity of G.

An **isomorphism** of topological groups $\varphi : G_1 \to G_2$ is an isomorphism of groups that is also a homeomorphism of topological spaces. A **locally compact topological group** is a Hausdorff topological group G with the property that every point (equivalently, the identity) has an open neighbourhood with compact closure.

Definition 8.5 Let G be a topological group. The functions

$$L_a : G \to G$$
$$x \mapsto ax$$

and

$$R_a : G \to G$$
$$x \mapsto xa$$

are called respectively the **left translation by** a and the **right translation by** a.

Remark 8.1 For any $a \in G$, the functions L_a and R_a are homeomorphisms from G to itself. Note that the inversion map $\iota : G \to G$ is also a homeomorphism.

Lemma 8.1 *Let G be a topological group and fix $a \in G$.*

1. *If $\mathcal{U}$ is a fundamental system of neighbourhoods for the identity e of G, then the collections $\{aU : U \in \mathcal{U}\}$ and $\{Ua : U \in \mathcal{U}\}$ are fundamental systems of neighbourhoods for a.*
2. *If K and L are compact subsets of G, then the sets aK, Ka, K^{-1} and KL are compact.*

Proof

1. We have already observed that L_a, R_a are homeomorphisms, hence they carry fundamental systems of neighbourhoods to fundamental systems of neighbourhoods.
2. The sets in question are images of K (respectively, $K \times L$) under the continuous maps L_a, R_a, ι (respectively, $m : G \times G \to G$). □

Lemma 8.2 *Let G be a topological group and U be an open neighbourhood of e.*

---

[1] More precisely: the underlying set of G is endowed.

1. There exists an open neighbourhood V of e such that $VV \subseteq U$.
2. There exists a symmetric open neighbourhood V of e such that $V \subseteq U$.

Proof

1. The set $m^{-1}(U) \subseteq G \times G$ is open, hence it contains an open set of the form $V_1 \times V_2$ with V_1, V_2 open in G. Setting $V := V_1 \cap V_2$ we then have $VV = m(V, V) \subseteq m(V_1, V_2) \subseteq U$.
2. The intersection $V := U \cap U^{-1}$ is open and symmetric.

□

Definition 8.6 Let X be a topological space and let $f : X \to \mathbb{C}$ be a continuous function. The **support of** f, denoted $\mathrm{supp}(f)$, is the closure of the set $\{x \in X : f(x) \neq 0\}$. If X is Hausdorff and locally compact, we write $\mathcal{K}(X)$ for the $\mathbb{C}$-vector space of all continuous functions $f : X \to \mathbb{C}$ with compact support.

Remark 8.2 For every function $f \in \mathcal{K}(X)$, the real-valued function $|f|$ is bounded. Moreover, if μ is a regular Borel measure on X, then f is μ-integrable. Indeed, f is Borel-measurable because it is continuous, and since μ is regular, $\mu(\mathrm{supp}(f)) < \infty$, so $\int_X |f| \, d\mu = \int_{\mathrm{supp}(f)} |f| \, d\mu \leq \mu(\mathrm{supp}(f)) \cdot \|f\|_\infty$ is certainly finite.

Proposition 8.1 *Let G be a topological group, let K be a compact subset of G, and let U be an open subset of G containing K. There exist open neighbourhoods V_R and V_L of e such that $KV_R \subseteq U$ and $V_L K \subseteq U$.*

Proof We give the construction for V_R, that for V_L being virtually identical. For every $x \in K$ consider the open neighbourhood $x^{-1}U$ of e. By Lemma 8.2 (2), there exists an open neighbourhood V_x of e such that $V_x V_x \subseteq x^{-1}U$. The set $\{xV_x\}_{x \in K}$ is an open cover of K. Let $x_1 V_{x_1}, \ldots, x_n V_{x_n}$ be a finite subcover, and set $V_R := \bigcap_{i=1}^n V_{x_i}$. We claim that $KV_R \subseteq U$. Indeed, for any $k \in K$ there exists $i \in \{1, \ldots, n\}$ such that $k \in x_i V_{x_i}$, hence (using $V_R \subseteq V_{x_i}$) we obtain

$$kV_R \in x_i V_{x_i} V_{x_i} \subseteq x_i \left(x_i^{-1} U \right) = U.$$

□

8.2 Haar Measure: Existence

Definition 8.7 (Haar Measure) Let G be a locally compact group. A **left Haar measure** on G is a nonzero regular Borel measure μ on G that is invariant under left-translations, in the sense that $\mu(gA) = \mu(A)$ for all $g \in G$ and all $A \in \mathcal{B}(G)$. Right Haar measures are defined similarly.

Remark 8.3 Note that, since L_g is a homeomorphism, $L_g(A) = gA$ is a Borel set if and only if A is. In particular, $\mu(gA)$ makes sense.

Remark 8.4 If G is commutative, a measure μ is a left Haar measure if and only if it is a right Haar measure. In general, μ is a left Haar measure if and only if $A \mapsto \mu(A^{-1})$ is a right Haar measure. From now on, following a well-established tradition, we will only consider the case of left Haar measures, leaving the case of right-invariant measures as a simple exercise.

Example 8.1 (Some Basic Haar Measures) The following are examples of (left and right) Haar measures.

1. Let G be a finite group endowed with the discrete topology. The counting measure on G (that is, $\mu(A) = |A|$) is a Haar measure.
2. Let $G = (\mathbb{R}, +)$ with its usual topology. The restriction of the Lebesgue measure to the σ-algebra $\mathcal{B}(\mathbb{R})$ is a Haar measure.
3. Let $G = (\mathbb{R}^+, \cdot)$ and let $\mu = \frac{1}{x} dx$, where dx is the standard Lebesgue measure (restricted to the σ-algebra $\mathcal{B}(\mathbb{R})$). Then μ is a Haar measure on G: this follows from the change-of-variables formula in elementary integration theory, since for any $g \in \mathbb{R}^+$ we have

$$\mu(gA) = \int_{gA} 1 \, d\mu = \int_{gA} \frac{dx}{x} = \int_A \frac{d(gx)}{gx} = \int_A \frac{dx}{x} = \mu(A).$$

The main theorem in this section shows that every locally compact topological group carries a Haar measure:

Theorem 8.1 (Existence of the Haar Measure) *Let G be a locally compact topological group. There exists a left Haar measure μ for G.*

The proof will occupy the rest of this section. We start with four simple lemmas in topology.

Lemma 8.3 *Let X be a Hausdorff topological space, and let K and L be disjoint compact subsets of X. There exist disjoint open subsets U and V of X such that $K \subseteq U$ and $L \subseteq V$.*

Proof Easy exercise. □

Lemma 8.4 *Let X be a locally compact Hausdorff topological space. Let $x \in X$ and let U be an open neighbourhood of x. There exists an open neighbourhood V of x with compact closure that satisfies $\overline{V} \subseteq U$.*

Proof Since X is locally compact, there is an open neighbourhood W of x with compact closure. Replacing W with $W \cap U$ if necessary, we may assume that $W \subseteq U$ (note that $\overline{W \cap U}$ is closed in $\overline{W}$, hence it is a closed subset of a compact set, and therefore it is itself compact). Consider the sets $\{x\}$ and $\overline{W} \setminus W$, which are compact and disjoint. Lemma 8.3 shows the existence of disjoint open subsets V_1 and V_2 such that $\{x\} \subseteq V_1$ and $\overline{W} \setminus W \subseteq V_2$. The set $V_1 \cap W$ an open neighbourhood of x whose closure is compact (proven as above) and satisfies $\overline{V_1 \cap W} \subseteq W \subseteq U$ (indeed, $\overline{V_1 \cap W}$ does not meet $V_2 = \overline{W} \setminus W$). □

8.2 Haar Measure: Existence

Corollary 8.1 *Let X be a locally compact Hausdorff topological space, let K be a compact subset of X, and let U be an open subset of X containing K. There exists an open subset V of X such that $\overline{V}$ is compact and satisfies $K \subseteq V \subseteq \overline{V} \subseteq U$.*

Proof For each point of K find a neighbourhood V_x as in the previous lemma. The union of the V_x covers K: extract a finite cover $V_{x_1}, \ldots, V_{x_n}$ and set $V = \cup V_{x_i}$. □

Lemma 8.5 *Let X be a locally compact Hausdorff topological space, let K be a compact subset of X, and let U_1 and U_2 be open subsets of X such that $K \subseteq U_1 \cup U_2$. There exist compact sets K_1 and K_2 such that $K = K_1 \cup K_2$, $K_1 \subseteq U_1$ and $K_2 \subseteq U_2$.*

Proof Let U, V be the open sets obtained from applying Lemma 8.3 to the disjoint compact sets $K \setminus U_1$ and $K \setminus U_2$. We can take $K_1 = K \setminus V$ and $K_2 = K \setminus U$: these are closed in K, hence compact, and $K_1 \subseteq K \setminus (K \cap U_2) \subseteq U_1$ (and similarly for K_2). Finally, their union is all of K since $U_1 \cap U_2 = \emptyset$. □

We now have all the ingredients to construct a Haar measure on any locally compact group. The key idea is that the 'size' of a subset X of G should be measured as follows: one takes a 'small' neighbourhood of the identity U and counts how many translates of U are necessary to cover X. In order to make sense of this, one should first work with *compact* subsets X (which ensures that finitely many translates of U suffice), and of course we also need to somehow normalise this counting (morally, we would like to divide by the size of U itself). Since we do not know how to assign a measure to U yet, we declare that its measure is inversely proportional to the number of translates needed to cover a *fixed* reference compact set K_0. Once we have a notion of size for compact sets, the rest of the construction is standard: we first obtain an outer measure on all Borel subsets of G and finally check that this is in fact already a measure. We now start putting this strategy in practice.

Remark 8.5 The construction sketched above is reminiscent of the Hausdorff measure in $\mathbb{R}^n$, which assigns a size to sets by counting how many balls of radius ε are needed to cover it, in the limit $\varepsilon \to 0$. The added difficulty we face here is that we don't know what volume to assign to (the analogue of) a ball of radius ε, because for general G there is no such simple fundamental system of neighbourhoods of the identity as in the case of $\mathbb{R}^n$.

Proof of Theorem 8.1 We start by defining a notion of 'index' $(K : V)$, whenever K is compact and V has non-empty interior V°: we set

$$(K : V) = \min\{n \in \mathbb{N} : \text{there exists a cover of } K \text{ by } n \text{ translates of } V^\circ\}.$$

It is clear that the definition is well-posed: $\{xV^\circ\}_{x \in G}$ is an open cover of K, hence we can extract a finite one. Notice that $(K : V) = 0$ if and only if $K = \emptyset$.

Now, using the assumption of local compactness, fix a neighbourhood of the identity with compact closure K_0. This is our 'reference compact set'; note that K_0° is non-empty, hence we can also measure the index $(K : K_0)$ for any compact set K.

Let C be the collection of all compact subsets of G and let $\mathcal{U}$ be the collection of all open neighbourhoods of the identity. For each $U \in \mathcal{U}$ we define a function

$$h_U : C \to \mathbb{R}$$
$$K \mapsto \frac{(K:U)}{(K_0:U)}.$$

The next lemma gives some basic properties of the functions h_U. □

Lemma 8.6 *Fix $U \in \mathcal{U}$, K, K_1, $K_2 \in C$ and $x \in G$. The following hold.*

1. $0 \leq h_U(K) \leq (K : K_0)$.
2. $h_U(K_0) = 1$.
3. $h_U(xK) = h_U(K)$.
4. *If $K_1 \subseteq K_2$, then $h_U(K_1) \leq h_U(K_2)$.*
5. $h_U(K_1 \cup K_2) \leq h_U(K_1) + h_U(K_2)$.
6. *If $K_1 U^{-1} \cap K_2 U^{-1} = \emptyset$, then $h_U(K_1 \cup K_2) = h_U(K_1) + h_U(K_2)$.*

Proof Parts (2) and (4) are clear, as is the lower bound in (1). For the upper bound in (1), let $\{x_i K_0^\circ\}_{i=1,\ldots,(K:K_0)}$ be an open cover of K with translates of K_0° and let $\{y_j U\}_{j=1,\ldots,(K_0:U)}$ be an open cover of K_0 with translates of U. Then, we have

$$K \subseteq \bigcup_{i=1}^{(K:K_0)} x_i K_0^\circ \subseteq \bigcup_{i=1}^{(K:K_0)} x_i K_0 \subseteq \bigcup_{i=1}^{(K:K_0)} x_i \left(\bigcup_{j=1}^{(K_0:U)} y_j U \right) = \bigcup_{i,j} x_i y_j U,$$

hence K can be covered by at most $(K : K_0)(K_0 : U)$ translates of U. This gives $(K : U) \leq (K : K_0)(K_0 : U)$, which is equivalent to the desired upper bound.

For (3), observe that translating every element of an open cover of K by x gives an open cover of xK with the same number of elements. For (5), use that the union of an open cover of K_1 and an open cover of K_2, consisting respectively of $(K_1 : U)$ and $(K_2 : U)$ translates of U, gives an open cover of $K_1 \cup K_2$ consisting of $(K_1 : U) + (K_2 : U)$ translates of U.

Finally, for (6), we need to show the inequality $h_U(K_1 \cup K_2) \geq h_U(K_1) + h_U(K_2)$, or equivalently, $(K_1 \cup K_2 : U) \geq (K_1 : U) + (K_2 : U)$. Let $\{xU\}$ be an open cover of $K_1 \cup K_2$. We claim that each set xU in the cover meets at most one of K_1, K_2: if we had $xU \cap K_1 \neq \emptyset$ and $xU \cap K_2 \neq \emptyset$, then x would be both in $K_1 U^{-1}$ and in $K_2 U^{-1}$, contradicting the assumption $K_1 U^{-1} \cap K_2 U^{-1} = \emptyset$. Hence, from $\{xU\}$ we can extract two disjoint subsets, one covering K_1 and the other covering K_2. The desired inequality follows immediately. □

We now construct a suitable limit of the functions h_U, arguing by compactness. Let

$$X := \prod_{K \in C} [0, (K : K_0)]$$

8.2 Haar Measure: Existence

and note that X, being a product of non-empty compact intervals of $\mathbb{R}$, is a non-empty compact space. We consider X as a subset of $\mathbb{R}^C$, the space of functions from C to $\mathbb{R}$. Lemma 8.6 shows that each h_U is an element of X.

We now construct the desired 'limit' of the functions h_U. For each open neighbourhood V of e, let $S(V)$ be the closure in X of the set $\{h_U : U \in \mathcal{U}, U \subseteq V\}$. If $V_1, \ldots, V_n$ are in $\mathcal{U}$ and V is their intersection, then $h_V \in \bigcap_{i=1}^n S(V_i)$. This implies that any finite intersection of sets $S(V_i)$ is nonempty; by compactness, $\bigcap_{V \in \mathcal{U}} S(V)$ is also non-empty. Let $h_\circ$ be an element of this intersection. The next lemma is the analogue of Lemma 8.6 for $h_\circ$.

Lemma 8.7 *Fix K, K_1, $K_2 \in C$ and $x \in G$. The following hold:*

1. $0 \leq h_\circ(K)$.
2. $h_\circ(\emptyset) = 0$.
3. $h_\circ(K_0) = 1$.
4. $h_\circ(xK) = h_\circ(K)$.
5. *If $K_1 \subseteq K_2$, then $h_\circ(K_1) \leq h_\circ(K_2)$.*
6. $h_\circ(K_1 \cup K_2) \leq h_\circ(K_1) + h_\circ(K_2)$.
7. *If $K_1 \cap K_2 = \emptyset$, then $h_\circ(K_1 \cup K_2) = h_\circ(K_1) + h_\circ(K_2)$.*

Proof The key point is that, by the definition of the product topology, for any fixed $K \in C$, the projection

$$X \to \mathbb{R}$$
$$h \mapsto h(K)$$

is continuous. This implies easily all statements (1) through (6): for example, for (6), fix K_1, K_2 and consider the function

$$X \to \mathbb{R}$$
$$h \mapsto h(K_1) + h(K_2) - h(K_1 \cup K_2).$$

As already argued, this function is continuous. By Lemma 8.6 (5) it is non-negative on every h_U, hence (by continuity) on every $S(V)$, and thus, in particular, on $h_\circ$.

Part (7) requires some more care. By Lemma 8.3, there exist open sets U_1, U_2 with $K_1 \subseteq U_1, K_2 \subseteq U_2$ and $U_1 \cap U_2 = \emptyset$. Proposition 8.1 gives two open neighbourhoods V_1, V_2 of e such that $K_i V_i \subseteq U_i$ for $i = 1, 2$. Letting $V = V_1 \cap V_2$, we have $K_1 U^{-1} \cap K_2 U^{-1} = \emptyset$ for any $U \subseteq V^{-1}$, and therefore $h_U(K_1 \cup K_2) = h_U(K_1) + h_U(K_2)$ for any such U (Lemma 8.6 (6)). By continuity, the same condition holds for all $h \in S(V^{-1})$, and in particular for $h_\circ$. □

The function $h_\circ$ is essentially the desired Haar measure. We now carry out the necessary measure-theoretic verifications. We define first an outer measure on open sets by setting

$$\mu^* : \mathcal{V} \to [0, \infty] \qquad (8.1)$$
$$U \mapsto \sup\{h_\circ(K) : K \subseteq U, K \in C\},$$

where $\mathcal{V}$ is the collection of all open sets in G. It is immediate to check that, if U is both open and compact, then $\mu^*(U) = h_\circ(U)$. Next, we extend μ^* to all subsets of G by setting

$$\begin{aligned} \mu^* : \mathcal{P}(G) &\to [0, \infty] \\ A &\mapsto \inf\{\mu^*(U) : A \subseteq U, U \in \mathcal{V}\}. \end{aligned} \quad (8.2)$$

Lemma 8.7 implies that μ^* is monotonic, non-negative, and satisfies $\mu^*(\emptyset) = 0$. It remains to check that μ^* is countably sub-additive and that all Borel sets are measurable.

It follows easily from the definition (8.2) that it suffices to check the sub-additivity on open sets. Let $\{U_i\}_{i \geq 1}$ be a countable collection of open sets in G and let $K \subseteq \bigcup_i U_i$ be a compact set. Since the U_i form an open cover of the compact set K, there is an index n such that $K \subseteq \bigcup_{i=1}^n U_i$. By Lemma 8.5 and a straightforward induction, we obtain compact sets $K_1, \ldots, K_n$ such that $K_i \subseteq U_i$ for $i = 1, \ldots, n$ and $K = K_1 \cup \cdots \cup K_n$. By Equation (8.1) we have $\mu^*(K_i) \leq \mu^*(U_i)$. Combining this with Lemma 8.7 (6) we obtain

$$\mu^*(K) \leq \sum_{i=1}^n \mu^*(K_i) \leq \sum_{i=1}^n \mu^*(U_i) \leq \sum_{i=1}^\infty \mu^*(U_i).$$

Since this holds for all compact sets $K \subseteq \bigcup_i U_i$, taking the supremum in K we get $\mu^*\left(\bigcup_i U_i\right) \leq \sum_{i=1}^\infty \mu^*(U_i)$, as desired.

We now turn to the measurability of Borel sets. Since the measurable sets form a σ-algebra, and since by definition $\mathcal{B}(G)$ is the σ-algebra generated by open sets, it suffices to show that open sets are measurable. Recall that, by definition, a set X is μ^*-measurable if and only if, for all subsets Y of G, we have

$$\mu^*(Y) = \mu^*(Y \cap X) + \mu^*(Y \setminus X). \quad (8.3)$$

We claim that it is enough to show this when Y is itself an open set. Indeed, suppose that the equality

$$\mu^*(V) = \mu^*(V \cap X) + \mu^*(V \setminus X) \quad (8.4)$$

holds for all open sets V. Consider the infimum of the above expression over all open sets V that contain Y. We have

1. $\inf_{V \supseteq Y} \mu^*(V) = \mu^*(Y)$, by definition;
2. $\inf_{V \supseteq Y} \mu^*(V \cap X) \geq \mu^*(Y \cap X)$: to see this, recall that we are assuming X to be open, so the intersection $V \cap X$ is open and contains $Y \cap X$. Thus, the left-hand side of the previous equality is the infimum of $\mu^*(U)$ over *certain* open subsets U containing $Y \cap X$. Thus, the left-hand side is at least as large as the right-hand side (which is the infimum of $\mu^*(U)$ over *all* open subsets U containing $Y \cap X$).

8.2 Haar Measure: Existence

3. $\inf_{V \supseteq Y} \mu^*(V \setminus X) = \mu^*(Y \setminus X)$: to see this, write the definition of both sides as

$$\inf_{\substack{V \supseteq Y \\ U_1 \text{ open} \\ U_1 \supseteq V \setminus X}} \mu^*(U_1) = \inf_{U_2 \supseteq Y \setminus X} \mu^*(U_2).$$

Given a set U_1 as on the left-hand side, this is in particular an open that contains $Y \setminus X$, hence it also appears in the infimum on the right-hand side. This proves that the LHS is greater than or equal to the LHS. Conversely, let U_2 be an open set that appears in the infimum on the right-hand side. Then $V = U_2 \cup X$ is an open set containing $(Y \setminus X) \cup X \supseteq Y$, and $U_1 = U_2$ is an open set containing $V \setminus X = (U_2 \cup X) \setminus X = U_2 \setminus X$. Thus, U_2 also appears in the infimum on the left-hand side, which establishes the opposite inequality.

Thus, if (8.4) holds for all open sets X and V, taking the infimum as above we obtain

$$\mu^*(Y) \geq \mu^*(Y \cap X) + \mu^*(Y \setminus X)$$

for all sets Y. On the other hand, the opposite inequality is true by sub-additivity (which we have already shown), hence we have obtained (8.3) for all open sets X and for all sets Y.

Hence, to finish the proof that Borel sets are measurable, it suffices to prove that if U, V are open sets with $\mu^*(V) < \infty$ we have

$$\mu^*(V) \geq \mu^*(V \cap U) + \mu^*(V \cap U^c).$$

Choose a compact subset K of $V \cap U$ with $h_\circ(K) \geq \mu^*(V \cap U) - \varepsilon$ and a compact subset L of $V \cap K^c$ with $h_\circ(L) \geq \mu^*(V \cap K^c) - \varepsilon$. Clearly K and L are disjoint and $V \cap U^c \subseteq V \cap K^c$. Since μ^* is monotonic, we have

$$h_\circ(L) \geq \mu^*(V \cap K^c) - \varepsilon \geq \mu^*(V \cap U^c) - \varepsilon$$

and from Lemma 8.7 (7) we obtain

$$h_\circ(K \cup L) = h_\circ(K) + h_\circ(L) \geq \big(\mu^*(V \cap U) - \varepsilon\big) + \big(\mu^*(V \cap U^c) - \varepsilon\big)$$
$$= \mu^*(V \cap U) + \mu^*(V \cap U^c) - 2\varepsilon.$$

Since $K \cup L \subseteq V$ we have obtained

$$\mu^*(V) \geq h_\circ(K \cup L) \geq \mu^*(V \cap U) + \mu^*(V \cap U^c) - 2\varepsilon$$

for all ε. Taking the limit $\varepsilon \to 0$ finishes the proof that V is μ^*-measurable. In particular, the restriction of μ^* to the Borel σ-algebra is a measure μ. By property (4) in Lemma 8.7, the measure μ satisfies $\mu(xA) = \mu(A)$ for every Borel set A and every $x \in G$. Furthermore, it follows from Lemma 8.7 (3) and the definition of

μ^* that μ is nonzero. To show that μ is the desired Haar measure it now suffices to check that it is regular (see Definition 8.2).

If K a compact set and U is an open set containing K, then by definition $h_\circ(K) \le \mu^*(U) = \mu(U)$. Taking the infimum over all U we obtain

$$h_\circ(K) \le \mu(K). \tag{8.5}$$

Suppose now that V is open with compact closure. Then for every compact subset L of V we have $h_\circ(L) \le h_\circ(\overline{V})$ by Lemma 8.7 (5). Taking the supremum over L we obtain that $\mu(V) \le h_\circ(\overline{V})$ is finite (notice that $h_\circ$ is finite on compact sets by definition). If K is an arbitrary compact subset of G, then by Corollary 8.1 (applied to K and $U = G$) there exists an open set V containing K whose closure is compact. Thus, by monotonicity we have that $\mu(K) \le \mu(V) \le h_\circ(\overline{V})$ is finite. We have checked property (1) in the definition of a regular measure. Outer regularity (property (2) in the definition) is an immediate consequence of the definition in Eq. (8.2), while inner regularity follows from Eqs. (8.1) and (8.5). □

We conclude this section by highlighting the key property of Haar measures from the standpoint of integration:

Theorem 8.2 (Invariance of the Haar Integral Under Translation) *Let G be a locally compact group and μ be a left Haar measure on G. Let furthermore f be a μ-integrable function on G and fix $x \in G$. We have*

$$\int_G f(t)\, d\mu(t) = \int_G f(x^{-1}t)\, d\mu(t).$$

Proof When f is the characteristic function of a (measurable) set A, the function $f(x^{-1}t)$ is the characteristic function of xA, so we have

$$\int_G f(t)\, d\mu(t) = \mu(A) = \mu(xA) = \int_G f(x^{-1}t)\, d\mu(t),$$

as desired. By linearity, the claim holds for all simple functions, and by passing to the limit we obtain it for all measurable functions. □

8.3 Haar Measure: Uniqueness (up to Constants)

In the previous section we have shown that every locally compact group carries at least one (left) Haar measure. It can be shown that this measure is in fact essentially unique:

Theorem 8.3 (Uniqueness of the Haar Measure up to Constants) *Let G be a locally compact group and let μ_1, μ_2 be two left Haar measures on G. There exists $c \in \mathbb{R}, c > 0$, such that $\mu_2 = c\mu_1$.*

8.3 Haar Measure: Uniqueness (up to Constants)

It wouldn't be too hard to prove this theorem in general, but since we only need it for the case of abelian groups, we limit ourselves to giving a (shorter and easier) proof for this special case. For the general case, the reader can refer to [2, pp. 16–19].

Proof in the Commutative Case Let K be a compact set with non-empty interior U (which exists, because by assumption G is locally compact). We first claim that $\mu_1(K)$ and $\mu_2(K)$ are non-zero. Indeed, suppose by contradiction that $\mu_j(K) = 0$ (for $j = 1$ or $j = 2$). Any compact set K' can be covered with finitely many translates $x_1 U, \ldots, x_n U$ of the interior U of K. Using the translation-invariance of Haar measures we then have

$$\mu_j(K') \leq \sum_{i=1}^{n} \mu_j(x_i U) = \sum_{i=1}^{n} \mu_j(U) \leq n\mu_j(K) = 0,$$

so every compact set has measure 0. By regularity, this implies that μ_j is identically 0, contradiction.

Take now $g = \mathbf{1}_{K^{-1}}$ to be the characteristic function of K^{-1} and $f = \mathbf{1}_V$ to be the characteristic function of any open set V with compact closure. Using the Fubini-Tonelli theorem together with Theorem 8.2 we compute

$$\int_G f(x)\, d\mu_1(x) \cdot \int_G g(y)\, d\mu_2(y) = \int_G \int_G f(x)g(y)\, d\mu_1(x)\, d\mu_2(y)$$

$$= \int_G \int_G f(xy)g(y)\, d\mu_1(x)\, d\mu_2(y)$$

$$\stackrel{y \mapsto x^{-1}y = yx^{-1}}{=} \int_G \int_G f(y)g(yx^{-1})$$

$$\times d\mu_1(x)\, d\mu_2(y)$$

$$= \int_G f(y) \left(\int_G g(yx^{-1})\, d\mu_1(x) \right) d\mu_2(y)$$

$$= \int_G f(y) \left(\int_G g(x^{-1})\, d\mu_1(x) \right) d\mu_2(y)$$

$$= \left(\int_G g(x^{-1})\, d\mu_1(x) \right) \int_G f(y)\, d\mu_2(y).$$

Rearranging, we have obtained

$$\int_G f(y)\, d\mu_2(y) = \frac{\int_G g(y)\, d\mu_2(y)}{\int_G g(x^{-1})\, d\mu_1(x)} \cdot \int_G f(x)\, d\mu_1(x),$$

where the ratio $c := \frac{\int_G g(y)\,d\mu_2(y)}{\int_G g(x^{-1})\,d\mu_1(x)}$ is independent of f (recall that we chose $g(x)$ so that the denominator is strictly positive). We have thus obtained

$$\mu_2(V) = \int_G f\,d\mu_2 = c\int_G f\,d\mu_1 = c\mu_1(V)$$

for all open sets V with compact closure. By outer regularity, this easily implies that μ_2 and $c\mu_1$ coincide on all compact sets; and by inner regularity, this finally implies $\mu_2 = c\mu_1$.

□

Definition 8.8 Let G be a locally compact group. We denote by $L^1(G)$ the $\mathbb{C}$-vector space of functions $f : G \to \mathbb{C}$ such that $\int_G |f|\,d\mu < \infty$, where μ is any left Haar measure on G. By Theorem 8.3, this definition is independent of the choice of the Haar measure.

We conclude this section by stating a form of the Riesz(-Markov-Kakutani) representation theorem. It will later help us obtain Haar measures on the multiplicative group of a field.

Theorem 8.4 (Riesz–Markov–Kakutani, [3, Theorem 2.14]) *Let X be a locally compact Hausdorff space. For any positive linear functional ψ on $\mathcal{K}(X)$, there is a unique Radon measure μ on X such that*

$$\psi(f) = \int_X f(x)\,d\mu(x) \qquad \forall f \in \mathcal{K}(X).$$

Problems

8.1 Let G be a locally compact topological group and let μ_G be a fixed choice of left-invariant Haar measure on G.

1. Prove that, for every $g \in G$, the map

$$A \mapsto \mu_G(Ag)$$

   induces a left-invariant Haar measure $\mu_{G,g}$.
2. Using the uniqueness statement, deduce that $\mu_{G,g} = \Delta(g)\mu_G$ for some $\Delta(g) \in \mathbb{R}^\times$.
3. Prove that $g \mapsto \Delta(g)$ is a homomorphism from G to $\mathbb{R}_{>0}$.
4. One can show that Δ (called the *modular function*) is continuous. Deduce that, if G is compact, any left-invariant Haar measure is also right-invariant.

8.2 Let G be a locally compact group that is homeomorphic to an open subset U of $\mathbb{R}^n$. Fix a homeomorphism $\varphi : G \to U$ and suppose that, under this identification, for every $g \in G$ the left-translation L_g is an affine map, that is, there exist $M_g \in \text{Mat}_{n \times n}(\mathbb{R})$ and $v_g \in \mathbb{R}^n$ such that

$$\varphi(L_g(h)) = M_g \varphi(h) + v_g \quad \forall h \in G.$$

Prove that $\varphi_*^{-1}\left(|\det M_g|^{-n} d\mathcal{L}\right)$, restricted to the Borel σ-algebra, is a Haar measure on G, where $\mathcal{L}$ denotes the usual Lebesgue measure on $\mathbb{R}^n$.

8.3 Compute $\int_{\mathbb{Z}_p} |x|_p \, dx$, where dx denotes the Haar measure on $\mathbb{Z}_p$.

References

1. Cohn, D.L.: Measure Theory, 2nd edn. Birkhäuser Advanced Texts: Basler Lehrbücher. [Birkhäuser Advanced Texts: Basel Textbooks]. Birkhäuser/Springer, New York (2013). https://doi.org/10.1007/978-1-4614-6956-8
2. Ramakrishnan, D., Valenza, R.J.: Fourier Analysis on Number Fields, Graduate Texts in Mathematics, vol. 186. Springer, New York (1999). https://doi.org/10.1007/978-1-4757-3085-2
3. Rudin, W.: Real and complex analysis, 3rd edn. McGraw-Hill Book Co., New York (1987)
4. Weil, A.: L'intégration dans les groupes topologiques et ses applications, Actualités Scientifiques et Industrielles [Current Scientific and Industrial Topics], vol. 869. Hermann & Cie, Paris (1940). [This book has been republished by the author at Princeton, N.J., 1941.]

Chapter 9
Abstract Fourier Analysis

Abstract In this chapter we briefly review the main tools of abstract Fourier analysis. We do not provide proofs and rely on the analogy with the finite case to guide the reader's intuition.

Our purpose in this chapter is to generalise Theorem 6.1 to an arbitrary locally compact abelian group. We will not give proofs: the statements should all look familiar from Sect. 6.1, but the detailed arguments get complicated. The interested reader may refer to any of the following sources: the original paper by Cartan and Godement [1]; Chapter 4 of [2]; Chapter 3 of [3] (Chapter 2 of the same book covers the relevant spectral theory prerequisites).

We mention right at the start that, when working with a general topological group G, one can consider both

- its **characters**, that is, the continuous homomorphisms $G \to \mathbb{S}^1 = \{z \in \mathbb{C} : |z| = 1\}$;
- its **quasi-characters**, that is, the continuous homomorphisms $G \to \mathbb{C}^\times$.

Note that the two notions coincide when G is finite, which is why we didn't have to worry about the distinction in Sect. 6.1. While in this section we are mostly concerned with the actual *characters* of a group, the more general notion of quasi-character will be central in Tate's thesis.

9.1 Pontryagin Duality: General Case

We start by stating Pontryagin duality in general (see Proposition 6.1 for the finite case). We omit the proof, for which the reader can refer to [3, Proposition 3-2 and Theorem 3-20].

Definition 9.1 (Topological Dual Group) Let G be a locally compact abelian group. We denote by

$$\hat{G} = \mathrm{Hom}_{\mathrm{cont}}(G, \mathbb{S}^1)$$

the group of continuous homomorphisms from G to $\mathbb{S}^1$. We endow $\hat{G}$ with the **compact-open** topology, defined as follows. For every compact neighbourhood K of id_G in G and every open neighbourhood V of $1 \in \mathbb{S}^1$, write

$$U(K, V) = \{\chi \in \hat{G} : \chi(K) \subseteq V\}.$$

The (compact-open) topology on $\hat{G}$ is by definition the topology having the sets $\{U(K, V)\}_{K,V}$ as a basis of neighbourhoods of the trivial character. We extend it to a topology on $\hat{G}$ by using the group structure as usual (that is, a basis of neighbourhoods around a general element $\chi \in \hat{G}$ is $\{\chi U(K, V)\}_{K,V}$). The topological group thus obtained is called the **Pontryagin dual** of G.

Theorem 9.1 (Pontryagin Duality) *For every locally compact abelian group G let $\hat{G}$ be the Pontryagin dual group as in Definition 9.1.*

1. *$\hat{G}$ is also a locally compact abelian group.*
2. *The map*

$$\Psi : G \to \hat{\hat{G}}$$
$$g \mapsto \Psi_g,$$

where Ψ_g is given by

$$\Psi_g : \hat{G} \to \mathbb{C}^\times$$
$$\chi \mapsto \chi(g),$$

is an isomorphism of topological groups.
3. *G is compact if and only if $\hat{G}$ is discrete.*

Remark 9.1 Every finite group G is a topological group when equipped with the discrete topology. Moreover, in this case, the compact-open topology on $\hat{G}$ is also the discrete topology. Thus, applying Theorem 9.1 to the case of finite groups recovers Proposition 6.1.

We also have the following analogue of Proposition 6.2 (see [2, Proposition 4.39, Theorem 4.40] for a proof):

Proposition 9.1 (Functoriality of $G \mapsto \hat{G}$, Locally Compact Case) *Let G be a locally compact abelian group. The following hold:*

1. *The association $G \mapsto \hat{G}$ can be extended to a contravariant functor from the category of locally compact abelian groups to itself by letting it act on arrows as*

9.1 Pontryagin Duality: General Case

follows: if $f : G \to H$ is a continuous group homomorphism between locally compact abelian groups, we define

$$\hat{f} : \hat{H} \to \hat{G}$$
$$\chi \mapsto \chi \circ f.$$

2. *This functor is exact: for every short exact sequence[1]*

$$0 \to H \xrightarrow{\iota} G \xrightarrow{\pi} G/H \to 0,$$

the dual sequence

$$0 \to \widehat{G/H} \xrightarrow{\hat{\pi}} \hat{G} \xrightarrow{\hat{\iota}} \hat{H} \to 0$$

is also exact in the category of locally compact abelian groups. In particular, $\widehat{G/H}$ is closed in $\hat{G}$, and can be identified with the subgroup

$$H^\perp = \{\chi \in \hat{G} : \chi|_H = 1\},$$

which is itself a closed subgroup of $\hat{G}$.

We also mention the following analogue of Proposition 6.3 for compact groups:

Proposition 9.2 *Let G be a compact abelian group and let $\chi \in \hat{G}$ be a character. We have*

$$\int_G \chi(g)\,dg = \begin{cases} 0, & \text{if } \chi \neq \text{id}_{\hat{G}} \\ \mu(G), & \text{otherwise} \end{cases}$$

Proof Exactly as in the finite case, if χ is nontrivial, there exists $a \in G$ such that $\chi(a) \neq 1$. Using the translation-invariance property of the Haar measure we obtain

$$\chi(a) \int_G \chi(g)\,dg = \int_G \chi(ag)\,dg = \int_G \chi(g)\,dg,$$

hence $\int_G \chi(g)\,dg = 0$. □

---

[1] In the category of locally compact abelian groups: in particular, H is assumed to be closed in G.

9.2 The Abstract Fourier Transform

Let G be a locally compact abelian group and let μ_G be a choice of Haar measure on G. (If G is compact, we can normalise our choices by taking as μ_G the unique *normalised* Haar measure, but this is not important for the discussion of this section.)

Recall from Definition 8.8 the $\mathbb{C}$-vector space $L^1(G)$ of complex-valued integrable functions on G. We now introduce a notion of Fourier transform in this generality, which will generalise both the usual notion of Fourier transform encountered in real and complex analysis and the Fourier transform of Definition 6.3.

Definition 9.2 (Abstract Fourier Transform) Let G be a locally compact topological group with a fixed choice μ_G of Haar measure and let $f \in L^1(G)$. We define the **(abstract) Fourier transform** of f as

$$\hat{f} : \hat{G} \to \mathbb{C}$$
$$\chi \mapsto \int_G f(g)\overline{\chi(g)}\, d\mu_G(g).$$

Remark 9.2 Note that the integral makes sense: we have

$$\int_G |f(g)\overline{\chi(g)}|\, d\mu_G(g) = \int_G |f(g)|\, d\mu_G(g) < \infty$$

since χ takes values in $\mathbb{S}^1$.

It will be useful to single out a class of well-behaved functions:

Definition 9.3 We denote by $\mathfrak{B}^1(G)$ the $\mathbb{C}$-vector space of functions $f : G \to \mathbb{C}$ that satisfy the following three conditions:

1. f is continuous;
2. f is in $L^1(G)$;
3. $\hat{f}$ is in $L^1(\hat{G})$.

Remark 9.3 In the previous definition, $L^1(G), L^1(\hat{G})$ are defined with respect to any choice of Haar measures on $G, \hat{G}$: since any two Haar measures only differ by a constant, the spaces $L^1(G), L^1(\hat{G})$ are independent of this choice.

The main theorem of abstract Fourier analysis can be stated as follows (for a proof see [2, Theorem 4.33] or [3, Theorem 3–9]; notice that in our statement the function f is assumed to be continuous, so that the point-wise equality in the statement makes sense).

Theorem 9.2 (Fourier Inversion for the Abstract Fourier Transform) *Let G be a locally compact topological group with a fixed choice μ_G of Haar measure. There is a unique Haar measure $\mu_{\hat{G}}$ on $\hat{G}$ such that the following holds: for all functions*

9.2 The Abstract Fourier Transform

$f \in \mathfrak{V}^1(G)$ we have

$$f(g) = \int_{\hat{G}} \hat{f}(\chi)\chi(g)\, d\mu_{\hat{G}}(\chi) \qquad \forall g \in G.$$

Definition 9.4 (Dual Measure) In the context of Theorem 9.2, we will say that $\mu_{\hat{G}}$ is the measure on $\hat{G}$ **dual** to the given Haar measure on G.

Remark 9.4 A crucial feature of Theorem 9.2 is the fact that the Haar measure $\mu_{\hat{G}}$ is *independent of the function* f. Notice that Haar measures are determined up to a constant, and therefore, one can determine $\mu_{\hat{G}}$ in the following way. Let μ be *any* Haar measure on $\hat{G}$: then $\mu_{\hat{G}} = \alpha\mu$ for some nonzero α. The inversion formula yields

$$f(g) = \alpha \int_{\hat{G}} \hat{f}(\chi)\chi(g)\, d\mu(\chi). \tag{9.1}$$

In particular, if one can compute $\int_{\hat{G}} \hat{f}(\chi)\chi(g)\, d\mu(\chi)$ for even a *single* function f and a *single* $g \in G$, the previous equation uniquely determines α (provided that $f(g) \neq 0$) and hence $\mu_{\hat{G}}$.

Example 9.1 (Recovering the Finite Case) Let G be a finite abelian group. We endow G with the counting measure, which is a Haar measure since it is obviously invariant under translation by any element of G. Notice that with this choice of Haar measure on G the Fourier transform coincides with that of Definition 6.3.

What is the measure $\mu_{\hat{G}}$ appearing in Theorem 9.2? Following Remark 9.4, we take $f = \mathbf{1}_e$, the characteristic function of the singleton $\{e\}$, where e is the identity of G. The Fourier transform is

$$\hat{f}(\chi) = \int_G f(g)\overline{\chi(g)}\, d\mu_G = \sum_{g \in G} \delta_{g,e}\overline{\chi(g)} = \overline{\chi(e)} = 1,$$

that is, the constant function 1. The dual measure $\mu_{\hat{G}}$ is some multiple α of the counting measure on $\hat{G}$. With reference to Eq. (9.1), we take μ to be the counting measure on $\hat{G}$, $f = \mathbf{1}_e$ and $g = e$. With these choices we obtain

$$1 = f(e) = \alpha \sum_{\chi \in \hat{G}} \hat{f}(\chi)\chi(e) = \alpha \sum_{\chi \in \hat{G}} 1 = \alpha \cdot |G|,$$

which yields $\alpha = \frac{1}{|G|}$. Thus, $\mu_{\hat{G}}$ is $\frac{1}{|G|}$ times the counting measure, and we recover Theorem 6.1.

References

1. Cartan, H., Godement, R.: Théorie de la dualité et analyse harmonique dans les groupes abéliens localement compacts. Ann. Sci. École Norm. Sup. (3) **64**, 79–99 (1947)
2. Folland, G.B.: A Course in Abstract Harmonic Analysis, 2nd edn. Textbooks in Mathematics. CRC Press, Boca Raton, FL (2016)
3. Ramakrishnan, D., Valenza, R.J.: Fourier Analysis on Number Fields, Graduate Texts in Mathematics, vol. 186. Springer, New York (1999). URL https://doi.org/10.1007/978-1-4757-3085-2

Chapter 10
Review of Local Fields

Abstract In this chapter we review the main properties of *local fields*, which for our applications we take to mean completions of number fields with respect to a non-Archimedean metric. We also discuss the relations between certain local and global invariants in extensions of local fields (resp. number fields).

In this chapter, we give a quick review of the basics of the theory of completions of number fields. The standard reference on the general theory of local fields is Serre's classical book [2].

Definition 10.1 (Norm on a Number Field, Place) Let K be a number field. A (multiplicative) **norm** on K is a function

$$d : K \to \mathbb{R}_{\geq 0}$$
$$x \to |x|$$

that satisfies the following:

1. $|xy| = |x| \cdot |y|$;
2. $|x + y| \leq |x| + |y|$;
3. $|x| = 0$ if and only if $x = 0$.

The norm is called **non-Archimedean** if it further satisfies $|x + y| \leq \max\{|x|, |y|\}$. Every norm induces a distance, hence a topology, on K. Two norms $|\cdot|_1, |\cdot|_2$ on a number field are called **equivalent** if they induce the same topology on K. An equivalence class of norms is called a **place** of K.

Remark 10.1 Since equivalent norms induce *the same* topology on K by definition, each place induces a topology on K.

The classification of places of a number field is known as Ostrowski's theorem. Before stating it, we describe a way to obtain a norm on a number field starting from a prime of its ring of integers.

Definition 10.2 (p-Adic Norm) Let K be a number field and let $\mathfrak{p}$ be a prime ideal of the ring of integers O_K. Let q be the size of the residue class field $\mathbb{F}_{\mathfrak{p}} := O_K/\mathfrak{p}$.

1. A **uniformiser** at $\mathfrak{p}$ is an element $\pi \in O_K$ such that the factorisation of (π) is of the form $\mathfrak{p}I$, with $(\mathfrak{p}, I) = (1)$. Equivalently, it is an element in $\mathfrak{p} \setminus \mathfrak{p}^2$ (the equivalence, and the fact that $\mathfrak{p} \setminus \mathfrak{p}^2$ is non-empty, follows from unique factorisation in ideals, Theorem 3.1).
2. The $\mathfrak{p}$-adic valuation $v_{\mathfrak{p}}$ on K is the function
$$v_{\mathfrak{p}} : K^{\times} \to \mathbb{Z}$$
$$x \mapsto v_{\mathfrak{p}}(x)$$
defined as follows. Given $x \in K^{\times}$, there exists a unique integer n such that x/π^n is in $O_K^{\times}$: we set $n = v_{\mathfrak{p}}(x)$. We further set, conventionally, $v_{\mathfrak{p}}(0) = \infty$.
3. The $\mathfrak{p}$-adic norm on K is then obtained by setting
$$\|x\|_{\mathfrak{p}} := q^{-v_{\mathfrak{p}}(x)},$$
with the natural convention $\|0\|_{\mathfrak{p}} = q^{-\infty} = 0$.

Theorem 10.1 (Ostrowski) *Let K be a number field. Denote by $\sigma_1, \ldots, \sigma_{n_1}$, $\tau_1, \overline{\tau_1}, \ldots, \tau_{n_2}, \overline{\tau_{n_2}}$ the embeddings of K in $\mathbb{C}$, as in Sect. 3.6 (in particular, the image of each σ_i is contained in $\mathbb{R}$, while the image of each τ_i is not). The following is a complete list of non-equivalent norms on K (that is, a complete list of places of K):*

1. *$\|x\|_{\sigma_i} := |\sigma_i(x)|$, for $i = 1, \ldots, n_1$, where $|\cdot|$ is the standard norm on $\mathbb{R}$; the corresponding places are called **real**;*
2. *$\|x\|_{\tau_j} := |\tau_j(x)|^2$, for $j = 1, \ldots, n_2$, where $|\cdot|$ is the standard Euclidean norm on $\mathbb{C}$; the corresponding places are called **complex**;*
3. *$\|x\|_{\mathfrak{p}}$ for each non-zero prime ideal $\mathfrak{p}$ of the ring of integers O_K; the corresponding places are called **finite**, and are precisely the non-Archimedean ones.*

Places that are not finite are called **Archimedean** or **infinite**. Each place is an *equivalence class* of norms. For each place, we will consistently take the representative given in the above statement of Ostrowski's theorem, with the normalisation of Definition 10.2. (We chose a normalisation when we set $\|\pi\|_{\mathfrak{p}} = q^{-1}$. One can replace q with any other real number greater than 1 and obtain an equivalent norm, but our choice has several technical advantages.)

One further piece of notation:

Definition 10.3 We will write Ω_K for the set of all places of K and Ω_K^{∞} for the subset of infinite places, that is, the Archimedean ones. As already mentioned in Theorem 10.1, a place is called *finite* if it is non-Archimedean, that is, if it lies in $\Omega_K \setminus \Omega_K^{\infty}$.

We will usually denote a place by v, or, if it comes from a prime $\mathfrak{p}$ of O_K, simply by $\mathfrak{p}$. By a slight abuse of notation, we will write $\|x\|_v$ for the norm of $x \in K$, as measured by our standard norm which represents the place v. When v is a finite place, corresponding to the prime ideal $\mathfrak{p}$ of O_K, we denote by q_v the size of the residue field $O_K/\mathfrak{p}$ and by p_v its characteristic.

Example 10.1 (Places of $\mathbb{Q}$) The places $\Omega_\mathbb{Q}$ of $\mathbb{Q}$ are in bijection with $\{p : p \text{ prime}\} \cup \{\infty\}$, with ∞ conventionally representing the place coming from the obvious embedding $\mathbb{Q} \hookrightarrow \mathbb{R}$. For every non-zero rational number $x = \frac{a}{b}$ and every prime p, write $x = p^{n_p} \frac{a'_p}{b'_p}$, where n_p is a (positive or negative) integer and $(a'_p, p) = (b'_p, p) = 1$. The different norms are then given by

$$\|x\|_p = p^{-n_p}, \quad \|x\|_\infty = |x|.$$

In Problem 10.1 you will be asked to show that $\prod_{v \in \Omega_\mathbb{Q}} \|x\|_v = 1$ for every $x \in \mathbb{Q}^\times$. This is no coincidence:

Theorem 10.2 (General Product Formula) *Let K be a number field. The equality*

$$\prod_{v \in \Omega_K} \|x\|_v = 1$$

holds for every $x \in K^\times$.

We will prove this result below in slightly different language, see Theorem 13.6.

Every place induces a corresponding completion of the number field:

Definition 10.4 (Completion of a Number Field) Let K be a number field and let v be a place of K. We denote by K_v the completion of K with respect to the topology induced by v. It is a topological field, that is, the operations $+ : K \times K \to K$, $\cdot : K \times K \to K$ and $^{-1} : K^\times \to K^\times$ are continuous. We will usually refer to K_v as the **completion of K at v**.

Remark 10.2 Let $\|\cdot\|$ be a norm corresponding to v. By general facts in topology, $\|\cdot\|$ extends to a norm on K_v (which we still denote by the same symbol) and makes K_v into a complete metric space.

Completions of number fields can be described fairly explicitly:

1. when v is a real place, the completion is isomorphic to $\mathbb{R}$;
2. when v is a complex place, the completion is isomorphic to $\mathbb{C}$;
3. when v is a finite place of characteristic p, the completion is a finite extension of the field $\mathbb{Q}_p$, the completion of $\mathbb{Q}$ with respect to the p-adic metric. Such fields are called *p*-**adic fields**.

In general, the completion of a number field at a (finite or infinite) place is called a **local field**. (There is a more general definition of local fields, but we will not need it here.)

The field $\mathbb{Q}_p$ of p-adic numbers, and more generally its finite extensions, have been studied extensively, and much is known about their structure. Here, we will limit ourselves to mentioning some of their fundamental properties, starting with the fact that, for every finite extension L of $\mathbb{Q}_p$, there exists a number field K and a place v of K of characteristic p such that L is isomorphic to K_v as a topological field.

Theorem 10.3 *Let L be a finite extension of $\mathbb{Q}_p$ (equivalently: let L be the completion of some number field K at a place v of characteristic p). Let $\|\cdot\|$ be the norm on L (see Remark 10.2 in case L is obtained as the completion of a number field). The following hold:*

1. $O_L = \{x \in L : \|x\| \leq 1\}$ *is a subring of L, called the **ring of integers**. When $L = \mathbb{Q}_p$, the ring of integers O_L is often denoted by $\mathbb{Z}_p$;*
2. $O_L^\times = \{x \in L : \|x\| = 1\}$ *is its group of units;*
3. O_L *is a local ring; its maximal ideal $\mathfrak{m}$ is principal, and any generator of $\mathfrak{m}$ is called a **uniformiser** π of L;*
4. *every ideal of O_L is a power of $\mathfrak{m}$; in particular, every element x of $L^\times$ can be written as $x = u \cdot \pi^n$ for some $n \in \mathbb{Z}$ and $u \in O_L^\times$; the integer n is called the **valuation** of x;*
5. *the quotient $\kappa_L := O_L/\mathfrak{m}$ is a finite field, of cardinality p^f; the integer f is called the **inertia degree** of L over $\mathbb{Q}_p$;*
6. *the ideal $(p)O_L$ is of the form (π^e); the integer e is called the **ramification index** of L over $\mathbb{Q}_p$;*
7. *let L be obtained as the completion of a number field K at a finite place $\mathfrak{p}$. Write $(p)O_K = \mathfrak{p}^{e'} I$, with $(\mathfrak{p}, I) = (1)$. The ramification index and inertia degree of L over $\mathbb{Q}_p$ coincide with the ramification index and inertia degree of $\mathfrak{p}$ over p.*

As in the case of number fields, if K is a p-adic field and I is a non-zero ideal of O_K, we denote by $N(I)$ the norm of I, that is, the cardinality of the quotient O_K/I. The ring O_K is a discrete valuation ring, and every ideal I of O_K is a power of the maximal ideal $\mathfrak{m} = (\pi)$. We define the π-adic valuation $v_\pi(I)$ of an ideal I as the unique integer n such that $I = \mathfrak{m}^n = (\pi^n)$. We have in particular the p-adic valuation v_p for the ideals of $\mathbb{Z}_p$.

We also mention the fact that, once the completions of $\mathbb{Q}$ have been constructed, the completions of an arbitrary number field can also be described in the following, more algebraic terms.

Theorem 10.4 (Completions and Tensor Products) *Let K be a number field of signature (n_1, n_2), and let $\mathfrak{p}_1^{e_1} \cdots \mathfrak{p}_r^{e_r}$ be the factorisation in O_K of the ideal (p), where p is a prime of $\mathbb{Z}$.*

1. *The tensor product $K \otimes_\mathbb{Q} \mathbb{R}$ is isomorphic (as a topological ring) to the product $\mathbb{R}^{n_1} \times \mathbb{C}^{n_2}$.*
2. *The tensor product $K \otimes_\mathbb{Q} \mathbb{Q}_p$ is isomorphic (as a topological ring) to the product $\prod_{i=1}^r K_{\mathfrak{p}_i}$. The field $K_{\mathfrak{p}_i}$ is a finite extension of $\mathbb{Q}_p$ of degree $e_i f_i$, where $e_i :=$*

$e(\mathfrak{p}_i \mid p)$ and $f_i := f(\mathfrak{p}_i \mid p)$. *Moreover, the ramification index and inertia degree of $K_{\mathfrak{p}_i}$ over $\mathbb{Q}_p$ are given by e_i and f_i, respectively.*

We will also make use of some fundamental topological properties of the completions K_v. The most important one is of course their completeness, which is true by construction. Another fact (well known in the real and complex case, and easy to prove in the p-adic setting) that we will need is the following:

Proposition 10.1 *Let K_v be a completion of a number field and let X be a subset of K_v. The topological closure $\overline{X}$ of X in K_v is compact if and only if X is bounded with respect to the norm on K_v. In particular, K_v is locally compact, and so is $K_v^\times$.*

In Part III we will need the notion of **different** of an extension of p-adic fields (in particular when the ground field is $\mathbb{Q}_p$ itself).

Definition 10.5 (Different) *Let $L/K/\mathbb{Q}_p$ be finite extensions of p-adic fields. The **different** of L over K is the fractional ideal of L given by*

$$\mathfrak{d}_{L/K}^{-1} = \{x \in L : \mathrm{tr}_{L/K}(xy) \in O_K \ \forall y \in O_L\}.$$

The **absolute different** of L, denoted simply by $\mathfrak{d}_L$, is the different of the extension $L/\mathbb{Q}_p$.

Remark 10.3 Notice that $\mathfrak{d}_{L/K}^{-1}$ contains the ring of integers O_L; it follows that its inverse $\mathfrak{d}_{L/K}$ is contained in O_L and is therefore an integral ideal.

It is not hard to show the following result:

Theorem 10.5 *Let L/K be a finite extension of p-adic fields. Define the **(relative) discriminant** of L over K as in the number field case, that is,*

$$d_{L/K} = \left(\det\left(\sigma_i(\alpha_j)\right)^2\right),$$

where the σ_i, for $i = 1, \ldots, [L:K]$ are the embeddings of L into $\overline{K}$ that fix K, and the α_j, for $j = 1, \ldots, [L:K]$, are a basis of O_L over O_K. Notice however that $d_{L/K}$ is only an integral ideal of O_K and not a well-defined element of O_K.

The ideal $(d_{L/K})$ is equal to $\left(N_{L/K}(\mathfrak{d}_{L/K})\right)$, the ideal generated by the norm from L to K of the different of L over K.

Using Theorem 10.4, the study of extensions of number fields can be reduced to a large extent to the local case. For example, one has the following:

Theorem 10.6 *Let K be a number field. For every prime p we have $K \otimes \mathbb{Q}_p = \prod_{\mathfrak{p} \mid p} K_\mathfrak{p}$, where $\mathfrak{p}$ ranges over the primes of O_K of characteristic p. We have*

$$(d_K) = \prod_p \prod_{\mathfrak{p} \mid p} N(d_{K_\mathfrak{p}/\mathbb{Q}_p}) :$$

the global discriminant is the product of the local ones.

Remark 10.4 Note that the equality in the previous theorem should be interpreted purely as an equality of ideals, not of numbers. In other words, for every prime p, the p-adic valuation of d_K is equal to $\sum_{\mathfrak{p}|p} v_p(N(d_{K_\mathfrak{p}/\mathbb{Q}_p}))$.

Since the different and the discriminant can be defined in terms of traces, Theorem 10.6 is related to the following statement (see e.g. [1, Proposition II.8.3 and Corollary II.8.4] for a proof):

Theorem 10.7 *Let K be a number field and p be a prime number. Write $K \otimes \mathbb{Q}_p$ as the direct product $\prod_{\mathfrak{p}|p} K_\mathfrak{p}$, where $\mathfrak{p}$ ranges over the primes of $\mathcal{O}_K$ of characteristic p. We have*

$$\mathrm{tr}_{K/\mathbb{Q}}(x) = \sum_{\mathfrak{p}|p} \mathrm{tr}_{K_\mathfrak{p}/\mathbb{Q}_p}(x)$$

for all $x \in K$. Similarly, for $p = \infty$ write $K \otimes \mathbb{R}$ as the direct product $\prod_{v \text{ infinite place}} K_v$, where v ranges over the infinite places of K. We have

$$\mathrm{tr}_{K/\mathbb{Q}}(x) = \sum_{v \text{ infinite place}} \mathrm{tr}_{K_v/\mathbb{R}}(x).$$

Problems

10.1 (Product Formula for $K = \mathbb{Q}$) Check that $\prod_{v \in \Omega_\mathbb{Q}} \|x\|_v = 1$ for every $x \in \mathbb{Q}^\times$.

Hint. Write $x = a/b$ with $(a, b) = 1$ and describe $\|x\|_v$ when v is a prime dividing a or b.

10.2 Let K be a (number) field and let $\|\cdot\|$ be a norm on K.

1. Prove that the subspace topology on $K^\times$ coincides with the subspace topology induced on $K^\times$ by the topology of $K \times K$, where $K^\times$ is embedded in $K \times K$ via $x \mapsto (x, x^{-1})$.
2. Prove that $x \mapsto x^{-1}$ is continuous.

10.3 Let R be the ring $\mathbb{Q}$ equipped with the topology for which a basis of open neighbourhoods of $q \in \mathbb{Q}$ is given by $\{q + m\mathbb{Z}\}_{m \in \mathbb{Z}_{>0}}$. Prove that R is a topological ring (that is, the operations $+$, $-$ and $\cdot$ are continuous), but $^{-1} : R^\times \to R^\times$ is not continuous for the subspace topology.

10.4 Prove Proposition 10.1.

10.5 Prove the case $p = \infty$ of Theorem 10.7.

Hint. Using properties of the trace, reduce to the Galois case. Treat this case explicitly.

10.6 Compute the different ideal of the extensions $\mathbb{Q}_2(\sqrt{17})/\mathbb{Q}_2$, $\mathbb{Q}_2(i)/\mathbb{Q}_2$, and $\mathbb{Q}_2(\sqrt{2})/\mathbb{Q}_2$.

References

1. Neukirch, J.: Algebraic Number Theory, Grundlehren der mathematischen Wissenschaften [Fundamental Principles of Mathematical Sciences], vol. 322. Springer, Berlin (1999). https://doi.org/10.1007/978-3-662-03983-0. Translated from the 1992 German original and with a note by Norbert Schappacher, With a foreword by G. Harder
2. Serre, J.P.: Local Fields, Graduate Texts in Mathematics, vol. 67. Springer, New York (1979). Translated from the French by Marvin Jay Greenberg

Chapter 11
Restricted Direct Products

Abstract In this chapter we develop the theory of restricted direct products of groups. We discuss the problem from several points of view: abstract group theory, topology, measure theory, and Fourier analysis.

In this chapter we discuss the notion of **restricted direct product** from several points of view: abstract group theory, topology, measure theory, Pontryagin duality, and Fourier analysis. Given a set of indices I, we say that a property holds for *almost all* $i \in I$ if it holds for all but finitely many i.

11.1 Abstract Group Theory

We start by introducing a general definition of **restricted product**:

Definition 11.1 (Restricted Product of Groups) Let $(G_i)_{i \in I}$ be a collection of groups and let $(H_i)_{i \in I}$ be a collection of subgroups, with $H_i < G_i$ for each i. The **restricted product of the groups** G_i **with respect to the subgroups** H_i is the subset of $\prod_{i \in I} G_i$ given by

$$\{(x_i) \in \prod_i G_i : \text{there exists a finite subset } S \subseteq I \text{ such that } x_i \in H_i \text{ for all } i \notin S\}.$$

Note that the subset S in the previous definition can depend on $(x_i)_{i \in I}$. The restricted product is often denoted by $\prod'_{i \in I}(G_i, H_i)$, $\prod_{i \in I}(G_i, H_i)$, or simply $\prod'_{i \in I} G_i$ if the H_i are clear from the context.

Remark 11.1 Note that the definition makes sense even if, for i in some finite subset S_0 of indices, the group H_i is not defined: indeed, we may always take S (as in the definition above) to contain S_0, and therefore, we don't need to know anything about the groups H_i for $i \in S_0$.

There is also an obvious variant of this definition where the G_i are replaced by rings R_i and the H_i by subrings:

Definition 11.2 (Restricted Product of Rings) Let $(R_i)_{i \in I}$ be a collection of rings and let $(T_i)_{i \in I}$ be a collection of subrings, with $T_i \subseteq R_i$ for every i. The **restricted product of the rings R_i with respect to the subrings T_i** is the subset of $\prod_{i \in I} R_i$ given by

$$\{(x_i) \in \prod_i R_i : \text{there exists a finite subset } S \subseteq I \text{ such that } x_i \in T_i \text{ for all } i \notin S\}.$$

For our applications, by far the most interesting examples of restricted products will be the group of idèles and the ring of adèles. We now introduce these objects, which we will then study in more detail in the next subsections. Let k be a number field and let Ω_k be the collection of its places. For each $v \in \Omega_k$, we can consider:

1. the completion k_v and its non-zero elements $k_v^\times$;
2. the "integers" O_v, which are defined in the usual way if v is a finite place, and as $O_v := k_v$ if v is an Archimedean place;
3. the "units" u_v, which are defined in all cases as $O_v^\times$.

Both the *ring* of adèles and the *group* of idèles are suitable restricted products:

Definition 11.3 (Adèles and Idèles) We call

$$\mathbb{A}_k := \prod{}' (k_v, O_v)$$

the **ring of adèles** of k, and

$$I_k := \prod{}' (k_v^\times, O_v^\times)$$

the **group of idèles** of k. We also define the **ring of finite adèles** by

$$\mathbb{A}_k^f := \prod_{v \text{ finite}}{}' (k_v, O_v).$$

Problem 11.2 describes the essential fact that I_k is the unit group of $\mathbb{A}_k$. The reader who is not familiar with this fact should definitely try to prove it.

11.2 Topological Groups

The definitions of the last section already cover the basic algebraic properties of adèles and idèles. However, matters become more complicated—and interesting—when we want to equip $\mathbb{A}_k$ and I_k with a topology. To this end, we give yet another definition of restricted product, this time in a topological setting.

11.2 Topological Groups

Definition 11.4 (Restricted Product of Locally Compact Abelian Groups) Let I be a set of indices and let $\{G_i\}_{i \in I}$ be a collection of locally compact abelian groups. Suppose that, for almost all $i \in I$ (meaning 'for all but a finite number of elements of I'), we are also given a subgroup $H_i \subset G_i$ which is open and compact. We define the topological group $G := \prod'_i (G_i, H_i)$ as follows:

- as a group, it is the restricted product of Definition 11.1 (with the interpretation of Remark 11.1);
- a fundamental system of neighbourhoods of 1 in G is given by the sets $\prod_{i \in I} N_i$, where each N_i is a neighbourhood of 1 in G_i for all i and $N_i = H_i$ for almost all i. We then define the topology as usual: U is an open neighbourhood of $g \in G$ if and only if $g^{-1}U$ is an open neighbourhood of $1 \in G$.

Remark 11.2

1. Naturally, we can and will identify each G_{i_0} to a subgroup of the restricted product (specifically, the subgroup of elements $x = (x_i)$ for which $x_i = 1$ for all $i \neq i_0$). The natural map is an isomorphism of topological groups.
2. Let S be a finite set of indices, including those for which H_i is not defined. The set $G_S := \{x = (x_i) \in G : x_i \in H_i \quad \forall i \notin S\}$ is a subgroup of G, and is an open neighbourhood of 1.
3. Let G_S be as above. Then, $G_S \cong \prod_{i \in S} G_i \times \prod_{i \notin S} H_i$ is a direct product of locally compact groups, almost all of which are compact. It follows that G_S is locally compact, hence (since G_S is a neighbourhood of 1 in G) that G is also locally compact.
4. By definition, $G = \bigcup_{\substack{S \subseteq I \\ S \text{ finite}}} G_S$.

The groups G_S, being direct products, are already easier to analyse than the restricted product G. An even simpler class of subgroups is given in the next definition:

Definition 11.5 Notation as in Definition 11.4. Let S be a finite subset of I. We denote by $G^S \subset G_S$ the subgroup of those elements $x = (x_i) \in G$ such that $x_i = 1$ for $i \in S$ and $x_i \in H_i$ for $i \notin S$.

A useful property of the topology on a restricted direct product is given in the following lemma:

Lemma 11.1 *Notation as in Definition 11.4. A subset $C \subseteq G$ is relatively compact (that is, has compact closure) if and only if it is contained in a product $\prod_{i \in I} B_i$ where each B_i is a compact subset of the corresponding G_i and $B_i = H_i$ for almost all i (recall that the H_i are compact by definition). Moreover, every compact subset of G is contained in some G_S.*

Proof Let K be a compact subset of G: we claim that there exists S such that $K \subseteq G_S$. To see this, recall that we have already observed that the groups G_S cover G, so $K \subseteq \bigcup G_S$, and by compactness we have $K \subseteq G_{S_1} \cup \cdots \cup G_{S_r}$ (recall that the G_{S_i} are open). Notice furthermore that $G_{S_1} \cup \cdots \cup G_{S_r} \subseteq G_{S_1 \cup \cdots \cup S_r}$, so—

setting $S = \bigcup S_i$—we have that K is a subset of $G_S = \prod_{i \in S} G_i \times \prod_{i \notin S} H_i$. This shows the last statement in the lemma. Finally, let $K_i := \pi_i(K)$ for $i \in S$: then K_i is compact (continuous image of a compact set), and by construction we have $K \subseteq \prod_{i \in S} K_i \times \prod_{i \notin S} H_i$, as desired.

Conversely, Tychonoff's theorem implies the compactness of any product $\prod_i B_i$ (where each B_i is compact and $B_i = H_i$ for almost all i); any such set is contained in some G_S, hence it is a compact subset of G. Hence, if K is contained in $\prod_i B_i$ with the B_i as above, its closure is contained in a compact set and is therefore compact. $\square$

11.3 (Quasi-)Characters of a Restricted Product

Let $G = \prod'(G_i, H_i)$ be a restricted direct product in the sense of Definition 11.4. We now wish to study the quasi-characters of G, that is, the continuous homomorphisms from G to $\mathbb{C}^\times$. Given a $c : G \to \mathbb{C}^\times$, we denote by c_i its restriction to G_i, that is,

$$c_i : G_i \to \mathbb{C}^\times$$
$$x_i \mapsto c((1, 1, \ldots, 1, x_i, 1, \ldots)).$$

It is clear that c_i is a homomorphism $G_i \to \mathbb{C}^\times$. The next two lemmas show how to factor any continuous homomorphism $c : G \to \mathbb{C}^\times$ as a product of c_i.

Lemma 11.2 *The homomorphism c_i is trivial on H_i for almost all i, and for every $x \in G$ we have*

$$c(x) = \prod_{i \in I} c_i(x_i),$$

where almost all of the factors of the product are equal to 1.

Proof Let U be a neighbourhood of 1 in the complex plane that contains no non-trivial multiplicative subgroup (Problem 12.6). Let $N = \prod_i N_i$ be a neighbourhood of the identity in G such that $c(N) \subseteq U$; such an N exists by continuity of c. By Definition 11.4, we may assume that $N_i = H_i$ for almost all i. If we let S be a finite set containing all the indices i for which $N_i \neq H_i$, then N contains G^S, and therefore $c(G^S) \subseteq U$ is a subgroup of U, hence is trivial. It follows that $c(N_i) = c(H_i) = \{1\}$ for $i \notin S$. Now, for a given $x \in G$, enlarging S if necessary we can assume $x \in G_S$. Identify $x_i \in G_i$ (the i-th component of x) to the element of G that coincides with x in position i and is 1 elsewhere, and denote by $x^S \in G^S$

11.3 (Quasi-)Characters of a Restricted Product

the element such that $x = \prod_{i \in S} x_i \cdot x^S$. We already know that $c(G^S) \subseteq c(N) = \{1\}$, hence

$$c(x) = \prod_{i \in S} c(x_i) \cdot c(x^S) = \prod_{i \in S} c_i(x_i) = \prod_i c_i(x_i)$$

as desired (the last equality holds since $c_i(x_i) = 1$ for $i \notin S$). □

Conversely, starting from a collection of homomorphisms $c_i : G_i \to \mathbb{C}^\times$ that are almost all trivial on the corresponding H_i we obtain a global homomorphism c:

Lemma 11.3 *For each $i \in I$ fix a continuous homomorphism $c_i : G_i \to \mathbb{C}^\times$. Suppose that c_i is trivial on H_i for all but finitely many i. The map*

$$c(x) = \prod_i c_i(x_i)$$

is a well-defined continuous homomorphism $G \to \mathbb{C}^\times$.

Proof It is clear that c is well-defined (almost all factors in the product are equal to 1) and multiplicative (it is by restriction to any G_S). To prove continuity, let U be an open neighbourhood of 1 in the complex plane and choose a finite set S containing all the i for which $c_i(H_i) \neq \{1\}$. Let furthermore V be a neighbourhood of 1 in $\mathbb{C}^\times$ such that $V^{\#S} \subseteq U$, where by $V^{\#S}$ we mean the image of the multiplication map $\underbrace{V \times \cdots \times V}_{\#S \text{ times}} \to \mathbb{C}^\times$. For each $i \in I$, define

$$N_i = \begin{cases} \text{an open neighbourhood of the identity in } G_i \text{ such that } c(N_i) \subseteq V, \text{ if } i \in S \\ H_i, \text{ if } i \notin S. \end{cases}$$

Letting $N = \prod_i N_i$, we have

$$c(N) \subseteq \prod_i c_i(N_i) \subseteq V^{\#S} \subseteq U.$$

To check continuity in general, consider now an arbitrary open subset V of $\mathbb{C}^\times$. Either $c^{-1}(V)$ is empty, in which case we are done, or it is not. If it is not, let g be a point in $c^{-1}(V)$. By definition of the topology on $\mathbb{C}^\times$, the open set V contains an open neighbourhood of $c(g)$ of the form $c(g)U_g$, where U_g is an open neighbourhood of 1 in $\mathbb{C}^\times$. By what we already showed, there is an open neighbourhood N_g of 1 in G such that $c(N_g) \subseteq U_g$. It follows that $c^{-1}(V)$ contains gN_g, which is an open neighbourhood of g. Since this holds for every g, the set $c^{-1}(V)$ is open and c is continuous. □

We summarise the above discussion as follows:

Proposition 11.1 *The quasi-characters c of G are in bijection with the collections $(c_i)_{i \in I}$, where each c_i is a quasi-character of G_i and $c_i|_{H_i}$ is trivial for almost all i.*

The same arguments apply verbatim to characters, and show that the characters of G are of the form $\prod_i c_i$, where each c_i is a character (and not just a quasi-character) of G_i and almost all c_i are trivial on H_i. This already suggests that the dual group of G is itself a restricted product. We now make this precise.

For each i where H_i is defined, let $H_i^\perp \subseteq \widehat{G_i}$ be the subgroup

$$H_i^\perp = \{c_i \in \widehat{G_i} \mid c_i(H_i) = \{1\}\}.$$

By Theorem 9.1 and Proposition 9.1, the fact that H_i is compact implies that its dual $\widehat{H_i} \cong \frac{\widehat{G_i}}{H_i^\perp}$ is discrete, and hence $H_i^\perp$ is open in $\widehat{G_i}$. Similarly, the fact that H_i is open implies that G_i/H_i is discrete, and hence $\widehat{G_i/H_i} \cong H_i^\perp$ is compact. Thus, the subgroups $H_i^\perp$ have all the necessary characteristics to form a restricted product, and we have:

Theorem 11.1 (Dual Group of a Topological Restricted Product) *The dual group of $\prod_i'(G_i, H_i)$ is $\prod_i'(\widehat{G_i}, H_i^\perp)$.*

Proof Restricting the bijection of Proposition 11.1 to characters yields an isomorphism φ of abstract groups $\prod_i'(\widehat{G_i}, H_i^\perp) \cong \widehat{G}$, sending $(c_i)_{i \in I}$ to $\prod_{i \in I} c_i$. We check that this isomorphism is also a homeomorphism.

Notice that, since φ and φ^{-1} are homomorphisms of topological groups, it suffices to check continuity at the origin, that is: a homomorphism of topological groups $f : G_1 \to G_2$ is continuous if and only if, for every open neighbourhood U_2 of the identity in G_2, there is an open neighbourhood U_1 of the identity in G_1 such that $f(U_1) \subseteq U_2$.

We begin by showing that φ is continuous. Fix a basis neighbourhood of 1 for the topology of $\hat{G}$, say $U(K, V)$, where K is a compact neighbourhood of the identity in G and V is an open neighbourhood of 1 in $\mathbb{S}^1$. Restricting V if necessary, we assume that V contains no non-trivial subgroup of $\mathbb{C}^\times$.

We want to find a neighbourhood $\prod_i N_i$ of 1 in $\prod_i'(\widehat{G_i}, H_i^\perp)$ such that $(c_i)_{i \in I} \in \prod_i'(\widehat{G_i}, H_i^\perp)$ implies $\prod_i c_i \in U(K, V)$. Since K is compact, by Lemma 11.1 it is contained in a product of the form $\prod_{i \in I} B_i$, where each B_i is compact and $B_i = H_i$ for almost all i. Let S be the (finite) set of indices for which $B_i \neq H_i$. Choose furthermore V' to be an open neighbourhood of 1 in $\mathbb{S}^1$ that satisfies $(V')^{\#S} \subseteq V$, where, as in the proof of Lemma 11.3, by $(V')^{\#S}$ we mean the image of the multiplication map

$$\underbrace{V' \times \cdots \times V'}_{\#S \text{ times}} \to \mathbb{S}^1.$$

Setting

$$N_i = \begin{cases} U(B_i, V') \text{ if } i \in S \\ H_i^\perp \text{ if } i \notin S \end{cases}$$

we obtain that $\prod N_i$ is a neighbourhood of 1 in $\prod'_i(\widehat{G_i}, H_i^\perp)$ and $\varphi(\prod_i N_i) \subseteq U(K, V)$. Indeed, if $(c_i)_{i \in I}$ is in $\prod_i N_i$ we have

$$(\prod_i c_i)(K) \subseteq (\prod_i c_i)(\prod_i B_i) = \prod_{i \in S} c_i(B_i) \subseteq (V')^{\#S} \subseteq V,$$

that is, $\varphi((c_i)_{i \in I}) \in U(K, V)$. This shows that φ is continuous.

Continuity of $\varphi^{-1} : \hat{G} \to \prod'_i(\widehat{G_i}, H_i^\perp)$ is similar. Let $\prod_i N_i$ be an open neighbourhood of 1 in $\prod'_i(\widehat{G_i}, H_i^\perp)$, where all but finitely many of the N_i coincide with $H_i^\perp$. Write S for the finite set of indices for which $N_i \neq H_i^\perp$. Restricting the neighbourhood (which we can certainly do), we may and do assume that, for $i \in S$, we have $N_i = U(K_i, V_i)$ for certain compact neighbourhoods K_i of the identity in G_i and open neighbourhoods V_i of 1 in $\mathbb{S}^1$. Let $V = \bigcap_{i \in S} V_i$. Shrinking V if necessary, we can assume that V contains no non-trivial subgroups of $\mathbb{S}^1$. Set $K_i = H_i$ for $i \notin S$ and consider the open subset $U(\prod K_i, V)$ of $\hat{G}$. We check that φ^{-1} sends $U(\prod K_i, V)$ inside $\prod_i N_i$. In other words, since φ^{-1} sends c to the collection $(c_i)_{i \in I}$ of the restrictions of c to each factor G_i, we have to check that for $c \in U(\prod K_i, V)$ we have $\varphi^{-1}(c)_i \in N_i$ for every i. Indeed:

- if $i \notin S$, then $c_i(K_i) = c_i(H_i) \subseteq V$ implies $c_i(H_i) = \{1\}$, hence $c_i \in H_i^\perp = N_i$.
- if $i \in S$, then $c_i(K_i) \subseteq V \subseteq V_i$ implies $c_i \in N_i$ by definition.

$\square$

11.4 Measure Theory

Consider again a collection of locally compact topological groups G_i and fix (for almost all i) an open, compact subgroup H_i of G_i. Fix furthermore a Haar measure dg_i on each G_i and suppose that $\int_{H_i} dg_i = 1$ for almost all i. Notice that we require dg_i to be defined for *all* indices i; it's only the condition $\int_{H_i} dg_i = 1$ that is allowed to fail (for finitely many indices).

We now define a Haar measure dg on $G = \prod'(G_i, H_i)$ which is morally the product of the measures dg_i. To do this, fix any finite subset S of the indices (containing those for which H_i is not defined) and write $G_S = (\prod_{i \in S} G_i) \times G^S$. Notice that $G^S \cong \prod_{i \notin S} H_i$ is compact. It carries a unique Haar measure dg^S

normalised in such a way that

$$\int_{G^S} dg^S = \prod_{i \notin S} \left(\int_{H_i} dg_i \right). \tag{11.1}$$

The product measure $dg_S = \left(\prod_{i \in S} dg_i\right) \times dg^S$ is therefore well-defined on $G_S = \left(\prod_{i \in S} G_i\right) \times G^S$. Furthermore, it is a Haar measure (one can check translation-invariance one coordinate at a time). Since G_S is open in G, there exists a unique measure Haar dg on G such that the restriction of dg to G_S coincides with dg_S. In principle, the measure dg thus constructed could depend on the set S. To show that it is well-defined, it suffices to check that, if T is another set of indices, the restriction of dg_S to $G_S \cap G_T$ coincides with the restriction of dg_T to $G_S \cap G_T$. Replacing T with $T \cup S$, we can and do assume that $S \subseteq T$, so that $G_S \subseteq G_T$. There is an obvious decomposition

$$G^S = \left(\prod_{i \in T \setminus S} H_i \right) \times G^T,$$

and we claim that

$$dg^S = \left(\prod_{i \in T \setminus S} dg_i \right) \times dg^T.$$

Indeed, both are Haar measures, so—in order to check that they coincide—it suffices to show that they give the same (non-zero) volume to some subset. In particular, evaluating the right-hand side of the previous (claimed) equality on G^S we obtain

$$\prod_{i \in T \setminus S} \int_{H_i} dg_i \cdot \int_{G^T} dg^T = \prod_{i \in T \setminus S} \int_{H_i} dg_i \cdot \prod_{i \notin T} \left(\int_{H_i} dg_i \right)$$

$$= \prod_{i \notin S} \left(\int_{H_i} dg_i \right) = \int_{G^S} dg^S,$$

as desired. Thus, upon restriction to G_S we have

$$dg_S = \left(\prod_{i \in S} dg_i \right) \times dg^S = \left(\prod_{i \in S} dg_i \right) \times \left(\prod_{i \in T \setminus S} dg_i \right) \times dg^T = dg_T,$$

as claimed.

Definition 11.6 (Haar Measure on a Restricted Product) We denote the measure dg just constructed by $\prod_i dg_i$.

11.4 Measure Theory

Given the nature of restricted products as increasing unions over subsets of a certain set of indices, it is useful to introduce the following general notion of *limit over sets of indices*:

Definition 11.7 (Limit over S) Let S be the collection of all finite subsets of I, let X be a topological space, and let $\varphi : S \to X$ be a function. We write $\lim_S \varphi(S) = x$ if the following holds: for every open subset U of X containing x, there exists a finite set $V(U) \in S$ such that $V(U) \subseteq S \Rightarrow \varphi(S) \in U$.

Equivalently: add to S a formal point ∞ and make $S \cup \{\infty\}$ into a topological space by declaring that a basis consists of the sets $W_V := \{\infty\} \cup \{S : S \supseteq V\}$. We then have $\lim_S \varphi(S) = x$ if and only if the extended function $\tilde{\varphi} : S \cup \{\infty\} \to V$ that sends ∞ to x is continuous at x.

In particular, if f is a function on indices, we define

$$\prod_{i \in I} f(i) = \lim_S \prod_{i \in S} f(i).$$

Intuitively, one can think of $\lim_S \varphi(S)$ as the limit of the values $\varphi(S)$ as the set S gets larger and larger.

Lemma 11.4 *Let $f : G \to \mathbb{C}$ satisfy one of the following:*

1. *f is measurable, real-valued and non-negative, or*
2. *f is in $L^1(G)$.*

Then, $\int_G f(g)\, dg = \lim_S \int_{G_S} f(g)\, dg$.

Proof Under either assumption, $\int_G f(g)\, dg$ is the limit of $\int_B f(g)\, dg$ for larger and larger compacts $B \subseteq G$. Since each compact is contained in some G_S by Lemma 11.1, the claim follows. □

The Haar measure dg constructed above lets us integrate arbitrary functions in $L^1(G)$. However, given the almost-product nature of G, it is natural to study more closely those functions that also decompose as a product:

Definition 11.8 (Product Functions) For each $i \in I$ fix a continuous function $f_i : G_i \to \mathbb{C}$ with $f_i \in L^1(G_i)$. Suppose that $f_i|_{H_i} = 1$ for almost all i. We define the function

$$f = \prod_{i \in I} f_i : \quad G \quad \to \quad \mathbb{C}$$
$$g = (g_i)_{i \in I} \mapsto \prod_{i \in I} f_i(g_i).$$

Notice that the product $\prod_{i \in I} f_i(g_i)$ contains only finitely many terms different from 1, for each $g = (g_i) \in G$.

Lemma 11.5 *Let $f = \prod_{i \in I} f_i$ be a product function as in Definition 11.8.*

1. *f is continuous on G.*

2. Let S be a finite subset of I containing the (finitely many) indices i for which $f_i(H_i) \neq \{1\}$ and those for which $\int_{H_i} dg_i \neq 1$. We have

$$\int_{G_S} f(g)\, dg = \prod_{i \in S} \left(\int_{G_i} f_i(g_i)\, dg_i \right).$$

Proof

1. Upon restriction to a set of the form G_S, the function f is the product of finitely many continuous functions, hence continuous. Since the G_S are open and cover G, f is continuous on G.
2. For $g \in G_S$, say $g = (g_i)_{i \in I}$, we have as above

$$f(g) = \prod_{i \in S} f_i(g_i).$$

Hence (recalling the defining property (11.1) of dg^S)

$$\int_{G_S} f(g)\, dg = \int_{G_S} f(g)\, dg_S = \int_{G_S} \left(\prod_{i \in S} f_i(g_i) \right) \left(\prod_{i \in S} dg_i \times dg^S \right)$$

$$= \prod_{i \in S} \left(\int_{G_i} f_i(g_i)\, dg_i \right) \times \int_{G^S} dg^S$$

$$= \prod_{i \in S} \left(\int_{G_i} f_i(g_i)\, dg_i \right) \times \prod_{i \notin S} \left(\int_{H_i} dg_i \right)$$

$$= \prod_{i \in S} \left(\int_{G_i} f_i(g_i)\, dg_i \right),$$

where in the last equality we used the fact that $\int_{H_i} dg_i = 1$ for every $i \notin S$. □

The previous two lemmas yield the following result:

Theorem 11.2 *Let f be a product function as in Definition 11.8. Assume*

$$\prod_i \left(\int_{G_i} |f_i(g_i)|\, dg_i \right) < \infty,$$

where the meaning of the product is as in Definition 11.7. Then f is in $L^1(G)$, and

$$\int_G f(g)\, dg = \prod_i \left(\int_{G_i} f_i(g_i)\, dg_i \right).$$

Proof Applying Lemmas 11.4 and 11.5 to $|f(g)| = \prod_i |f_i(g_i)|$ shows that $f \in L^1(G)$, at which point the same lemmas (applied to $f(g)$ itself) yield the result. □

11.5 Fourier Analysis

We have seen in Theorem 11.1 that for a restricted direct product $G = \prod'_{i \in I}(G_i, H_i)$ we have

$$\hat{G} = \prod_{i \in I}{}' (\widehat{G_i}, H_i^\perp).$$

Denote by $c = (c_i)_{i \in I}$ an element of $\hat{G}$, that is, a continuous homomorphism $G \to \mathbb{S}^1$. Let dc_i be the measure on $\widehat{G_i}$ dual to the measure dg_i on G_i (see Definition 9.4).

Lemma 11.6 *The following hold for every $i \in I$ such that H_i is an open, compact subgroup of G_i.*

1. *Let f_i be the characteristic function of H_i. Its Fourier transform is $\int_{H_i} dg_i$ times the characteristic function of $H_i^\perp$.*
2. $\left(\int_{H_i} dg_i\right)\left(\int_{H_i^\perp} dc_i\right) = 1.$

Proof

1. By definition,

$$\hat{f}_i(c_i) = \int_{G_i} f_i(g_i)\overline{c_i(g_i)}\, dg_i = \int_{H_i} \overline{c_i(g_i)}\, dg_i.$$

By Proposition 9.2, this integral is 0 if the character c_i is nontrivial on H (that is, if $\mathbf{1}_{H_i^\perp}(c_i) = 0$), and is $\int_{H_i} dg_i$ otherwise.

2. Since $f = \mathbf{1}_{H_i}$ is in $L^1(G_i)$, is continuous, and has its Fourier transform in $L^1(\hat{G}_i)$ (by part (1) of the lemma), Fourier inversion (Theorem 9.2) gives

$$\mathbf{1}_{H_i}(g) = \int_{\hat{G}_i} \hat{f}(c_i) c_i(g)\, dc_i$$

$$= \int_{\hat{G}_i} \left(\int_{H_i} dg_i\right) \mathbf{1}_{H_i^\perp}(c_i) c_i(g)\, dc_i$$

$$= \left(\int_{H_i} dg_i\right)\left(\int_{H_i^\perp} c_i(g) dc_i\right).$$

Evaluating at $g \in H_i$ yields

$$1 = \left(\int_{H_i} dg_i\right)\left(\int_{H_i^\perp} dc_i\right)$$

as claimed. □

Thus, we see that the collection $\{dc_i\}_{i \in I}$ of the dual measures satisfies the condition to give a measure $\prod dc_i$ on $\hat{G} = \prod'_{i \in I}(\widehat{G_i}, H_i^\perp)$. We denote this measure by dc.

Lemma 11.7 (Product Decomposition for the Fourier Transform) *If f_i belongs to $\mathfrak{B}^1(G_i)$ for all $i \in I$ and $f_i(g_i) = \mathbf{1}_{H_i}$ for almost all i, then the function $f(g) = \prod_i f_i(g_i)$ belongs to $\mathfrak{B}^1(G)$ and has Fourier transform $\hat{f}(c) = \prod_i \hat{f}_i(c_i)$.*

Proof Apply Theorem 11.2 to the function $f(g)\bar{c}(g) = \prod_i f_i(g_i)\overline{c_i(g_i)}$: it implies that $\hat{f}(c) = \prod_i \hat{f}_i(c_i)$.

Since $f_i \in \mathfrak{B}^1(G_i)$ for all i, we have $\hat{f}_i \in L^1(\widehat{G_i})$ for all i. Moreover, by Lemma 11.6, $\hat{f}_i$ is the characteristic function of $H_i^\perp$ for almost all i. From this, Lemmas 11.5 and 11.6 (2) we then obtain $\hat{f} \in L^1(\hat{G})$. Since f is continuous and in $L^1(G)$ (again by Lemma 11.5), we get that f is in $\mathfrak{B}^1(G)$. □

Corollary 11.1 *The measure $dc = \prod_i dc_i$ is dual to $dg = \prod_i dg_i$.*

Proof The previous lemma implies that the Fourier inversion formula

$$f(g) = \int_{\hat{G}} \hat{f}(c)c(g)\, dc$$

holds at least for the product functions considered in the lemma. Since the dual measure of (G, dg) is unique, it must be dc. □

Problems

11.1 Check that the restricted product $\prod'(G_i, H_i)$ is a subgroup of $\prod G_i$.

11.2 The group of units of $\mathbb{A}_k$ (as a ring) is $\mathbb{I}_k$.

11.3 Let $(G_i)_{i \in I}$ be a family of abelian groups and let H_i be the trivial subgroup of G_i for every i. Prove that $\prod'(G_i, H_i) \cong \bigoplus_{i \in I} G_i$.

Part III
Tate's Thesis

Chapter 12
The Local Theory

Abstract In this chapter we develop the theory of *local* zeta functions. Specifically, given a local field k, we make compatible choices of Haar measures for the additive and multiplicative group of k, and we discuss Fourier analysis on these groups. We further show that local zeta functions, defined in terms of Fourier analysis, satisfy a very general functional equation, which we describe in detail for all the cases of interest, namely when k is isomorphic to $\mathbb{R}$, $\mathbb{C}$, or a p-adic field.

We have now developed all the tools necessary to discuss Tate's *local and global ζ functions* [3]. Their definition and the verification of their properties will require a certain amount of technical work, but before we go into any calculations we pause to describe the general content of this and the next chapter and try to motivate where the constructions come from. The reader will hopefully forgive the length of this introduction, which we think is helpful in understanding the relevance of the results of this chapter.

For the purposes of this discussion, it is useful to briefly recall some facts of local and global class field theory. Let k be a perfect field, fix an algebraic closure $\overline{k}$, and let k^{ab} be the maximal abelian extension of k contained in $\overline{k}$. Consider the Galois group $\Gamma_k^{\mathrm{ab}} := \mathrm{Gal}(k^{\mathrm{ab}}/k)$. Class field theory aims to describe this group purely in terms of data *internal* to the field k itself. The simplest example is provided by local fields of characteristic 0: in this case, Γ_k^{ab} is isomorphic to the profinite completion $\varprojlim_n k^\times/k^{\times n}$. In particular, we can compose the natural map $k^\times \to \varprojlim_n k^\times/k^{\times n}$ (which has dense image) with the isomorphism $\varprojlim_n k^\times/k^{\times n} \cong \Gamma_k^{\mathrm{ab}}$ to obtain a map $k^\times \to \Gamma_k^{\mathrm{ab}}$ with dense image. When k is a number field, we instead have a canonical surjective map $I_k \to \Gamma_k^{\mathrm{ab}}$ (see Theorem 14.1 for more details). In both cases, the upshot is that every character (resp. quasi-character) of Γ_k^{ab} extends to a character (resp. quasi-character) of a group that is in many ways simpler: $k^\times$ in the local case and I_k in the global case.

We saw in Part I that an L-function is determined by a representation of a Galois group—hence, in the abelian case, by a character χ of a group of the form $\mathrm{Gal}(L/k)$, where k is a number field and L is a finite extension of k. Since every abelian Galois group $\mathrm{Gal}(L/k)$ is a quotient of Γ_k^{ab}, by the above discussion we can consider the composition

$$I_k \to \Gamma_k^{ab} \twoheadrightarrow \operatorname{Gal}(L/k) \xrightarrow{\chi} \mathbb{C}^\times,$$

which ultimately gives us a character of I_k. A similar argument works in the local case: any character of $\operatorname{Gal}(L/k)$, where L/k is a finite extension of local fields, induces a character $k^\times \to \mathbb{C}^\times$. Thus, in this light, it is very natural to investigate the (quasi-)characters of groups of the form $k^\times$ or I_k.

To fix ideas, consider the global case, namely, abelian Artin L-functions: for a fixed character χ, they are functions of a complex variable s. One of Tate's many contributions is the realisation that in this context the variable s can be used to parametrise all quasi-characters that are equivalent to χ in a suitable sense. Concretely, if we define the **idèlic norm** by the formula[1]

$$\begin{aligned} \|\cdot\|: \quad I_k &\to \mathbb{R}_{>0} \\ (a_v)_{v \in \Omega_k} &\mapsto \prod_v \|a_v\|_v, \end{aligned}$$

the quasi-characters equivalent to the character χ are precisely those of the form $a \mapsto \chi(a)\|a\|^s$. One could then imagine attaching an 'L-number' $L(\psi)$ to each *quasi*-character ψ of I_k, thus defining a function on the seemingly wild space of all quasi-characters. However, the above considerations show that each *equivalence class* can be parametrised by a complex variable s: for each character we have an equivalence class of quasi-characters, hence an induced function of a complex variable s (specifically, the function that sends s to $L(\chi(a)\|a\|^s)$). The hope is now to arrange matters so that these functions of a complex variable are precisely the L-functions we are interested in: we would like to have $L(s, \chi) = L(\chi(a)\|a\|^s)$, where the L on the left is the Hecke L-function of χ and the L on the right is the function just defined on the space of quasi-characters.

Tate makes this vague vision concrete by defining the 'L-numbers' $L(\psi)$ (which he calls *local* and *global zeta functions*) on the space of quasi-characters, and by showing that—upon restriction to the equivalence class of a character χ—they give exactly the Hecke L-function corresponding to χ (this is not quite accurate, since Tate's ζ functions also depend on the choice of an auxiliary function on k or $\mathbb{A}_k$. However, as we will see, we shall mostly *fix* the right auxiliary functions to get back Hecke's L-functions).

The other important point, of course, is how to actually define these local and global zeta functions: they will arise as modified Fourier transforms on the spaces $k^\times$ and I_k, allowing us to bring to bear the machinery of abstract Fourier analysis that we described in Part II.

From this discussion, it should now be clear that (both in the local and global setting) we need to address the following questions:

---

[1] The product is well-defined, because by definition all but finitely many a_v have v-adic norm 1.

12 The Local Theory

1. what do the quasi-characters of $k^\times$ and I_k look like?
2. how do we choose meaningful Haar measures on $k^\times$ and I_k?
3. what sort of generalised Fourier transforms on $k^\times$ (respectively I_k) produce the local factors of L-functions (respectively, global Hecke L-functions)?
4. do these zeta functions satisfy a functional equation and analytic continuation?

In this chapter and the next we will answer these questions, first in the local case and then in the global one. Naturally, the local study will be an essential input for the global case, because—as a consequence of Theorem 11.1—the characters of the idèles are products of characters of the multiplicative groups of various local fields. In Tate's approach, this fact, which holds by the general theory of restricted products in Chap. 11, is the source of all Euler product expansions of L-functions.

It turns out that in order to answer question 2 in the local case (namely, to pin down an arithmetically meaningful Haar measure on $k^\times$) it is important to fix a suitable Haar measure also on the additive group $(k, +)$. This additive Haar measure will be determined by the property of being self-dual with respect to the Fourier transform, in the following sense. We will show that the dual group of $(k, +)$ can be identified with $(k, +)$ itself. The theory of Fourier inversion then gives, for any Haar measure on $(k, +)$, a dual measure on the dual group $(k, +)$: it is then not surprising that the technically superior choice of Haar measure is the unique one that coincides with its dual.

Remark 12.1 The self-dual Haar measure on $(k, +)$ depends on the specific identification of $(k, +)$ with its dual. As we will see in Proposition 12.1, this identification in turn depends on the choice of a non-trivial character of k^+. There is, therefore, a degree of arbitrariness in the construction. We spell out in Definition 12.1 the choice we make for the non-trivial additive character $k^+ \to \mathbb{S}^1$; especially in the p-adic case, it is made so as to obtain nicer formulas in the computations.

This chapter is structured as follows. In Sect. 12.1 we discuss the characters of the additive group of a local field, prove that the dual group of $(k, +)$ is isomorphic to $(k, +)$ itself, and show how the choice of an isomorphism between $(k, +)$ and its dual leads to a corresponding choice of Haar measure on $(k, +)$. In Sect. 12.2 we describe the quasi-characters of $k^\times$ and we make a choice of Haar measure on this group that is compatible in an appropriate sense with the choice of Haar measure on $(k, +)$. In Sect. 12.3 we define Tate's local zeta functions as generalised Fourier transforms and prove that they satisfy a general functional equation. This functional equation will lead us to introduce a certain function $\rho(c)$ on characters. Finally, in Sect. 12.4 we explicitly determine $\rho(c)$ for each character c of a local field, distinguishing cases according to whether k is real, complex, or p-adic.

Notation In this chapter we let $k := K_v$ be the completion of a number field K at a place v. If v is a finite place, so that k is a p-adic field, we denote by O its ring of integers. We denote by ξ the variable in the additive group $(k, +)$, and by α the variable in $(k^\times, \cdot)$. In particular, we will eventually denote by $d\xi$ and $d\alpha$ certain Haar measures on $(k, +)$ and on $(k^\times, \cdot)$, respectively.

12.1 The Additive Group

Recall our choice of norm on $k = K_v$:

1. the ordinary absolute value if $k \cong \mathbb{R}$;
2. the square of the ordinary absolute value if $k \cong \mathbb{C}$;
3. $\|\alpha\| = (N\mathfrak{p})^{-v\mathfrak{p}(\alpha)}$, if v is the finite place corresponding to the prime $\mathfrak{p}$.

We begin by studying the group of characters of the locally compact group $k^+ := (k, +)$.

Proposition 12.1 *Let χ be any non-trivial character of k^+.*

1. *For each $\eta \in k$, the map $\xi \mapsto \chi(\eta\xi)$ is a continuous character.*
2. *The map $\alpha_\chi : k^+ \to \widehat{k^+}$ given by $\eta \mapsto \chi_\eta$, where $\chi_\eta(\xi) := \chi(\eta\xi)$, is an isomorphism (of topological groups) between k^+ and its character group.*

Proof Since $\xi \mapsto \eta\xi$ is a continuous homomorphism of k^+ into itself, by composition with χ we see that χ_η is indeed a continuous character of k^+. One checks easily that $\eta \mapsto \chi_\eta$ is a homomorphism. Moreover, α_χ is injective, because χ_η is the trivial character 1 if and only if $\chi(\eta\xi) = 1$ for all $\xi \in k$. Since χ is non-trivial, there exists $y \in k$ such that $\chi(y) \neq 1$. If $\eta \neq 0$, setting $\xi = y/\eta$ gives a contradiction, so χ_η is trivial only for $\eta = 0$.

Next we show that α_χ is a topological isomorphism between k and its image (and in particular, that α_χ is open). We start by recalling the topology on $\widehat{k^+}$. By definition, a basis of neighbourhoods of the identity in $\widehat{k^+}$ is given by the $U(K, V)$ (see Definition 9.1), where K is a compact neighbourhood of $0 \in k$ and V is a neighbourhood of $1 \in \mathbb{S}^1$. Clearly, it suffices to let K and V range over a basis of neighbourhoods of 0 and 1 in k and $\mathbb{S}^1$, respectively. Thus, we may only consider

$$K = C_m = \{\xi \in F : \|\xi\| \leq m\} \text{ and } V = V_\varepsilon = \{z \in \mathbb{S}^1 : \|z - 1\| < \varepsilon\}.$$

Since χ is continuous, for all $\varepsilon > 0$ there exists a $\delta > 0$ such that

$$\|\chi(\xi) - 1\| < \varepsilon \text{ whenever } \|\xi\| < \delta. \tag{12.1}$$

Since α_χ is a group homomorphism, to show that it is continuous it suffices to show that it is continuous at the identity. Explicitly, we have to show that for every $U(C_m, V_\varepsilon)$ there exists an open neighbourhood W of 0 in k^+ such that

$$\alpha_\chi(W) \subseteq U(C_m, V_\varepsilon),$$

that is, we must choose W in such a way that for every $\xi \in W$

$$\|\alpha_\chi(\xi)(y) - 1\| < \varepsilon \quad \forall y \in C_m,$$

12.1 The Additive Group

or equivalently,

$$\|\chi(\xi y) - 1\| < \varepsilon \quad \forall y \in C_m.$$

Thus, it suffices to take as W the open set $\{\xi \in F : \|\xi\| < \frac{\delta}{m}\}$: for $\xi \in W$ and $y \in C_m$ one has $\|\xi y\| < \frac{\delta}{m} m = \delta$, and hence $\|\chi(\xi y) - 1\| < \varepsilon$ by (12.1).

Now we show continuity of $\alpha_\chi^{-1} : \alpha_\chi(k) \to k$. Again by definition of the respective topologies, we have to show that, for every $\delta > 0$, there exist $\varepsilon > 0$, $m \in \mathbb{R}$ such that

$$\alpha_\chi(\xi) \in \alpha_\chi(k) \cap U(C_m, V_\varepsilon) \implies \|\xi\| < \delta.$$

Since χ is nontrivial, there exists $\xi_0 \in k$ such that $\chi(\xi_0) \neq 1$. Set

$$\varepsilon = \|\chi(\xi_0) - 1\|, \quad m = \frac{2\|\xi_0\|}{\delta}.$$

Suppose now that ξ is such that $\alpha_\chi(\xi) \in U(C_m, V_\varepsilon)$, that is to say,

$$\|\chi(\xi y) - 1\| < \varepsilon \quad \forall y \text{ with } \|y\| < m.$$

Note that $y = \xi^{-1}\xi_0$ does *not* satisfy $\|\chi(\xi y) - 1\| < \varepsilon$, hence $y = \xi^{-1}\xi_0$ does *not* satisfy $\|y\| < m$. This implies

$$\|\xi^{-1}\xi_0\| \geq m = \frac{2\|\xi_0\|}{\delta} \implies \|\xi\|^{-1} \geq \frac{2}{\delta} \implies \|\xi\| \leq \frac{\delta}{2} < \delta,$$

as desired.

Since α_χ is a topological isomorphism, $\alpha_\chi(k)$ is locally compact, hence closed (see Problem 12.1). Thus, in order to show that $\alpha_\chi(k) = \widehat{k^+}$, it suffices to prove that $H := \alpha_\chi(k)$ is everywhere dense in $\widehat{k^+}$. We now observe that

$$H^\perp = \{\xi \in k : \psi(\xi) = 1 \quad \forall \psi \in H\} = \{\xi \in k : \chi(\xi y) = 1 \quad \forall y \in k\} = \{0\},$$

which, by Proposition 9.1, yields

$$0 = \widehat{H^\perp} = \widehat{k^+}/H,$$

hence $H = \widehat{k^+}$, as desired. □

Using Proposition 12.1, we may identify k^+ with its dual provided that we fix a non-trivial character. We start by defining a certain function $\lambda : \mathbb{Q}_p \to \mathbb{R}/\mathbb{Z}$ for each $p \in \{\text{primes}\} \cup \{\infty\}$, where by convention we set $\mathbb{Q}_\infty = \mathbb{R}$.

1. if $p = \infty$, we let $\lambda(\xi)$ be the class of $-\xi$ in the quotient $\mathbb{R}/\mathbb{Z}$. **Note the choice of sign!**

2. if p is a prime number, one can prove (see Problem 12.2) that $\mathbb{Q}_p/\mathbb{Z}_p$ is isomorphic to the subgroup of $\mathbb{Q}/\mathbb{Z}$ given by torsion elements of order a power of p. Identifying $\mathbb{Q}/\mathbb{Z}$ to a subset of $\mathbb{R}/\mathbb{Z}$, we take λ to be the projection

$$\lambda : \mathbb{Q}_p \to \frac{\mathbb{Q}_p}{\mathbb{Z}_p} \hookrightarrow \frac{\mathbb{Q}}{\mathbb{Z}} \hookrightarrow \frac{\mathbb{R}}{\mathbb{Z}}.$$

Concretely, $\lambda(\xi)$ can be described as follows: let v be an integer such that $p^v \xi$ is in $\mathbb{Z}_p$, and let n be an integer such that $n \equiv p^v \xi \pmod{p^v}$. We then have $\lambda(\xi) = \frac{n}{p^v} \pmod{1}$. In particular, $\xi - \lambda(\xi)$ is a p-adic integer.

Finally, if $k = K_v$ is an arbitrary completion of a number field and if p is the prime of $\mathbb{Q}$ 'lying under v' (that is, $p = \infty$ if v is Archimedean, and $p = p_v$ if v is finite), we have a natural inclusion $\mathbb{Q}_p \subseteq k$. We then give the following definition.

Definition 12.1 (Fundamental Character of the Additive Group) We set $\Lambda(\xi) := \lambda\left(\operatorname{tr}_{k/\mathbb{Q}_p}(\xi)\right).$

Since the trace map is continuous, Λ is a non-trivial, continuous map from k to $\mathbb{R}/\mathbb{Z}$. Using Proposition 12.1, we obtain

Theorem 12.1 (Dual Group of k^+) *The additive group k^+ is isomorphic to its dual group via the isomorphism $\eta \mapsto \chi_\eta$, where*

$$\chi_\eta(\xi) = e^{2\pi i \Lambda(\eta \xi)}.$$

For later use, we record a lemma connecting properties of characters with the arithmetic of k.

Lemma 12.1 *Let v be a finite place of characteristic p. The character $e^{2\pi i \Lambda(\eta \xi)}$ corresponding to η is trivial on O if and only if η belongs to the inverse different ideal $\mathfrak{d}_k^{-1}$ (see Definition 10.5).*

Proof The character $e^{2\pi i \Lambda(\eta \xi)}$ is trivial if and only if $\Lambda(\eta \xi)$ is an integer for every $\xi \in O$, if and only if $\operatorname{tr}_{k/\mathbb{Q}_p}(\eta \xi)$ is in $\mathbb{Z}_p$ for every $\xi \in O$, if and only if η is in the inverse different. □

12.1.1 Choice of Haar Measure

Let μ be a Haar measure on k^+. We now investigate the interactions between μ and the multiplicative structure of k, and describe the measure on the dual group which appears in Theorem 9.2 (abstract Fourier inversion).

Lemma 12.2 *For every $\alpha \in k^\times$ and for every measurable set M in k we have $\mu(\alpha M) = \|\alpha\| \mu(M)$ for the choice of norm $\|\cdot\|$ recalled at the beginning of this chapter.*

12.1 The Additive Group

Proof Note that $M \mapsto \mu(\alpha M)$ is a Haar measure on k^+, so $\mu(\alpha M) = \varphi(\alpha)\mu(M)$ for some constant $\varphi(\alpha) > 0$ which may depend on α but not on M. To identify this constant, note that in the real and complex case this is precisely $\|\alpha\|$, as follows from the change-of-variables formula for integration that is familiar from analysis (in the complex case, note that $\alpha = u + iv$ acts on $\mathbb{C} \cong \mathbb{R}^2$ as the linear transformation

$$\begin{pmatrix} x \\ y \end{pmatrix} \mapsto \begin{pmatrix} u & v \\ -v & u \end{pmatrix} \begin{pmatrix} x \\ y \end{pmatrix},$$

whose determinant is $u^2 + v^2 = \|\alpha\|$). Finally, in the p-adic case, consider the set $M = O$. Suppose first that α is integral: then $O/\alpha O$ has $N(\alpha)$ elements, which means that $O = \bigsqcup_{i=1}^{N(\alpha)} (\xi_i + \alpha O)$ for some collection $\xi_1, \ldots, \xi_{N(\alpha)}$ of points in O. Since the Haar measure is translation-invariant, we obtain

$$\mu(O) = \sum_{i=1}^{N(\alpha)} \mu(\xi_i + \alpha O) = N(\alpha)\mu(\alpha O).$$

If we let π be a uniformiser of k, we have $\alpha = \pi^v u$ with $v \in \mathbb{N}$ and $u \in O^\times$, and $N(\alpha) = N(\pi)^v = \|\alpha\|^{-1}$. Thus, we conclude that

$$\mu(O) = \|\alpha\|^{-1}\mu(\alpha O),$$

as desired. Finally, if α has negative valuation, the same argument with α replaced by α^{-1} gives the desired statement. $\square$

By standard results in measure theory, Lemma 12.2 implies the following equality for integrable functions f on k^+ and for $\alpha \in k^\times$:

$$\int_{k^+} f(\xi) \, d\mu(\xi) = \int_{k^+} f(\alpha\xi) \, d\mu(\alpha\xi) = \|\alpha\| \int_{k^+} f(\alpha\xi) \, d\mu(\xi).$$

Remark 12.2 (Measure of a Fractional Ideal) Every fractional ideal of k is principal, generated by a power of the uniformiser π. Writing $I = (\pi^v)$ and using Lemma 12.2, we obtain

$$\mu(I) = \mu(\pi^v O) = \|\pi\|^v \mu(O) = (N\pi)^{-v}\mu(O) = (N(I))^{-1}\mu(O).$$

The identification of k^+ with its dual group provided by Theorem 12.1 allows us to interpret the abstract Fourier transform of a function on k^+ (which would formally be a function on the dual group of k^+) as another function on k^+ itself. We now look for a choice of Haar measure on k^+ that is 'natural' with respect to Fourier inversion. More specifically, for every Haar measure on k^+, Theorem 9.2 yields the existence of a corresponding Haar measure on the dual. However, since we have identified k^+ with its dual, this means that from every Haar measure on

k^+ we obtain a 'Fourier-dual Haar measure' on k^+ itself. The most natural choice is then to require that the Fourier-dual measure coincide with the original measure. This is achieved by taking μ_{k^+} as in the following definition.

Definition 12.2 (Choice of Haar Measure) We define μ_{k^+} to be

1. the ordinary Lebesgue measure on the real line, if k is real;
2. twice the ordinary Lebesgue measure in the plane, if k is complex;
3. the unique Haar measure for which $\mu_{k^+}(O) = N(\mathfrak{d}_k)^{-1/2}$, if k is p-adic.

We will sometimes simply write $d\xi$ for $d\mu_{k^+}(\xi)$.

We summarise the above discussion in the next theorem.

Theorem 12.2 *Let $d\xi$ denote the measure on k^+ introduced in Definition 12.2. If we define the Fourier transform $\hat{f}$ of a function $f \in L^1(k^+)$ by*

$$\hat{f}(\eta) = \int_{k^+} f(\xi) e^{-2\pi i \Lambda(\eta \xi)} \, d\xi, \tag{12.2}$$

then the inversion formula

$$f(\xi) = \int_{k^+} \hat{f}(\eta) e^{2\pi i \Lambda(\xi \eta)} \, d\eta = \hat{\hat{f}}(-\xi)$$

holds for $f \in \mathfrak{B}^1(k^+)$ (see Definition 9.3 for the notation $\mathfrak{B}^1$).

Proof Theorem 9.2 implies that the identity

$$f(\xi) = c \int_{k^+} \hat{f}(\eta) e^{2\pi i \Lambda(\xi \eta)} \, d\eta$$

holds for some nonzero constant c independent of f, because the Fourier transform defined in the statement is equivalent to the general, abstract Fourier transform of Definition 9.2 under the isomorphism between k^+ and its dual provided by Theorem 12.1. Thus, it suffices to check that $c = 1$ for a single function f. We distinguish three cases, according to the nature of k:

1. if k is real, we take $f(\xi) = e^{-\pi \|\xi\|^2}$. The result then reduces to the classical calculation of the Fourier transform of a Gaussian, see Problem 12.3.
2. if k is complex, we similarly take $f(\xi) = e^{-2\pi \|\xi\|}$ (recall that our norm on complex fields is the *square* of the usual absolute value), see again Problem 12.3.
3. if k is p-adic, we take as f the characteristic function of O. We compute its Fourier transform: by definition,

$$\hat{f}(\eta) = \int_O e^{-2\pi i \Lambda(\eta \xi)} d\xi.$$

12.2 The Multiplicative Group

We are integrating a character of O on the whole group: since O is compact, by Proposition 9.2, the result is either 0 (if this character is nontrivial) or $\mu_{k^+}(O)$. Using Lemma 12.1, we obtain

$$\hat{f}(\eta) = \begin{cases} \mu_{k^+}(O), & \text{if } \eta \in \mathfrak{d}_k^{-1} \\ 0, & \text{otherwise}, \end{cases}$$

and therefore $\hat{f}(\eta) = \mu_{k^+}(O)\mathbf{1}_{\mathfrak{d}_k^{-1}} = N(\mathfrak{d}_k)^{-1/2}\mathbf{1}_{\mathfrak{d}_k^{-1}}$, where the last equality follows from our normalisation $\mu_{k^+}(O) = N(\mathfrak{d}_k)^{-1/2}$. Next, we plug $\hat{f}(\eta)$ in the Fourier inversion formula:

$$\int_{k^+} \hat{f}(\eta) e^{2\pi i \Lambda(\xi\eta)} \, d\eta = \int_{\mathfrak{d}_k^{-1}} N(\mathfrak{d}_k)^{-1/2} e^{2\pi i \Lambda(\xi\eta)} \, d\eta.$$

Fix a uniformiser π of k and write $\mathfrak{d}_k = \pi^r$. The change of variables $\eta = \pi^{-r} y$ leads to

$$\int_{\mathfrak{d}_k^{-1}} N(\mathfrak{d}_k)^{-1/2} e^{2\pi i \Lambda(\xi\eta)} \, d\eta = \int_O N(\mathfrak{d}_k)^{-1/2} e^{2\pi i \Lambda(\xi\pi^{-r}y)} \|\pi^{-r}\| \, dy,$$

where we have used Lemma 12.2. By definition, $\|\pi^{-r}\| = N(\pi^r) = N(\mathfrak{d}_k)$. Applying Lemma 12.1 again we get

$$\int_{k^+} \hat{f}(\eta) e^{2\pi i \Lambda(\xi\eta)} \, d\eta = N(\mathfrak{d}_k)^{1/2} \mu_{k^+}(O) \mathbf{1}_O(\xi) \, d\xi = \mathbf{1}_O(\xi),$$

which concludes the proof.

Finally, we check the equality $\hat{\hat{f}}(\xi) = f(-\xi)$. By definition, $\hat{\hat{f}}(\xi)$ is the Fourier transform of (12.2), hence it is given by

$$\hat{\hat{f}}(\eta) = \int_{k^+} \hat{f}(\xi) e^{-2\pi i \Lambda(\eta\xi)} \, d\xi = \int_{k^+} \hat{f}(\xi) e^{2\pi i \Lambda((-\eta)\xi)} \, d\xi = f(-\eta)$$

by what we already showed. $\square$

12.2 The Multiplicative Group

We now turn to the study of the quasi-characters of $k^\times$. A special such homomorphism is given by the norm itself, $\alpha \mapsto \|\alpha\| \in \mathbb{R}^\times$. Of particular importance will be its kernel:

Definition 12.3 We denote by $\mathfrak{u}$ the subgroup of $k^\times$ of elements of norm 1, that is, the kernel of $\|\cdot\| : k^\times \to \mathbb{R}^\times$. Note that, if k is a p-adic field, $\mathfrak{u}$ is nothing else than the group of units $O^\times$. We say that a quasi-character $\chi : k^\times \to \mathbb{C}^\times$ is **unramified** if $\chi(\mathfrak{u}) = 1$.

The unramified quasi-characters are easy to classify:

Lemma 12.3 *The unramified quasi-characters of $k^\times$ are the maps of the form $c(\alpha) = |\alpha|^s := e^{s \log \|\alpha\|}$, where s is any complex number. When v is Archimedean, s is uniquely determined by c; when v is finite, corresponding to a prime $\mathfrak{p}$, the number s is determined modulo $\frac{2\pi i}{\log N(\mathfrak{p})}$.*

Proof Clearly, it suffices to classify the continuous homomorphisms from $\tilde{c} : k^\times/\mathfrak{u}$ to $\mathbb{C}^\times$. Since $\mathfrak{u}$ is the kernel of the norm, $k^\times/\mathfrak{u}$ is isomorphic to the image of $\|\cdot\|$ (the 'value group'), which is the group of positive real numbers when v is Archimedean and $\langle N(\mathfrak{p}) \rangle \cong \mathbb{Z}$ when v is finite and corresponds to $\mathfrak{p}$. The claim follows easily: when v is finite, the value group $\|k^\times\|$ is isomorphic to $\mathbb{Z}$, so a homomorphism from $\|k^\times\|$ to $\mathbb{C}^\times$ is determined by its value on a generator. For the Archimedean case, see Problem 12.4. □

We now give a partial description of all the quasi-characters of $k^\times$. This will be an immediate consequence of Lemma 12.3 once we make the following observations:

1. if v is Archimedean, every element α of $k^\times$ can be written uniquely as

$$\alpha = \tilde{\alpha}\rho \tag{12.3}$$

with $\tilde{\alpha} \in \mathfrak{u}$ and $\rho > 0$;

2. if v is finite, letting π be a uniformiser of k, every element α of $k^\times$ can be written uniquely as $\tilde{\alpha}\rho$ with $\tilde{\alpha} \in \mathfrak{u}$ and ρ a power of π.

In either case, the map $\alpha \mapsto \tilde{\alpha}$ is a continuous homomorphism $k^\times \to \mathfrak{u}$ which is the identity on $\mathfrak{u}$.

The following classification of quasi-characters is now immediate:

Theorem 12.3 *The quasi-characters of $k^\times$ are the maps of the form $c : \alpha \mapsto \tilde{c}(\tilde{\alpha})\|\alpha\|^s$, where $\tilde{c}$ is any (continuous) character of $\mathfrak{u}$. The character $\tilde{c}$ is uniquely determined by c (it is its restriction to $\mathfrak{u}$), while s is determined as in Lemma 12.3.*

Thus, the classification of the quasi-characters of $k^\times$ reduces to the classification of those of $\mathfrak{u}$.

Proposition 12.2 (Classification of Quasi-Characters of $\mathfrak{u}$) *The quasi-characters $\tilde{c}$ of $\mathfrak{u}$ can be described as follows:*

1. *if k is real, $\mathfrak{u} = \{\pm 1\}$ and the quasi-characters are $x \mapsto x^n$ for $n = 0, 1$;*
2. *if k is complex, $\mathfrak{u} \cong \mathbb{S}^1$ and the quasi-characters are of the form $x \mapsto x^n$ for $n \in \mathbb{Z}$;*

12.2 The Multiplicative Group

3. *if k is p-adic, with uniformiser π, there is an integer $n \geq 1$ such that $\tilde{c}$ factors via the finite set $\mathfrak{u}/(1+(\pi)^n)$.*

Proof The real case is obvious. The complex case requires us to classify all continuous homomorphisms from $\mathbb{S}^1$ to $\mathbb{C}^\times$; the answer is well-known, but we re-derive it here.

Since $\mathbb{S}^1$ is compact, the image of any continuous homomorphism $c : \mathbb{S}^1 \to \mathbb{C}^\times$ is compact, hence contained in $\mathbb{S}^1$. Indeed, if z is an element of the image with $|z| \neq 1$, then either z or $1/z$ is in the image and has absolute value greater than one. This easily implies that the image of c is unbounded, hence not compact. Thus, it suffices to classify the homomorphisms $c : \mathbb{S}^1 \to \mathbb{S}^1$.

We use the map $x \mapsto \exp(2\pi i x)$ to identify $\mathbb{R}/\mathbb{Z}$ with $\mathbb{S}^1$. Pre-composing with this map, we obtain a character $c \circ \exp(2\pi i x) : \mathbb{R} \to \mathbb{S}^1$. By additive duality (Proposition 12.1), every such character is of the form $x \mapsto \exp(2\pi i \eta x)$ for some $\eta \in \mathbb{R}$. In order for c to descend to the quotient $\mathbb{R}/\mathbb{Z} \cong \mathbb{S}^1$, a necessary and sufficient condition is that $\eta \in \mathbb{Z}$. Writing $n = \eta$, it follows that $c(x) = x^n$, as claimed.

Finally, the p-adic case follows essentially from topological considerations. Specifically, since $\tilde{c} : \mathfrak{u} \to \mathbb{C}^\times$ is continuous, the inverse image of an open subset of $\mathbb{C}^\times$ is an open subset of $\mathfrak{u}$. Choose an open subset V of $\mathbb{C}^\times$ that does not contain any non-trivial multiplicative subgroup of $\mathbb{C}^\times$ (see Problem 12.6). Then, $\tilde{c}^{-1}(V)$ is an open neighbourhood of 1 in $\mathfrak{u}$.

By definition of the p-adic topology, the subgroups $1 + (\pi)^n$ form a basis of open neighbourhoods of the identity of $\mathfrak{u}$. Hence, there exists $n \geq 1$ such that $H := 1 + (\pi)^n \subseteq \tilde{c}^{-1}(V)$. However, this implies $\tilde{c}(H) \subseteq V$, and $\tilde{c}(H)$ is a subgroup of $\mathbb{C}^\times$, so $\tilde{c}(H)$ must be trivial, which proves that the kernel of $\tilde{c}$ contains $1 + (\pi)^n$, hence that $\tilde{c}$ factors via $\mathfrak{u}/(1+(\pi)^n)$, as desired. □

Remark 12.3 Let K be a number field. The second part of Problem 12.6, together with the fact that $\mathrm{Gal}(\overline{K}/K)$ is a profinite group, shows that the definition of an Artin L-function may be reformulated by considering—formally more generally—any complex representation of $\mathrm{Gal}\left(\overline{K}/K\right)$. Indeed, any such representation factors via a finite quotient of $\mathrm{Gal}(\overline{K}/K)$, hence via $\mathrm{Gal}(L/K)$ for some finite extension L/K. This topological obstruction prevents one from fully understanding $\mathrm{Gal}(\overline{K}/K)$ by only looking at continuous complex representations: to get a more complete picture, one should also consider continuous representations with values in $\mathrm{GL}_n(\overline{\mathbb{Q}_\ell})$, where ℓ is any prime. These constitute a very rich area of research, but we won't discuss them here, since this would take us too far afield. The interested reader can start by consulting Serre's book [2], where the case of abelian ℓ-adic representations is treated in detail.

Theorem 12.3 justifies the following definition.

Definition 12.4 (Exponent of a Quasi-Character) Let $c : k^\times \to \mathbb{C}^\times$ be a quasi-character. By Theorem 12.3, we may write $c(\alpha) = \tilde{c}(\tilde{\alpha})\|\alpha\|^s$, hence $|c(\alpha)| = \|\alpha\|^\sigma$, where $\sigma = \Re(s)$ is uniquely determined by the character χ. We call σ the **exponent** of c and denote it by $\sigma(c)$.

Remark 12.4 Notice that a quasi-character $c : k^\times \to \mathbb{C}^\times$ is a character if and only if its exponent is zero.

12.2.1 Choice of Haar Measure

We would now like to choose a Haar measure on $k^\times$ in a way that is compatible with the Haar measure on k^+. If g is a function in $\mathcal{K}(k^\times)$, then by definition there is an open neighbourhood B of 0 (for the topology of k) such that $g|_B$ vanishes. Since $\|\xi\|$ is bounded away from 0 on the complement of B, we conclude that $h(\xi) := \frac{g(\xi)}{\|\xi\|}$ is a continuous function with bounded support on all of k (where of course we set $h(0) = 0$). In particular, we may consider the (positive) linear functional

$$\psi : \mathcal{K}(k^\times) \to \mathbb{R}$$
$$g \mapsto \int_{k^+} g(\xi) \frac{d\xi}{\|\xi\|}.$$

By the representation theorem (Theorem 8.4), there exists a unique Radon measure $\mu_{k^\times}$ on $k^\times$ such that

$$\psi(g) = \int_{k^\times} g(\alpha) \, d\mu_{k^\times}(\alpha).$$

Moreover, $\mu_{k^\times}$ is invariant under translation by elements of $k^\times$: it suffices to check that $\psi(g(\beta \cdot)) = \psi(g(\cdot))$ for every $g \in \mathcal{K}(k^\times)$ and $\beta \in k^\times$, and this follows from the identities

$$\psi(g(\beta \cdot)) = \int_{k^+} g(\beta \xi) \frac{d\xi}{\|\xi\|} = \int_{k^+} g(y) \frac{d(\beta^{-1} y)}{\|\beta^{-1} y\|}$$
$$= \int_{k^+} g(y) \frac{\|\beta^{-1}\| dy}{\|\beta^{-1} y\|} = \int_{k^+} g(y) \frac{dy}{\|y\|},$$

where we have used Lemma 12.2. Hence, $\mu_{k^\times}$ is a Haar measure on $k^\times$. We now select an appropriate multiple that is more suitable to our arithmetic applications:

Definition 12.5 (Haar Measure on $k^\times$) We denote by $d\alpha$ the Haar measure given by

1. $d\alpha = d\mu_{k^\times} = \frac{d\mu_{k^+}(\alpha)}{\|\alpha\|}$, if v is Archimedean;
2. $d\alpha = \frac{N\mathfrak{p}}{N\mathfrak{p}-1} d\mu_{k^\times} = \frac{N\mathfrak{p}}{N\mathfrak{p}-1} \frac{d\mu_{k^+}(\alpha)}{\|\alpha\|}$, if v is finite and corresponds to the prime $\mathfrak{p}$.

The next lemma gives the measure of the group of units:

Lemma 12.4 *If v is finite, we have*

$$\int_{\mathfrak{u}} d\alpha = (N\mathfrak{d})^{-1/2}.$$

Proof Let π be a uniformiser. We have

$$O = \bigsqcup_{n \geq 0} \pi^n \mathfrak{u},$$

hence

$$\mu_{k^+}(O) = \sum_{n \geq 0} \mu_{k^+}\left(\pi^n \mathfrak{u}\right) = \sum_{n \geq 0} \|\mathfrak{p}\|^n \mu_{k^+}(\mathfrak{u})$$

$$= \frac{1}{1 - N(\mathfrak{p})^{-1}} \mu_{k^+}(\mathfrak{u}) = \frac{N(\mathfrak{p})}{N(\mathfrak{p}) - 1} \mu_{k^+}(\mathfrak{u}).$$

Recalling that we have defined μ_{k^+} so that $\mu_{k^+}(O) = N(\mathfrak{d})^{-1/2}$ (see Definition 12.2), we obtain

$$\int_{\mathfrak{u}} d\alpha = \int_{\mathfrak{u}} \frac{N\mathfrak{p}}{N\mathfrak{p} - 1} d\mu_{k^\times}(\alpha) = \frac{N\mathfrak{p}}{N\mathfrak{p} - 1} \int_{\mathfrak{u}} \frac{d\xi}{\|\xi\|} = \frac{N\mathfrak{p}}{N\mathfrak{p} - 1} \int_{\mathfrak{u}} d\xi$$

$$= \frac{N\mathfrak{p}}{N\mathfrak{p} - 1} \mu_{k^+}(\mathfrak{u}) = \mu_{k^+}(O) = N(\mathfrak{d})^{-1/2}.$$

□

12.3 Local Zeta Functions I: The General Functional Equation

Fix a function $f : k \to \mathbb{C}$. Continuing with our notation from the previous sections, we will denote by $f(\xi)$ the function on the whole of k, and by $f(\alpha)$ its restriction to $k^\times$.

Definition 12.6 (Class of $\mathfrak{z}$-functions) We denote by $\mathfrak{z}$ the class of all functions $f : k \to \mathbb{C}$ that satisfy

1. $f(\xi) \in \mathfrak{V}_1(k^+)$ (that is, $f(\xi)$ is continuous and in $L^1(k^+)$, and $\hat{f}$ is in $L^1(k^+)$, see both Definition 9.3 and Theorem 12.1);
2. $f(\alpha)\|\alpha\|^\sigma$ and $\hat{f}(\alpha)\|\alpha\|^\sigma$ are in $L^1(k^\times)$ for all $\sigma > 0$.

For each function f of class $\mathfrak{z}$ we can introduce a generalised Fourier transform where—instead of integrating $f(\xi)$ only against characters—we more generally consider all quasi-characters of positive exponent. This is made precise in the following definition.

Definition 12.7 (Tate's Local ζ Function) Let $f \in \mathfrak{z}$ and let c be a quasi-character of $k^\times$ with strictly positive exponent. We set

$$\zeta(f, c) = \int_{k^\times} f(\alpha) c(\alpha) \, d\alpha$$

and call such a function a ζ-function of k.

At least 'locally' (to be defined shortly), we can consider $\zeta(f, c)$ as a holomorphic function. More precisely:

Definition 12.8 We say that two quasi-characters c_1, c_2 are **equivalent** if there exists an unramified quasi-character χ such that $c_2(\alpha) = c_1(\alpha) \chi(\alpha)$.

By Lemma 12.3, the equivalence class of the quasi-character c is given by the set of all quasi characters of the form

$$\alpha \mapsto c(\alpha) \|\alpha\|^s.$$

This allows us to consider a ζ function of k as a collection of many functions of a complex variable s: for each quasi-character c of positive exponent, we can consider the function

$$s \mapsto \zeta(f, c \cdot \|\alpha\|^s).$$

Notice that, by Lemma 12.3, this function can (and should) be considered as being defined

1. on the whole complex plane, if v is Archimedean;
2. on the cylinder $\dfrac{\mathbb{C}}{\mathbb{Z} \cdot \frac{2\pi i}{\log \|N\mathfrak{p}\|}}$, if v is finite and corresponds to $\mathfrak{p}$.

Each of these functions turns out to be holomorphic, in the following sense.

Lemma 12.5 For every quasi-character c of positive exponent, the function $s \mapsto \zeta(f, c \cdot \|\alpha\|^s)$ is well-defined and holomorphic in $\{\Re s > 0\}$.

Proof Convergence of the integral is guaranteed by the fact that $f(\xi)$ is a function of class $\mathfrak{z}$.

To show holomorphicity, it suffices to check that we can differentiate under the integral sign. By definition,

$$\zeta(f, c \cdot \|\alpha\|^s) = \int_{k^\times} f(\alpha) c(\alpha) \|\alpha\|^s \, d\alpha = \int_{k^+} f(\xi) c(\xi) \|\xi\|^{s-1} \, d\xi.$$

12.3 Local Zeta Functions I: The General Functional Equation

Fix a compact subset K of $\{\Re s > 0\}$. We will check that we can differentiate under the integral at a point s in the interior of K. The derivative (in s) of the function being integrated is $f(\xi)c(\xi)\log\|\xi\|\|\xi\|^{s-1}$. For $s \in K$, this function is uniformly absolutely integrable: denoting by e the exponent of c, the integral of the absolute value is

$$\int_{k^+} |f(\xi)| \cdot |c(\xi)| \cdot |\log\|\xi\|| \cdot \|\xi\|^{s-1} \, d\xi = \int_{k^+} |f(\xi)| \cdot |\log\|\xi\|| \cdot \|\xi\|^{e+\Re s-1} \, d\xi$$

with $\Re s$ bounded above and below. Convergence can only fail around 0 and as $\|\xi\| \to \infty$. Let $C = \{\xi \in k^+ : \|\xi\| \leq 1\}$ be a compact neighbourhood of 0 in k^+. Splitting the integral as $\int_C + \int_{k^+ \setminus C}$, we have

$$\int_C |f(\xi)| \cdot |\log\|\xi\|| \cdot \|\xi\|^{e+\Re s-1} \, d\xi \leq \|f|_C\|_\infty \int_C |\log\|\xi\|| \cdot \|\xi\|^{e+\Re s-1} \, d\xi.$$

The exponent $e + \Re s - 1$ is bounded below by a constant κ_0 strictly larger than -1 (since $e > 0$ and $\Re s \geq \min_{s \in K} \Re s > 0$). The integral $\int_C |\log\|\xi\|| \cdot \|\xi\|^{\kappa_0} \, d\xi$ converges (Problem 12.7).

As for the integral on $k^+ \setminus C$, we have

$$\int_{k^+ \setminus C} |f(\xi)| \cdot |\log\|\xi\|| \cdot \|\xi\|^{e+\Re s-1} \, d\xi \leq \int_{k^+ \setminus C} |f(\xi)| \cdot \|\xi\|^{e+\Re s} \, d\xi,$$

and for $s \in K$ the exponent $e + \Re s$ is bounded above by some $\kappa_1 > 1$, so that the function being integrated is dominated by the L^1 function $|f(\xi)| \cdot \|\xi\|^{\kappa_1}$. Thus, thanks to the dominated convergence theorem we may in fact differentiate under the integral sign at any s in the interior of K. As K is arbitrary, this proves the desired holomorphicity. □

Remarkably, all these ζ functions satisfy a functional equation of a very general type. To state it, we introduce the following notation:

Definition 12.9 For a quasi-character c we set $\hat{c}(\alpha) = \|\alpha\|c(\alpha)^{-1}$.

Remark 12.5 It is clear from the definitions that $\sigma(\hat{c}) = 1 - \sigma(c)$.

The following proposition, while comparatively easy to prove, will be the key to all subsequent results about analytic continuation and functional equations.

Proposition 12.3 (Functional Equation of Local ζ Functions) Let $f, g \in \mathfrak{z}$. For every quasi-character c with $\sigma(c) \in (0, 1)$ we have

$$\zeta(f, c)\zeta(\hat{g}, \hat{c}) = \zeta(\hat{f}, \hat{c})\zeta(g, c). \tag{12.4}$$

Remark 12.6 An equivalent (but perhaps easier to remember) way to state the proposition is that the 'pairing' $(f, g) \mapsto \zeta(f, c)\zeta(\hat{g}, \hat{c})$ is symmetric in f, g.

Proof The condition $\sigma(c) \in (0, 1)$ guarantees that both sides are well-defined. By definition, $\hat{c}(\alpha) = \|\alpha\| c(\alpha)^{-1}$, and so

$$\zeta(f, c) \zeta(\hat{g}, \hat{c}) = \int_{k^\times \times k^\times} c(\alpha) f(\alpha) \hat{g}(\beta) \hat{c}(\beta) \, d\alpha \, d\beta$$

$$= \int_{k^\times \times k^\times} c(\alpha) f(\alpha) \hat{g}(\beta) \|\beta\| c(\beta)^{-1} \, d\alpha \, d\beta$$

$$= \int_{k^\times \times k^\times} c(\alpha \beta^{-1}) f(\alpha) \hat{g}(\beta) \|\beta\| \, d\alpha \, d\beta.$$

Replacing $(\alpha, \beta) \to (\alpha, \alpha\beta)$, which (by properties of the Haar measure) does not change $d\alpha \, d\beta$, we rewrite the above as

$$\int_{k^\times \times k^\times} c(\beta^{-1}) f(\alpha) \hat{g}(\alpha\beta) \|\alpha\beta\| \, d\alpha \, d\beta.$$

We now express everything in terms of the additive measure on k^+. Recall from Definition 12.5 that $d\alpha = \frac{d\mu_{k^+}(\alpha)}{\|\alpha\|}$, up to multiplicative constants. Thus, again up to multiplicative constants independent of f, g, the above integral is equal to

$$\int_{k \times k} c(\beta^{-1}) f(\alpha) \hat{g}(\alpha\beta) \|\alpha\beta\| \frac{1}{\|\alpha\| \|\beta\|} \, d\mu_{k^+}(\alpha) \, d\mu_{k^+}(\beta).$$

We finally replace $\hat{g}$ with its definition (Theorem 12.2) to obtain

$$\int_{k \times k} \int_k g(\xi) e^{-2\pi i \Lambda(\alpha\beta\xi)} c(\beta^{-1}) f(\alpha) \, d\mu_{k^+}(\xi) \, d\mu_{k^+}(\alpha) \, d\mu_{k^+}(\beta),$$

which is manifestly symmetric in f and g. $\square$

The crucial remark is now the following: provided that $\zeta(\hat{f}, \hat{c})$ is not identically zero (as a function of c), Proposition 12.3 shows that the ratio

$$\rho(c) := \frac{\zeta(f, c)}{\zeta(\hat{f}, \hat{c})}$$

is independent of f. We will compute $\rho(c)$ when f is a particularly simple function. We will then find that in all cases $\rho(c)$ can be expressed in terms of elementary functions, and in particular that $\rho(c)$ admits meromorphic continuation to the space of all characters. We do this in the next section, but first we establish some formal properties of the function $\rho(c)$ that follow directly from the functional equation.

12.3 Local Zeta Functions I: The General Functional Equation

Proposition 12.4

1. $\rho(\hat{c}) = \frac{c(-1)}{\rho(c)}$
2. $\rho(\bar{c}) = c(-1)\overline{\rho(c)}$
3. $|\rho(c)| = 1$ for c of exponent $1/2$.

Proof

1. $\zeta(f,c) = \rho(c)\zeta(\hat{f},\hat{c}) = \rho(c)\rho(\hat{c})\zeta(\hat{\hat{f}},\hat{\hat{c}}) = \rho(c)\rho(\hat{c})\zeta(f(-\alpha),c)$, where in the last step we used $\hat{\hat{c}} = c$ (by definition) and $\hat{\hat{f}} = f(-\alpha)$ (by Theorem 12.2). On the other hand, by definition,

$$\zeta(f,c) = \int_{k^\times} f(\alpha)c(\alpha)d\alpha = \int_{k^\times} f(-\alpha)c(-\alpha)d(-\alpha)$$

$$= c(-1)\int_{k^\times} f(-\alpha)c(\alpha)d(\alpha) = c(-1)\zeta(f(-\alpha),c).$$

Comparing the two expressions for $\zeta(f,c)$ we get $\rho(c)\rho(\hat{c}) = c(-1)$.

Remark 12.7 There seems to be a typo in Tate's proof of this relation.

2. $\overline{\zeta(f,c)} = \zeta(\bar{f},\bar{c}) = \rho(\bar{c})\zeta(\hat{\bar{f}},\hat{\bar{c}})$. Now observe that $\hat{\bar{c}}(\alpha) = \|\alpha\|\overline{c(\alpha)}^{-1} = \overline{\hat{c}}(\alpha)$, while

$$\hat{\bar{f}}(\eta) = \int_{k^+} \overline{f(\xi)}e^{-2\pi i \Lambda(\eta\xi)}d\xi = \overline{\int_{k^+} f(\xi)e^{2\pi i \Lambda(\eta\xi)}d\xi} = \overline{\hat{f}(-\eta)}.$$

Replacing in $\overline{\zeta(f,c)} = \rho(\bar{c})\zeta(\hat{\bar{f}},\hat{\bar{c}})$ we get

$$\overline{\zeta(f,c)} = \rho(\bar{c})\zeta(\overline{\hat{f}(-\eta)},\overline{\hat{c}}) = \rho(\bar{c})c(-1)\zeta(\overline{\hat{f}},\overline{\hat{c}}) = \rho(\bar{c})c(-1)\overline{\zeta(\hat{f},\hat{c})},$$

where the last equality follows immediately from the definition of $\zeta(f,c)$. On the other hand,

$$\overline{\zeta(f,c)} = \overline{\rho(c)\zeta(\hat{f},\hat{c})} = \overline{\rho(c)}\,\overline{\zeta(\hat{f},\hat{c})}.$$

Comparing the two expressions yields the result.

3. If c has exponent $1/2$, then $c(\alpha)\overline{c(\alpha)} = \|c(\alpha)\|^2 = \|\alpha\| = c(\alpha)\hat{c}(\alpha)$, and therefore $\bar{c} = \hat{c}$. Comparing the expressions for $\rho(\bar{c})$ and $\rho(\hat{c})$ given in (1) and (2) yields $\rho(c)\overline{\rho}(c) = 1$. □

We will check in the next section that $\rho(c)$ is a 'familiar' function, for all quasi-characters c, and in particular it trivially admits meromorphic continuation to $\mathbb{C}$. As a consequence, the local functional equation of Proposition 12.3 yields the following important theorem:

Theorem 12.4 *Any ζ-function of k has meromorphic continuation to the domain of all quasi-characters given by a functional equation of the form*

$$\zeta(f, c) = \rho(c)\zeta(\hat{f}, \hat{c}),$$

where $\rho(c)$ is a meromorphic function of c.

Proof By Lemma 12.5, the function $\zeta(f, c)$ is defined and holomorphic for c of positive exponent. The function $\rho(c)\zeta(\hat{f}, \hat{c})$ is similarly defined and meromorphic (since $\rho(c)$ is only known to be meromorphic) for $\hat{c}$ of positive exponent, that is, for c of exponent strictly less than 1.

In particular, both functions are defined and meromorphic for all $0 < \sigma(c) < 1$, and they coincide in this domain by Proposition 12.3. Thus, we get meromorphic continuation of $\zeta(f, c)$ to the domain of all quasi-characters. □

12.4 Local Zeta Functions II: Computation of the Local Factors

Our objective in this section is to compute the function

$$\rho(c) := \frac{\zeta(f, c)}{\zeta(\hat{f}, \hat{c})}$$

when f is a particularly simple function taken in class $\mathfrak{z}$. We will organise the computation according to the equivalence class of the quasi-character $c(\alpha) = c_0(\alpha)\|\alpha\|^s$. For c in a fixed equivalence class, we will find that $\rho(c)$, seen as a function of the complex variable s, is holomorphic and non-vanishing for $\Re s \in (0, 1)$. These functions $\rho(c)$ will form the basis of all our discussion of the functional equations satisfied by the ζ functions.

It will be necessary to distinguish cases according to whether k is real, complex, or p-adic. Following Tate, we begin each section by recalling our choices for the map Λ, the norm on k, and the Haar measures on k^+ and on $k^\times$.

12.4.1 Real Case

12.4.1.1 Conventions

1. ξ is a real variable
2. $\Lambda(\xi) = -\xi \pmod 1$
3. $\|\alpha\|$ is the ordinary absolute value
4. $d\mu_{k^+}(\xi) = d\xi$ is the ordinary Lebesgue measure
5. $d\alpha = \frac{d\mu_{k^+}(\alpha)}{\|\alpha\|}$

12.4 Local Zeta Functions II: Computation of the Local Factors

12.4.1.2 Equivalence Classes of Quasi-Characters

According to Lemma 12.3 and Proposition 12.2, there are two equivalence classes:

$$\alpha \mapsto \|\alpha\|^s \quad \text{and} \quad \alpha \mapsto (\operatorname{sign}\alpha)\|\alpha\|^s.$$

We denote the former by $\|\cdot\|^s$ and the latter by $\pm \|\cdot\|^s$.

12.4.1.3 Choice of f

We correspondingly take

$$f(\xi) = e^{-\pi\xi^2} \quad \text{and} \quad f_\pm(\xi) = \xi e^{-\pi\xi^2}. \tag{12.5}$$

12.4.1.4 Fourier Transforms

We have

$$\hat{f}(\xi) = f(\xi) \quad \text{and} \quad \hat{f}_\pm(\xi) = i f_\pm(\xi).$$

Before we check these equalities, we pause to recall a classical lemma in real Fourier analysis:

Lemma 12.6 *Let $a, b \in \mathbb{R}$. We have*

$$\int_\mathbb{R} e^{-2\pi y^2 + 4\pi i a y} \, dy = \frac{e^{-2\pi a^2}}{\sqrt{2}}$$

and more generally

$$\int_\mathbb{R} e^{-2b\pi y^2 + 4\pi i a y} \, dy = \frac{1}{\sqrt{2b}} e^{-2\pi(a^2/b)}.$$

Proof We have

$$\int_\mathbb{R} e^{-2\pi y^2 + 4\pi i a y} \, dy = \int_\mathbb{R} e^{-2\pi(y^2 - 2iay - a^2 + a^2)} \, dy$$

$$= \int_\mathbb{R} e^{-2\pi(y-ia)^2 - 2\pi a^2} \, dy$$

$$= e^{-2\pi a^2} \int_\mathbb{R} e^{-2\pi(y-ia)^2} \, dy.$$

Since the function $y \mapsto e^{-2\pi(y-ia)^2}$ is holomorphic, we can shift the integration contour from the real line $\mathbb{R}$ to $ia + \mathbb{R}$, thus rewriting the above integral as

$$e^{-2\pi a^2} \int_{\mathbb{R}} e^{-2\pi y^2}\, dy.$$

The value $I := \int_{\mathbb{R}} e^{-2\pi y^2} = \frac{1}{\sqrt{2}}$ is well-known, and can be obtained by the standard trick

$$I^2 = \int_{\mathbb{R}^2} e^{-2\pi(x^2+y^2)}\, dx\, dy = \int_0^\infty \int_0^{2\pi} e^{-2\pi r^2} r\, dr\, d\vartheta$$
$$= \frac{1}{2} \int_0^\infty e^{-2\pi r^2}\, d(2\pi r^2) = \frac{1}{2} \int_0^\infty e^{-t}\, dt = \frac{1}{2}.$$

This proves the first formula in the statement. The second follows: upon writing $y = \frac{z}{\sqrt{b}}$ we obtain

$$\int_{\mathbb{R}} e^{-2b\pi y^2 + 4\pi i a y}\, dy = \int_{\mathbb{R}} e^{-2\pi z^2 + 4\pi i \frac{a}{\sqrt{b}} z}\, \frac{dz}{\sqrt{b}} = \frac{1}{\sqrt{2b}} e^{-2\pi(a^2/b)}.$$

$\square$

Let us now compute $\hat{f}(\xi)$. By definition, recalling from Eq. (12.2) the definition of the Fourier transform in our setting, we have

$$\hat{f}(\xi) = \int_{\mathbb{R}} f(y) e^{-2\pi i \Lambda(y\xi)}\, dy = \int_{\mathbb{R}} e^{-\pi y^2} e^{2\pi i y\xi}\, dy = e^{-\pi \xi^2},$$

where in the last step we used Lemma 12.6 with $a = \frac{1}{2}\xi, b = \frac{1}{2}$. For the sake of completeness, we also derive the expression for $\hat{f}_\pm$:

$$\hat{f}_\pm(\xi) = \int_{\mathbb{R}} f(y) e^{-2\pi i \Lambda(y\xi)}\, dy = \int_{\mathbb{R}} y e^{-\pi y^2} e^{2\pi i y\xi}\, dy$$
$$= \int_{\mathbb{R}} e^{-\pi \xi^2} y e^{-\pi(y-i\xi)^2}\, dy = e^{-\pi \xi^2} \int_{\mathbb{R}} y e^{-\pi(y-i\xi)^2}\, dy$$
$$= e^{-\pi \xi^2} \int_{\mathbb{R}} (y + i\xi) e^{-\pi y^2}\, dy = i\xi e^{-\pi \xi^2} + \int_{\mathbb{R}} y e^{-\pi y^2}\, dy = i\xi e^{-\pi \xi^2},$$

where the integral $\int_{\mathbb{R}} y e^{-\pi y^2}\, dy$ vanishes since we are summing a (rapidly decaying) odd function over a symmetric domain.

Remark 12.8 The same result can also be derived more cleanly by recalling that, writing $g(\xi) = 2\pi i \xi f(\xi)$, one has $\frac{d\hat{f}(\xi)}{d\xi} = \hat{g}$. Applying this to $f(\xi)$ and $g(\xi) =$

12.4 Local Zeta Functions II: Computation of the Local Factors

$2\pi i f_\pm(\xi)$ immediately yields

$$2\pi i \hat{f}_\pm = \frac{d}{d\xi}\left(e^{-\pi\xi^2}\right) = -2\pi\xi e^{-\pi\xi^2} \Rightarrow \hat{f}_\pm = i\xi e^{-\pi\xi^2} = if_\pm(\xi).$$

12.4.1.5 The ζ-Functions

From the definitions we obtain

$$\zeta\left(f, \|\cdot\|^s\right) = \int_{\mathbb{R}^*} f(\alpha)\|\alpha\|^s \, d\alpha = \int_{\mathbb{R}} e^{-\pi\xi^2} |\xi|^s \frac{d\xi}{|\xi|}$$

$$= 2\int_0^\infty \xi^{s-1} e^{-\pi\xi^2} \, d\xi \stackrel{y=\xi^2}{=} \int_0^\infty y^{(s-1)/2} e^{-\pi y} y^{-1/2} dy$$

$$\stackrel{z=\pi y}{=} \frac{1}{\pi} \int_0^\infty \left(\frac{z}{\pi}\right)^{s/2-1} e^{-z} \, dz = \pi^{-s/2} \Gamma(s/2).$$

Remark 12.9 Note that we have essentially interpreted the Γ function as a ζ function of the local field $\mathbb{R}$. Furthermore, notice that this exact function $\pi^{-s/2}\Gamma(s/2)$ is precisely the 'missing factor' from the completed ζ function of Remark 1.2. This already strongly suggests that Tate's local ζ functions should have something to do with the global Dedekind ζ function we met in Sect. 1.2. We will clarify this connection in the next chapters.

Remark 12.10 Since the function $\Gamma(s/2)$ is meromorphic and has no zeroes (see Problem 1.1), the function $\zeta(f, \|\cdot\|^s)$ is meromorphic and nowhere zero. We will see that this holds for all the ζ functions we compute in this chapter; we will state this result formally in Proposition 12.5.

A similar calculation leads to

$$\zeta\left(f_\pm, \pm\|\cdot\|^s\right) = \int_{-\infty}^0 -\xi e^{-\pi\xi^2} |\xi|^s \frac{d\xi}{|\xi|} + \int_0^\infty \xi e^{-\pi\xi^2} |\xi|^s \frac{d\xi}{|\xi|}$$

$$= 2\int_0^\infty \xi^s e^{-\pi\xi^2} d\xi = \pi^{-\frac{s+1}{2}} \Gamma\left(\frac{s+1}{2}\right).$$

As above, we note that this is a meromorphic and nowhere vanishing function.

Now note that $\widehat{\|\cdot\|^s} = \|\cdot\|^{1-s}$ and $\widehat{\pm\|\cdot\|^s} = \pm\|\cdot\|^{1-s}$. Combined with the linearity of $\zeta(f, c)$ in the first argument, this shows

$$\zeta(\hat{f}, \widehat{\|\cdot\|^s}) = \zeta(f, \|\cdot\|^{1-s}) = \pi^{-\frac{1-s}{2}} \Gamma\left(\frac{1-s}{2}\right),$$

$$\zeta\left(\hat{f}_\pm, \widehat{\pm\|\cdot\|^s}\right) = \zeta(if_\pm, \pm\|\cdot\|^{1-s}) = i\pi^{-\frac{(1-s)+1}{2}} \Gamma\left(\frac{(1-s)+1}{2}\right).$$

12.4.1.6 The Function $\rho(c)$

We have obtained

$$\rho(\|\cdot\|^s) = \frac{\pi^{-s/2}\Gamma\left(\frac{s}{2}\right)}{\pi^{-\frac{1-s}{2}}\Gamma\left(\frac{1-s}{2}\right)} = 2^{1-s}\pi^{-s}\cos\left(\frac{\pi s}{2}\right)\Gamma(s)$$

$$\rho(\pm\|\cdot\|^s) = i\frac{\pi^{-\frac{s+1}{2}}\Gamma\left(\frac{s+1}{2}\right)}{\pi^{-\frac{(1-s)+1}{2}}\Gamma\left(\frac{(1-s)+1}{2}\right)} = -i2^{1-s}\pi^{-s}\sin\left(\frac{\pi s}{2}\right)\Gamma(s).$$

Both of these functions admit meromorphic continuation to $\mathbb{C}$ by well-known properties of the Γ function.

12.4.2 Complex Case

12.4.2.1 Conventions

1. $\xi = x + iy = re^{i\vartheta}$
2. $\Lambda(\xi) = -2\mathfrak{R}(\xi) = -2x$
3. $\|\alpha\| = \alpha\bar{\alpha} = |\alpha|^2 = r^2$ is the square of the ordinary absolute value
4. $d\mu_{k^+}(\xi) = 2\,dx\,dy$ is twice the ordinary Lebesgue measure
5. $d\alpha = \frac{d\mu_{k^+}(\alpha)}{|\alpha|^2}$

12.4.2.2 Equivalence Classes of Quasi-Characters

According to Lemma 12.3 and Proposition 12.2, the equivalence classes are parametrised by $n \in \mathbb{Z}$. Representatives for each equivalence class are given by

$$c_n(re^{i\vartheta}) = e^{in\vartheta},$$

and the corresponding equivalence class consists of all characters of the form $c_n(\alpha)\|\alpha\|^s$.

12.4.2.3 Choice of f

We take

$$f_n(\xi) = \begin{cases} (x - iy)^{|n|}e^{-2\pi(x^2+y^2)}, & \text{for } n \geq 0 \\ (x + iy)^{|n|}e^{-2\pi(x^2+y^2)}, & \text{for } n < 0 \end{cases} \qquad (12.6)$$

12.4 Local Zeta Functions II: Computation of the Local Factors

Notice that, writing $\alpha = re^{i\vartheta} \neq 0$, the value $f_n(\alpha)$ (for $n \geq 0$) can also be written as

$$(r(\cos\vartheta - i\sin\vartheta))^n e^{-2\pi r^2} = r^n e^{-in\vartheta} e^{-2\pi r^2},$$

while for $n \leq 0$ we have

$$f_n(\alpha) = (r(\cos\vartheta + i\sin\vartheta))^{-n} e^{-2\pi r^2} = r^{-n} e^{-in\vartheta} e^{-2\pi r^2}.$$

We may thus write the uniform formula $f_n(re^{i\vartheta}) = r^{|n|} e^{-in\vartheta} e^{-2\pi r^2}$.

12.4.2.4 Fourier Transforms

We have

$$\hat{f}_n(\xi) = i^{|n|} f_{-n}(\xi)$$

for all n. To prove this we proceed as follows:

1. for $n = 0$ we compute

$$\hat{f}_0(\xi) = \int_{\mathbb{C}} f(\eta) e^{-2\pi i \Lambda(\eta\xi)} \, dk^+(\eta) = 2\int_{\mathbb{R}^2} e^{-2\pi(x^2+y^2)} e^{4\pi i \Re(\xi(x+iy))} \, dx\, dy.$$

Writing $\xi = u + iv$, the previous integral becomes

$$\hat{f}_0(\xi) = 2\int_{\mathbb{R}^2} e^{-2\pi(x^2+y^2)} e^{4\pi i(ux-bv)} \, dx\, dy$$

$$= 2\left(\int_{\mathbb{R}} e^{-2\pi x^2 + 4\pi i u x} \, dx\right)\left(\int_{\mathbb{R}} e^{-2\pi y^2 - 4\pi i v y} \, dy\right)$$

$$= 2\left(\frac{e^{-2\pi u^2}}{\sqrt{2}}\right)\left(\frac{e^{-2\pi v^2}}{\sqrt{2}}\right) = e^{-2\pi(u^2+v^2)}$$

$$= f_0(u+iv) = f_0(\xi),$$

where the single real integrals are treated using the first part of Lemma 12.6 (with $a = u$ and $a = -v$, respectively).

2. for $n \geq 0$ we proceed by induction. Assume that we know $\hat{f}_n(\xi) = i^{|n|} f_{-n}(\xi)$, that is,

$$2\int_{\mathbb{R}}\int_{\mathbb{R}} (u-iv)^n e^{-2\pi(u^2+v^2)+4\pi i(xu-yv)} \, du\, dv = i^n (x+iy)^n e^{-2\pi(x^2+y^2)}.$$

(12.7)

Introduce the differential operator $D = \frac{1}{4\pi i}\left(\frac{\partial}{\partial x} + i\frac{\partial}{\partial y}\right)$ and observe that $D(x+iy) = 0$, hence $D(x+iy)^n = 0$ (this is essentially a consequence of the Cauchy-Riemann equation for the analytic function $z \mapsto z^n$). Applying D to both sides of (12.7) we obtain

$$2\int_{\mathbb{R}}\int_{\mathbb{R}} (u-iv)^{n+1} e^{-2\pi(u^2+v^2)+4\pi i(xu-yv)}\, du\, dv = i^{n+1}(x+iy)^{n+1} e^{-2\pi(x^2+y^2)},$$

which is nothing but the equality $\widehat{f_{n+1}}(\xi) = i^{n+1} f_{-(n+1)}(\xi)$.

3. finally, to handle the case $n < 0$, we write $n = -m$ with $m \geq 0$ and consider the equality (consequence of the previous case)

$$\widehat{f_m}(\xi) = i^m f_{-m}(\xi).$$

Taking the Fourier transform of both sides, we get

$$f_m(-\xi) = \widehat{\widehat{f_m}}(\xi) = i^m \widehat{f_{-m}}(\xi),$$

which, using $f_m(-\xi) = (-1)^m f_m(\xi)$, yields the desired equality.

12.4.2.5 The ζ-Functions

Recall the formula $f_n(re^{i\vartheta}) = r^{|n|} e^{-in\vartheta} e^{-2\pi r^2}$. From this, we obtain

$$\zeta(f_n, c_n \|\cdot\|^s) = \int_{\mathbb{C}^\times} f(\alpha) c_n(\alpha) \|\alpha\|^s\, d\alpha = \int_{\mathbb{C}} r^{|n|} e^{-in\vartheta} e^{-2\pi r^2} e^{in\vartheta} r^{2s} \frac{2dx\, dy}{r^2}$$

$$= \int_0^\infty \int_0^{2\pi} r^{|n|+2s-2} e^{-2\pi r^2} 2r\, dr\, d\vartheta = 2\pi \int_0^\infty (r^2)^{\frac{|n|+2s-2}{2}} e^{-2\pi r^2}\, d(r^2)$$

$$= 2\pi \int_0^\infty t^{\frac{|n|}{2}+(s-1)} e^{-2\pi t}\, dt = 2\pi \int_0^\infty \left(\frac{u}{2\pi}\right)^{\frac{|n|}{2}+(s-1)} e^{-u} \frac{du}{2\pi}$$

$$= (2\pi)^{-\frac{|n|}{2}+(1-s)} \int_0^\infty u^{\frac{|n|}{2}+(s-1)} e^{-u}\, du = (2\pi)^{-\frac{|n|}{2}+(1-s)} \Gamma\left(s + \frac{|n|}{2}\right).$$

As in the real case (see Remark 12.10), this is a meromorphic and nowhere vanishing function.

We highlight two aspects of this calculation that are easy to get wrong: on the one hand, recall that our Haar measure is *twice* the standard Lebesgue measure, which justifies the factor of 2 in front of $dx\, dy$, hence the factor of 2 in $2r\, dr\, d\vartheta$. On the other hand, note $\|\alpha\|^s = r^{2s}$ (and not r^s).

12.4 Local Zeta Functions II: Computation of the Local Factors

Remark 12.11 Tate finds $(2\pi)^{\frac{|n|}{2}+(1-s)}\Gamma\left(s+\frac{|n|}{2}\right)$ instead (note the sign change in $\frac{|n|}{2}$). This seems to be a typo on Tate's part. Luckily, the sign in question is irrelevant for the computation of $\rho(c)$.

Now note that $\widehat{c_n\|\cdot\|^s} = c_{-n}\|\cdot\|^{1-s}$, so

$$\zeta(\hat{f}_n, \widehat{c_n\|\cdot\|^s}) = \zeta(i^{|n|}f_{-n}, c_{-n}\|\cdot\|^{1-s}) = i^{|n|}(2\pi)^{-\frac{|n|}{2}+s}\Gamma\left(1-s+\frac{|n|}{2}\right).$$

12.4.2.6 The Function $\rho(c)$

We simply take the ratio of the functions computed above to get

$$\rho(c_n\|\cdot\|^s) = (-i)^{|n|}\frac{(2\pi)^{1-s}\Gamma\left(s+\frac{|n|}{2}\right)}{(2\pi)^s\Gamma\left((1-s)+\frac{|n|}{2}\right)},$$

which admits meromorphic continuation to $\mathbb{C}$ by well-known properties of the Γ function.

12.4.3 p-Adic Case

12.4.3.1 Conventions

1. ξ is a p-adic variable
2. $\Lambda(\xi) = \lambda(\mathrm{tr}_{k/\mathbb{Q}_p}(\xi))$
3. $d\mu_{k^+}(\xi)$ is normalised in such a way that O has measure $(N\mathfrak{d})^{-1/2}$
4. $\alpha = \tilde{\alpha}\pi^n$, where α is a variable in $k^\times$, π is a uniformiser, and $\tilde{\alpha}$ is a unit
5. $\|\alpha\| = (N\mathfrak{p})^{-n}$
6. $d\mu_{k^\times}(\alpha) = \frac{N\mathfrak{p}}{N\mathfrak{p}-1}\frac{d\mu_{k^+}(\alpha)}{\|\alpha\|}$, so that $\mathfrak{u}$ gets measure $(N\mathfrak{d})^{-1/2}$

12.4.3.2 Equivalence Classes of Quasi-Characters

The classification of quasi-characters is complicated (see Proposition 12.2). Luckily, for the computation of the local ζ function we only need to know the **conductor** of our character (that is, the minimal $n \geq 0$ such that c is trivial on the subgroup[2] $(1+\mathfrak{p}^n)\cap \mathfrak{u}$ of $\mathfrak{u}$). Also note that each equivalence class of characters contains a

---
[2] Note that for $n \geq 1$ the set $1+\mathfrak{p}^n$ is contained in $\mathfrak{u}$ and is a subgroup. For $n=0$, we set conventionally $1+\mathfrak{p}^0 = \mathfrak{u}$.

representative for which $c(\pi) = 1$. Let then c_n be a character such that

$$c(\pi) = 1, \quad c(1 + \mathfrak{p}^n) = \{1\}, \quad c(1 + \mathfrak{p}^{n-1}) \neq \{1\} \text{ if } n \geq 1.$$

12.4.3.3 Choice of f

We take f to depend only on the conductor. Precisely, we set

$$f_n(\xi) = \begin{cases} e^{2\pi i \Lambda(\xi)}, & \text{if } \xi \in \mathfrak{d}^{-1}\mathfrak{p}^{-n} \\ 0, & \text{otherwise} \end{cases} \quad (12.8)$$

Notice that for $n = 0$ the function $f_0(\xi)$ is the characteristic function of $\mathfrak{d}^{-1}$ (for $\xi \in \mathfrak{d}^{-1}$ we have $\Lambda(\xi) = 0$).

12.4.3.4 Fourier Transforms

We claim that

$$\hat{f}_n(\xi) = \begin{cases} (N\mathfrak{d})^{1/2}(N\mathfrak{p})^n, & \text{if } \xi \equiv 1 \pmod{\mathfrak{p}^n} \\ 0, & \text{if } \xi \not\equiv 1 \pmod{\mathfrak{p}^n}. \end{cases} \quad (12.9)$$

By definition,

$$\hat{f}_n(\xi) = \int_k f_n(\eta) e^{-2\pi i \Lambda(\xi\eta)} \, d\eta$$

$$= \int_k e^{2\pi i \Lambda(\eta)} \mathbf{1}_{\eta \in \mathfrak{d}^{-1}\mathfrak{p}^{-n}} e^{-2\pi i \Lambda(\xi\eta)} \, d\eta$$

$$= \int_{\mathfrak{d}^{-1}\mathfrak{p}^{-n}} e^{-2\pi i \Lambda((\xi-1)\eta)} \, d\eta.$$

If $\xi \equiv 1 \pmod{\mathfrak{p}^n}$, then $(\xi - 1)\eta$ lies in $\mathfrak{d}^{-1}$ for every $\eta \in \mathfrak{d}^{-1}\mathfrak{p}^{-n}$. By definition of the different, we have $\text{tr}_{k/\mathbb{Q}_p}(\mathfrak{d}^{-1}) \subseteq \mathbb{Z}_p$, which implies $\Lambda((\xi - 1)\eta) = 0$. Thus, if $\xi \equiv 1 \pmod{\mathfrak{p}^n}$, the function that is integrated is identically equal to 1, hence we get the measure of the set over which we are integrating, that is,

$$\mu_{k^+}\left(\mathfrak{d}^{-1}\mathfrak{p}^{-n}\right) = (N\mathfrak{p})^n (N\mathfrak{d})^{1/2},$$

where we have used Remark 12.2.

On the other hand, suppose that $\xi \not\equiv 1 \pmod{\mathfrak{p}^n}$. Then (by definition of conductor) the map $\eta \mapsto \Lambda((\xi - 1)\eta)$ is a non-trivial character of the group

12.4 Local Zeta Functions II: Computation of the Local Factors

$\mathfrak{d}^{-1}\mathfrak{p}^{-n}$, hence its integral over this (compact) group vanishes by Proposition 9.2. This completes the verification of Eq. (12.9).

12.4.3.5 The ζ-Functions

Let again c be a character of conductor π^n that satisfies $c(\pi) = 1$. We begin by computing the ζ function in case c is unramified, that is, $n = 0$. In this case, the conditions $c(\pi) = 1$ and $c(\mathfrak{u}) = \{1\}$ force c to be trivial, and its equivalence class is the class of the powers of the norm, $\|\cdot\|^s$. The local ζ function is then

$$\zeta(f_0, \|\cdot\|^s) = \int_{k^\times} f_0(\alpha)\|\alpha\|^s \, d_{k^\times}(\alpha) = \int_{\mathfrak{d}^{-1}} e^{2\pi i \Lambda(\alpha)} \|\alpha\|^s \, d_{k^\times}(\alpha).$$

Next, we observe that $\Lambda(\alpha) \in \mathbb{Z}$ for $\alpha \in \mathfrak{d}^{-1}$, hence the integral reduces to

$$\int_{\mathfrak{d}^{-1}} \|\alpha\|^s \, d_{k^\times}(\alpha).$$

Writing $\mathfrak{d} = \pi^d$, we have

$$\mathfrak{d}^{-1} = \bigsqcup_{v=-d}^{\infty} \{\xi \in k : \|\xi\| = (N\mathfrak{p})^{-v}\}.$$

Further set $A_v := \{\xi \in k : \|\xi\| = (N\mathfrak{p})^{-v}\}$.

Remark 12.12 Since $\mu_{k^\times}$ is invariant under rescaling by elements of $k^\times$ (that's the point of the Haar measure!), we have $\mu_{k^\times}(A_v) = \mu_{k^\times}(\pi^v \mathfrak{u}) = \mu_{k^\times}(\mathfrak{u}) = (N\mathfrak{d})^{-1/2}$ for all v.

Using this remark we easily obtain

$$\int_{\mathfrak{d}^{-1}} \|\alpha\|^s \, d\mu_{k^\times}(\alpha) = \sum_{v=-d}^{\infty} \int_{A_v} \|\alpha\|^s d\mu_{k^\times}(\alpha) = \sum_{v=-d}^{\infty} \int_{A_v} (N\mathfrak{p})^{-vs} d\mu_{k^\times}(\alpha)$$

$$= \sum_{v=-d}^{\infty} (N\mathfrak{p})^{-vs} \mu_{k^\times}(A_v) = (N\mathfrak{d})^{-1/2} \frac{(N\mathfrak{p})^{ds}}{1-(N\mathfrak{p})^{-s}}.$$

Finally, recalling that $\mathfrak{d} = \mathfrak{p}^d$ has norm $(N\mathfrak{p})^d$, the result may be rewritten as

$$\zeta(f_0, \|\cdot\|^s) = \frac{(N\mathfrak{d})^{s-1/2}}{1-(N\mathfrak{p})^{-s}}. \tag{12.10}$$

We note that this function is everywhere meromorphic and has no zeroes in $\mathbb{C}$.

Now for the Fourier transform $\hat{f}_0$: we have computed above that this is $(N\mathfrak{d})^{1/2}\mathbf{1}_\mathcal{O}$, hence

$$\zeta(\hat{f}_0, \widehat{\|\cdot\|^s}) = \zeta(\hat{f}_0, \|\cdot\|^{1-s}) = (N\mathfrak{d})^{1/2}\int_\mathcal{O} \|\alpha\|^{1-s}\, d\mu_{k^\times}(\alpha)$$

$$= (N\mathfrak{d})^{1/2}\sum_{v\geq 0}(N\mathfrak{p})^{-v(1-s)}\mu_{k^\times}(A_v)$$

$$= (N\mathfrak{d})^{1/2}\sum_{v\geq 0}(N\mathfrak{p})^{-v(1-s)}\mu_{k^\times}(A_v) = \frac{1}{1-(N\mathfrak{p})^{s-1}},$$

where we have used again Remark 12.12. More generally, essentially by the same computation one shows:

Theorem 12.5 *Let χ be an unramified character of $k^\times$ (so that $\chi(u\cdot\pi^v) = \chi(\pi)^v$ for all v and all $u \in \mathfrak{u}$; we may then evaluate χ on any fractional ideal (π^v)). Let f be the characteristic function of the fractional ideal $I = (\pi^n)$. The local ζ function $\zeta(f, \chi)$ is given by*

$$\zeta(f, \chi) = \frac{(N\mathfrak{d})^{-1/2}\chi(I)(NI)^{-s}}{1 - \chi(\mathfrak{p})(N\mathfrak{p})^{-s}}.$$

Remark 12.13 If we take $k = \mathbb{Q}_p$ and $I = (1)$ in Theorem 12.5 we get that the local zeta function $\zeta(f, \chi)$ looks very much like the local factor at p of the classical Dirichlet L-functions (see Eq. (1.1)). We will clarify the connection later, when we discuss global zeta functions.

We are left with computing the local zeta functions of ramified characters. Write the different as $\mathfrak{d} = \mathfrak{p}^d$. We start by decomposing the integral defining ζ as a sum over the annuli A_v:

$$\zeta(f_n, c_n\|\cdot\|^s) = \int_{\mathfrak{d}^{-1}\mathfrak{p}^{-n}} e^{2\pi i\Lambda(\alpha)}c_n(\alpha)\|\alpha\|^s\, d\mu_{k^\times}(\alpha)$$

$$= \sum_{v=-d-n}^{\infty}(N\mathfrak{p})^{-vs}\int_{A_v} e^{2\pi i\Lambda(\alpha)}c_n(\alpha)\, d\mu_{k^\times}(\alpha).$$

Next, we observe that all but one of the terms in this sum actually vanish:

Lemma 12.7 *For every $v > -d - n$ we have $\int_{A_v} e^{2\pi i\Lambda(\alpha)}c_n(\alpha)\, d\mu_{k^\times}(\alpha) = 0$.*

12.4 Local Zeta Functions II: Computation of the Local Factors

Proof We distinguish two cases:

1. $v \geq -d$. In this case $A_v \subseteq \mathfrak{d}^{-1}$, hence $\Lambda(\alpha) \in \mathbb{Z}$ for all $\alpha \in A_v$ by definition of the different. It follows that $e^{2\pi i \Lambda(\alpha)} = 1$ on all of A_v, and the integral in question is

$$\int_{A_v} c_n(\alpha)\, d\mu_{k^\times}(\alpha) = \int_{\pi^v \mathfrak{u}} c_n(\alpha)\, d\mu_{k^\times}(\alpha) = \int_{\mathfrak{u}} c_n(\pi^v \alpha)\, d\mu_{k^\times}(\alpha)$$

$$= c_n(\pi)^v \int_{\mathfrak{u}} c_n(\alpha)\, d\mu_{k^\times}(\alpha) = 0$$

by Proposition 9.2 (notice that c_n, being ramified, is by definition nontrivial on $\mathfrak{u}$).

2. $-d > v > -d-n$. We write A_v as the disjoint union of sets of the form $\alpha_0 + \mathfrak{p}^{-d} = \alpha_0(1 + \mathfrak{p}^{-d-v})$. On each such set, Λ is constant and equal to $\Lambda(\alpha_0)$. It follows that

$$\int_{\alpha_0 + \mathfrak{d}^{-1}} e^{2\pi i \Lambda(\alpha)} c_n(\alpha)\, d\mu_{k^\times}(\alpha) = e^{2\pi i \Lambda(\alpha_0)} \int_{\alpha_0 + \mathfrak{d}^{-1}} c_n(\alpha)\, d\mu_{k^\times}(\alpha).$$

We now prove that the last integral is zero. Translating (multiplicatively) by α_0 shows that

$$\int_{\alpha_0 + \mathfrak{d}^{-1}} c_n(\alpha)\, d\mu_{k^\times}(\alpha) = \int_{\alpha_0(1+\mathfrak{p}^{-v-d})} c_n(\alpha)\, d\mu_{k^\times}(\alpha)$$

$$= \int_{1+\mathfrak{p}^{-v-d}} c_n(\alpha_0 \alpha)\, d\mu_{k^\times}(\alpha)$$

$$= c_n(\alpha_0) \int_{1+\mathfrak{p}^{-v-d}} c_n(\alpha)\, d\mu_{k^\times}(\alpha).$$

Since $-v-d > 0$ by assumption, $1 + \mathfrak{p}^{-v-d}$ is a (compact) subgroup of $\mathfrak{u}$. On the other hand, $c_n(\alpha)$ is nontrivial on it, because by definition of the conductor the smallest exponent k such that c_n is trivial on $1 + \mathfrak{p}^k$ is $k = n$, and $-v-d < n$ by assumption. Once again, we conclude by applying Proposition 9.2. □

Thus, the local zeta function is given simply by

$$\zeta(f_n, c_n \|\cdot\|^s) = (N\mathfrak{p})^{(d+n)s} \int_{A_{-d-n}} e^{2\pi i \Lambda(\alpha)} c_n(\alpha)\, d\mu_{k^\times}(\alpha).$$

As is usual in the p-adic setting, we can use the fact that the functions we integrate are locally constant to rewrite the remaining integral as a finite sum. More precisely, fix a set $\{\varepsilon\}$ of representatives for the quotient $\mathfrak{u}/(1+\mathfrak{p}^n)$. Then

$$A_{-d-n} = \mathfrak{u}\pi^{-d-n} = \bigsqcup_\varepsilon \varepsilon\pi^{-d-n}(1+\mathfrak{p}^n) = \bigsqcup_\varepsilon (\varepsilon\pi^{-d-n} + \mathfrak{d}^{-1}).$$

On each set $\varepsilon\pi^{-d-n}(1+\mathfrak{p}^n)$ the character c_n is constant by definition of n, and its value is $c_n(\varepsilon)c_n(\pi)^{-d-n} = c_n(\varepsilon)$ (recall that we chose our representatives c_n to satisfy $c_n(\pi) = 1$). Similarly, Λ is also constant and equal to $\Lambda(\varepsilon\pi^{-d-n})$. Thus,

$$\zeta(f_n, c_n \|\cdot\|^s) = (N\mathfrak{p})^{s(d+n)} \sum_\varepsilon c_n(\varepsilon) e^{2\pi i \Lambda(\varepsilon\pi^{-d-n})} \int_{1+\mathfrak{p}^n} d\mu_{k^\times}(\alpha).$$

This expression makes it clear that, once more, this ζ function is everywhere meromorphic (even holomorphic) and has no zeroes in $\mathbb{C}$.

Finally, we compute the local zeta functions attached to the Fourier transforms of the f_n for $n > 0$. We have already seen in Eq. (12.9) that the Fourier transform of f_n is $(N\mathfrak{d})^{1/2}(N\mathfrak{p})^n \mathbf{1}_{1+\mathfrak{p}^n}$. On the set $1 + \mathfrak{p}^n$, both $c_n(\alpha)^{-1}$ and $\|\alpha\|^{1-s}$ are equal to 1 (here we use $n > 0$), and therefore

$$\zeta(\widehat{f_n}, \widehat{c_n\|\cdot\|^s}) = (N\mathfrak{d})^{1/2}(N\mathfrak{p})^n \int_{1+\mathfrak{p}^n} d\mu_{k^\times}(\alpha),$$

which is simply a constant.

12.4.3.6 The Function $\rho(c)$

For the unramified character c_0 we get

$$\rho(c\|\cdot\|^s) = (N\mathfrak{d})^{s-1/2} \frac{1-(N\mathfrak{p})^{s-1}}{1-(N\mathfrak{p})^{-s}}.$$

The situation is slightly more complicated for a ramified character c of conductor $\mathfrak{f} = \mathfrak{p}^n$. Let $\{\varepsilon\}$ be a set of representatives for the quotient $\mathfrak{u}/(1+\mathfrak{f})$ and set

$$\rho_0(c) = (N\mathfrak{f})^{-1/2} \sum_\varepsilon c(\varepsilon) e^{2\pi i \Lambda(\varepsilon/\pi^{v(\mathfrak{d}\mathfrak{f})})}.$$

The function $\rho(c)$ is then given by

$$\rho(c\|\cdot\|^s) = \frac{(N\mathfrak{p})^{s(d+n)} \sum_\varepsilon c_n(\varepsilon) e^{2\pi i \Lambda(\varepsilon\pi^{-d-n})} \int_{1+\mathfrak{p}^n} d\mu_{k^\times}(\alpha)}{(N\mathfrak{d})^{1/2}(N\mathfrak{p})^n \int_{1+\mathfrak{p}^n} d\mu_{k^\times}(\alpha)}$$

$$= \rho_0(c) \frac{(N\mathfrak{p})^{s(d+n)}(N\mathfrak{f})^{1/2}}{(N\mathfrak{d})^{1/2}(N\mathfrak{p})^n} = \rho_0(c)(N\mathfrak{p})^{(s-1/2)(d+n)}$$

$$= \rho_0(c)(N\mathfrak{d}\mathfrak{f})^{s-1/2},$$

12.4 Local Zeta Functions II: Computation of the Local Factors

where we have used the definitions $\mathfrak{d} = \mathfrak{p}^d, \mathfrak{f} = \mathfrak{p}^n$. Setting $s = \frac{1}{2}$ we obtain $\rho(c\|\cdot\|^{1/2}) = \rho_0(c)$, which—using Proposition 12.4 (3)—shows the nontrivial fact that $|\rho_0(c)| = 1$.

Remark 12.14 Even though the notation may not make it clear, the quantities $\rho_0(c)$ may be familiar to the reader from elementary number theory, at least when $k = \mathbb{Q}_p$ and $n = 1$. Indeed, in this case, the quotient $\mathfrak{u}/(1 + \mathfrak{f})$ is $\mathbb{Z}_p^\times/(1 + (p)) \cong \mathbb{F}_p^\times$, so c is simply a character of the finite group $\mathbb{F}_p^\times$. Taking the representatives ε to be the integers between 1 and $p - 1$, the factor $e^{2\pi i \Lambda(\varepsilon/p)}$ gives simply ζ_p^ε, and $\rho_0(c)$ reduces to

$$\frac{1}{\sqrt{p}} \sum_{j \in \mathbb{F}_p^\times} c(j) \zeta_p^j.$$

This is an example of a so-called *Gauss sum*, extensively studied in elementary and algebraic number theory. The fact that $|\rho_0(c)| = 1$ implies in particular the interesting relation $|\sum_{j \in \mathbb{F}_p^\times} c(j) \zeta_p^j| = \sqrt{p}$ for every non-trivial character c. We will use this fact in Chap. 16.

For a modern and very accessible treatment of this classical theory, we recommend the book [1] by Ireland and Rosen. See in particular Proposition 8.2.2 in Chapter 6 of *op. cit.* for an elementary proof of the statement that Gauss sums have absolute value 1.

12.4.4 Non-vanishing of the Standard Local ζ Functions

We conclude this section with a remark that will be useful when we discuss the global theory over number fields. For each character c of the multiplicative group $k^\times$ of a local field k, we have constructed a 'standard' function f_c (depending on c) and computed $\zeta(f_c, c)$. Collecting the results of the previous sections, we have obtained:

Proposition 12.5 (Non-vanishing of the Standard Local ζ Functions) *Let c be a character of the local field $k^\times$. The function $\zeta(f_c, c)$ is meromorphic and everywhere non-vanishing. In particular, $1/\zeta(f_c, c)$ is everywhere holomorphic.*

Problems

12.1 Let H be a locally compact subgroup of a topological group G. Prove that H is closed in G.

Sketch of Solution. Let K be the closure of H. It suffices to prove that $K = H$. Clearly, H is dense in K (which is Hausdorff, since G is Hausdorff by assumption). Given a point $h \in H$, let U be an open neighbourhood of h in H whose closure C in H is compact. Write $U = V \cap H$ for some open V in K. Since C is compact and G (hence K) is Hausdorff, C is also closed in K. Now observe that $V \setminus C$ is open in K and does not meet H (since $V \cap H = U \subseteq C$). As H is dense in K but does not meet the open set $V \setminus C$, we must have $V \setminus C = \emptyset$, that is, $V \subseteq C \subseteq H$, so H contains a neighbourhood of h. As h was arbitrary, we see that H is open in K. Every open subgroup of a topological group is closed (consider the partition given by its cosets), so in particular H is closed in K. Since K is the closure of H, we have $K = H$ as desired.

12.2 Let p be a prime number.

1. Describe an isomorphism between $\mathbb{Q}_p/\mathbb{Z}_p$ and the p-power torsion of $\mathbb{Q}/\mathbb{Z}$.
2. Check the description of λ given above.
3. Check that λ is continuous.

12.3 Fill in the details of the proof of Theorem 12.2 in the real and complex cases. It can be useful to recall (and prove, if necessary) the classical formula

$$\int_{-\infty}^{\infty} e^{-ax^2} e^{-2\pi i k x}\, dx = \sqrt{\frac{\pi}{a}} e^{-\pi^2 k^2/a}.$$

Indication. It is important that you try this exercise if you haven't seen it before. A solution is given in the proof of Lemma 12.6.

12.4 Show that every continuous homomorphism from the positive reals to $\mathbb{C}^\times$ is of the form $x \mapsto x^s$, and that different values of s give different homomorphisms.

12.5 Let $\tilde{\alpha}$ be the map of Equation (12.3). Check that $\alpha \mapsto \tilde{\alpha}$ is continuous when v is a finite place.

Hint. ρ is locally constant.

12.6

1. Show that every sufficiently small neighbourhood of 1 in $\mathbb{C}^\times$ contains no non-trivial subgroup of $\mathbb{C}^\times$.
2. Mimicking the proof of Proposition 12.2 in the p-adic case, show that if G is a profinite group and $\rho : G \to \mathrm{GL}_n(\mathbb{C})$ is a continuous representation, then the image of ρ is finite.

12.7 Prove that $\int_{\|\xi\| \leq 1} |\log \|\xi\|| \cdot \|\xi\|^\kappa \, d\xi$ converges for all local fields k and all $\kappa > -1$.

Hint. The cases $k = \mathbb{R}$ and $k = \mathbb{C}$ are easy exercises in analysis (but do pay attention to the fact that $\|\xi\|$ is the *square* of the usual complex absolute value). For the p-adic case, reduce to summing over certain annuli A_v (see Remark 12.12 if necessary).

12.8 Check the identities of Sect. 12.4.1.6 using the results of Problem 1.1 (including Euler's reflection formula).

12.9 Prove Theorem 12.5.
Hint. We will essentially show this later as Equation (15.6).

References

1. Ireland, K., Rosen, M.: A Classical Introduction to Modern Number Theory, Graduate Texts in Mathematics, vol. 84, 2nd edn. Springer, New York (1990). https://doi.org/10.1007/978-1-4757-2103-4
2. Serre, J.P.: Abelian ℓ-adic Representations and Elliptic Curves, Research Notes in Mathematics, vol. 7. A K Peters, Wellesley, MA (1998). With the collaboration of Willem Kuyk and John Labute, Revised reprint of the 1968 original
3. Tate, J.T.: Fourier analysis in number fields, and Hecke's zeta-functions. In: Algebraic Number Theory (Proc. Instructional Conf., Brighton, 1965), pp. 305–347. Thompson, Washington, D.C. (1967)

Chapter 13
The Global Theory

Abstract In this chapter we extend the notion of zeta function to the global setting, replacing the multiplicative group of a local field with the idèle group of a global field. We prove the fundamental arithmetic Riemann-Roch theorem and use it to deduce the functional equation satisfied by global zeta functions. As an immediate corollary, we obtain that global zeta functions admit meromorphic, and usually holomorphic, continuation to the whole complex plane.

Following the philosophy outlined at the beginning of the previous chapter, we now turn to the study of ζ functions of number fields. This requires us to replace the multiplicative group $k^\times$ of a local field with the idèles group I_k of a number field, and correspondingly the additive group k is replaced by the adèles. The organisation of this chapter closely parallels that of the previous one: we first describe the dual group of the adèles and make a choice of Haar measure, then we do the same for I_k, and finally we introduce global ζ functions as modified Fourier transforms and study their general properties (functional equation and analytic continuation).

Notation In this chapter we let k be a number field. We denote by v a place of k (Definition 10.1) and by k_v the corresponding completion. For each v we then have all the analogues of the quantities defined in the previous chapter, which we decorate with a subscript v: the ring of integers O_v, the units $\mathfrak{u}_v$, the norm $\|\cdot\|_v$, the character Λ_v, the different $\mathfrak{d}_v$ if v is finite, etc.

13.1 The Additive Group: The Adèles

Recall from Definition 11.3 the notion of ring of adèles, which we consider in the topological sense of Definition 11.4. Thus, as a topological group, $\mathbb{A}_k$ is $\prod'_v (k_v^+, O_v)$. The ring structure is provided by the component-wise multiplication. We will denote the generic adèle by $x = (x_v)$, where $x_v \in k_v$ for every $v \in \Omega_k$.

Theorem 11.1, combined with Theorem 12.1 and Lemma 12.1, shows that the dual group of $\mathbb{A}_k$ is the restricted direct product of the groups $\widehat{k_v^+} \cong k_v^+$ with respect to the subgroups $\mathfrak{d}_v^{-1}$ (for v finite). Since $\mathfrak{d}_v^{-1} = O_v$ for almost all v, this shows that the dual group is simply $\mathbb{A}_k$ itself. More precisely, an adèle $(\eta_v) \in \mathbb{A}_k$ corresponds to the character

$$(x_v) \mapsto \prod_v e^{2\pi i \Lambda_v(\eta_v x_v)} = e^{2\pi i \sum_v \Lambda_v(\eta_v x_v)}.$$

It is then useful to define the **standard adèlic character**

$$\Lambda((x_v)) = \sum_v \Lambda_v(x_v). \tag{13.1}$$

Since we also have a local measure $d\mu_{k_v^+}$ for each place v, from the discussion in Sect. 11.4 we get a product measure $d\mu_{\mathbb{A}_k} = \prod_v d\mu_{k_v^+}$ on $\mathbb{A}_k$. Moreover, since each $d\mu_{k_v^+}$ is self-dual for the Fourier transform (see Theorem 12.2), we obtain from Corollary 11.1 that $d\mu_{\mathbb{A}_k}$ is also self-dual. Thus, the abstract general theory leads to the following:

Theorem 13.1 (Fourier Inversion on the Adèles) *The additive group of adèles $\mathbb{A}_k$ is its own character group. An isomorphism is obtained by identifying the adèle $\eta := (\eta_v)$ with the character $x := (x_v) \mapsto e^{2\pi i \Lambda(\eta x)}$. If for a function $f(x) \in L^1(\mathbb{A}_k)$ we define the Fourier transform by the formula*

$$\hat{f}(\eta) = \int_{\mathbb{A}_k} f(x) e^{-2\pi i \Lambda(\eta x)} d\mu_{\mathbb{A}_k}(x),$$

then for $f \in \mathfrak{V}^1(\mathbb{A}_k)$ we have the inversion formula

$$f(x) = \int_{\mathbb{A}_k} \hat{f}(\eta) e^{2\pi i \Lambda(x\eta)} d\mu_{\mathbb{A}_k}(\eta).$$

We also recall the following fact (see Problem 11.2):

Lemma 13.1 *The unit group of $\mathbb{A}_k$ is I_k, the group of idèles of k. In particular, for $\eta = (\eta_v) \in \mathbb{A}_k$, the map $x \mapsto \eta x$ of $\mathbb{A}_k$ into itself is an automorphism if and only if η is an idèle.*

The following is the global analogue of Lemma 12.2.

Lemma 13.2 (Rescaling the Adèlic Haar Measure) *Let a be an idèle of k. We have*

$$d\mu_{\mathbb{A}_k}(ax) = \|a\| d\mu_{\mathbb{A}_k}(x),$$

where $\|a\|$ is the product $\prod_v \|a_v\|_v$ (the product is finite, in the sense that all but finitely many terms are equal to 1).

13.1 The Additive Group: The Adèles

Proof Since $d\mu_{\mathbb{A}_k}(x)$ is a Haar measure and $x \mapsto ax$ is a ring automorphism, $d\mu_{\mathbb{A}_k}(ax)$ is another Haar measure. Thus, it suffices to compare the measures of any set of positive measure N. We take $N = \prod_v N_v$, where $N_v = O_v$ for v finite, and N_v is a compact neighbourhood of 1 if v is infinite. Applying Lemma 12.2 to each place we obtain

$$\int_{aN} d\mu_{\mathbb{A}_k}(x) = \prod_v \int_{a_v N_v} d\mu_{k_v^+}(x_v)$$

$$= \prod_v \|a_v\|_v \int_{N_v} d\mu_{k_v^+}(x_v)$$

$$= \left(\prod_v \|a_v\|_v\right) \int_N d\mu_{\mathbb{A}_k}(x).$$

□

13.1.1 The Field as a Subring of the Adèles

We consider k as embedded in $\mathbb{A}_k$ via the map $\xi \mapsto (\xi, \xi, \ldots, \xi, \ldots)$ which sends an element of k to the adèle whose components are all equal to ξ. The next lemma shows that k acts as a sort of complement for the subring of 'integral adèles'. To state it, let $S_\infty = \Omega_k^\infty$ denote the set of Archimedean places of k, and observe that $\mathbb{A}_{k,S_\infty} := (\mathbb{A}_k)_{S_\infty}$ is by definition the set of adèles (x_v) such that x_v is in O_v for every finite v.

Lemma 13.3 *The following hold.*
1. $k \cap \mathbb{A}_{k,S_\infty} = O_k$.
2. $k + \mathbb{A}_{k,S_\infty} = \mathbb{A}_k$.

Proof

1. This is simply the statement that a field element that has non-negative valuation at each finite place is an algebraic integer.
2. We need to show that, given an adèle $x = (x_v)$, there is $\xi \in k$ such that $x + \xi$ is integral at every finite place $\mathfrak{p}$. Let m be an integer divisible by all the primes $\mathfrak{p}$ such that $x_\mathfrak{p} \notin O_\mathfrak{p}$. Replacing m by m^N for some $N \gg 0$ we may assume that $mx_\mathfrak{p}$ has non-negative valuation at $\mathfrak{p}$ for all $\mathfrak{p}$.

   Denote by S the finite set of places dividing m; note that S contains all the places at which $x_\mathfrak{p}$ is not integral.

   We look for a field element ξ of the form $\frac{a}{m}$ with $a \in O_k$. Since both $x_\mathfrak{p}$ and $\frac{a}{m}$ are integral at $\mathfrak{p}$ for $\mathfrak{p} \notin S$, it suffices to show that we can choose a in such a way that $mx_\mathfrak{p} + a \equiv 0 \pmod{\mathfrak{p}^{v_\mathfrak{p}(m)}}$ for all $\mathfrak{p} \in S$. Such an a exists by the Chinese remainder theorem. Note that, with a slight abuse of notation, we have identified

$m x_\mathfrak{p} \in O_\mathfrak{p}/\mathfrak{p}^{v_\mathfrak{p}(m)}$ with an element of $O_k/\mathfrak{p}^{v_\mathfrak{p}(m)}$. This identification is possible since the canonical map

$$\frac{O_k}{\mathfrak{p}^{v_\mathfrak{p}(m)}} \to \frac{O_v}{O_v \mathfrak{p}^{v_\mathfrak{p}(m)}}$$

is an isomorphism (it is injective between groups with the same cardinality).

□

We now introduce the following notation:

Definition 13.1 (Infinite Part of the Adèles) We denote by $\mathbb{A}_k^\infty$ the product $\prod_{v \in S_\infty} k_v$ of the Archimedean completions of k. If (r_1, r_2) is the signature of k (see Definition 3.4), then $\mathbb{A}_k^\infty$ is isomorphic to $\mathbb{R}^{r_1} \times \mathbb{C}^{r_2}$. Given $x \in \mathbb{A}_k$, we denote by x^∞ its projection on $\mathbb{A}_k^\infty$.

Lemma 13.4 *Let $\omega_1, \ldots, \omega_n$ be a $\mathbb{Z}$-basis of O_k (so that in particular $[k : \mathbb{Q}] = n$ and $r_1 + 2r_2 = n$).*

1. *$\omega_1^\infty, \ldots, \omega_n^\infty$ is an $\mathbb{R}$-basis of $\mathbb{A}_k^\infty$.*
2. *Let $D^\infty = \{\sum_{i=1}^n x_i \omega_i : x_i \in [0, 1)\}$ be the 'fundamental parallelotope' spanned by the given basis. The volume of D^∞ with respect to the measure $\prod_{v \in S_\infty} dx_v$ is $\sqrt{|d_k|}$, where d_k is the discriminant of k.*

Proof Denote by $\sigma_1, \ldots, \sigma_{r_1}$ the r_1 real embeddings of k and by $\sigma_{r_1+1}, \ldots, \sigma_{r_1+r_2}$ a choice of r_2 non-equivalent non-real embeddings of k into $\mathbb{C}$ (here, by *non-equivalent* we mean that no two of them are complex conjugate of each other). An isomorphism $\mathbb{A}_k^\infty \cong \mathbb{R}^{r_1} \times \mathbb{R}^{2r_2}$ is given by

$$\xi \mapsto \left((\sigma_i(\xi))_{i=1,\ldots,r_1}, (\Re \sigma_{r_1+i}(\xi), \Im \sigma_{r_1+i}(\xi))_{i=1,\ldots,r_2}\right).$$

Here we use the fact that $\mathbb{C}$ (with its **standard** Lebesgue measure) is isomorphic as a measure space to $\mathbb{R} \times \mathbb{R}$ (with its standard Lebesgue measure) via the map $z \mapsto (\Re z, \Im z)$. Via this isomorphism, the elements ω_m are sent to the vectors

$$\omega_m^\infty = \begin{pmatrix} \sigma_1(\omega_m) \\ \vdots \\ \sigma_{r_1}(\omega_m) \\ \Re \sigma_{r_1+1}(\omega_m) \\ \Im \sigma_{r_1+1}(\omega_m) \\ \vdots \\ \Re \sigma_{r_1+r_2}(\omega_m) \\ \Im \sigma_{r_1+r_2}(\omega_m) \end{pmatrix} = \begin{pmatrix} \sigma_1(\omega_i) \\ \vdots \\ \sigma_{r_1}(\omega_m) \\ \frac{\sigma_{r_1+1}(\omega_m) + \overline{\sigma_{r_1+1}(\omega_m)}}{2} \\ \frac{\sigma_{r_1+1}(\omega_m) - \overline{\sigma_{r_1+1}(\omega_m)}}{2i} \\ \vdots \\ \frac{\sigma_{r_1+r_2}(\omega_m) + \overline{\sigma_{r_1+r_2}(\omega_m)}}{2} \\ \frac{\sigma_{r_1+r_2}(\omega_m) - \overline{\sigma_{r_1+r_2}(\omega_m)}}{2i} \end{pmatrix}.$$

13.1 The Additive Group: The Adèles

Consider the matrix $\boldsymbol{\omega}^\infty$ having ω_m^∞ as columns. We want to compute the absolute value of the determinant of this matrix; in order to do so, we can consider $\boldsymbol{\omega}^\infty$ as a matrix with complex coefficients.

Summing $1/i$-times each row $\Re\sigma_{r_1+j}(\omega_m)$ to the following one (which does not change the determinant of $\boldsymbol{\omega}^\infty$) we replace the m-th column of ω^∞ with

$$\begin{pmatrix} \sigma_1(\omega_m) \\ \vdots \\ \sigma_{r_1}(\omega_m) \\ \frac{\sigma_{r_1+1}(\omega_m)+\overline{\sigma_{r_1+1}(\omega_m)}}{2} \\ \sigma_{r_1+1}(\omega_m)/i \\ \vdots \\ \frac{\sigma_{r_1+r_2}(\omega_m)+\overline{\sigma_{r_1+r_2}(\omega_m)}}{2} \\ \sigma_{r_1+r_2}(\omega_m)/i \end{pmatrix}.$$

We now subtract $i/2$-times each row $\sigma_{r_1+r_2}(\omega_m)/i$ from the previous one, obtaining a matrix (with the same determinant as ω^∞) having as m-th column the vector

$$\begin{pmatrix} \sigma_1(\omega_m) \\ \vdots \\ \sigma_{r_1}(\omega_m) \\ \frac{\sigma_{r_1+1}(\omega_m)}{2} \\ \sigma_{r_1+1}(\omega_m)/i \\ \vdots \\ \frac{\overline{\sigma_{r_1+r_2}(\omega_m)}}{2} \\ \sigma_{r_1+r_2}(\omega_m)/i \end{pmatrix}.$$

Pulling out a factor of $1/2$ from each row $\frac{\overline{\sigma_{r_1+j}(\omega_m)}}{2}$ and a factor $1/i$ from each row $\sigma_{r_1+j}(\omega_m)$, we obtain

$$\det \boldsymbol{\omega}^\infty = i^{-r_2} 2^{-r_2} \det (\sigma_i(\omega_m))_{i,m},$$

hence in absolute value we have

$$|\det(\boldsymbol{\omega}^\infty)| = 2^{-r_2} |\det (\sigma_i(\omega_m))_{i,m}| = 2^{-r_2} \sqrt{|d_k|},$$

see Definition 3.1.

1. The previous determinant computation shows in particular that the vectors ω_m^∞ are linearly independent.
2. Note that (up to a set of measure zero) D^∞ is the image of $[0, 1]^n$ under the linear map that sends a vector $\mathbf{x} = \begin{pmatrix} x_1 \\ x_2 \\ \vdots \\ x_n \end{pmatrix}$ to $\omega^\infty \mathbf{x}$, where ω^∞ is as above. Thus, one of the basic properties of the determinant shows that the volume of D^∞ is $|\det(\omega^\infty)| \cdot \mathrm{vol}([0, 1]^n)$. The claim follows from the fact that $|\det(\omega^\infty)| = 2^{-r_2}\sqrt{|d_k|}$ and $\mathrm{vol}([0, 1]^n) = 2^{r_2}$, since our choice of Haar measure on $\mathbb{R}^{r_2}$ is **twice** the standard Lebesgue measure.

□

Definition 13.2 (Additive Fundamental Domain) The **additive fundamental domain** D of $\mathbb{A}_k$ is the set $\{x \in \mathbb{A}_k : x \in \mathbb{A}_{k,S_\infty} \text{ and } x^\infty \in D^\infty\}$. Equivalently, $D = \mathbb{A}_k^{S_\infty} \times D^\infty$.

Theorem 13.2 (Properties of the Additive Fundamental Domain) *The following hold:*

1. *$\mathbb{A}_k$ is the disjoint union $\bigsqcup_{\xi \in k} (\xi + D)$*
2. *The measure of D is 1.*

Proof

1. We first prove that $\xi_1 + D$ and $\xi_2 + D$ intersect trivially if ξ_1, ξ_2 are elements of k with $\xi_1 \neq \xi_2$. Equivalently, we need to show that $\xi_1 - \xi_2 \in D$ implies $\xi_1 = \xi_2$. Since D is contained in $\mathbb{A}_{k,S_\infty}$, by Lemma 13.3 we see that $\xi_1 - \xi_2$ is an algebraic integer, hence an integral linear combination of $\omega_1, \ldots, \omega_n$ (notation as in Lemma 13.4). However, by projecting on $\mathbb{A}_k^\infty$ we obtain that its coordinates in the $\mathbb{Z}$-basis $\omega_1, \ldots, \omega_n$ are all strictly less than 1, hence they are all equal to zero, that is, $\xi_1 - \xi_2 = 0$.

    Now we show that every adèle x is in some $\xi + D$. By Lemma 13.3, there exists ξ_1 such that $y := x - \xi_1$ is in $\mathbb{A}_{k,S_\infty}$. Consider y^∞: by Lemma 13.4 the set $\omega_1, \ldots, \omega_n$ is a basis of this vector space, so we can write $y^\infty = \sum_{i=1}^n c_i \omega_i + \sum_{i=1}^n \delta_i \omega_i$ with $c_i \in \mathbb{Z}$ and $\delta_i \in [0, 1)$. The field element $\xi_2 = \sum_{i=1}^n c_i \omega_i$ is in O_k, so $y - \xi_2$ is still in $\mathbb{A}_{k,S_\infty}$, and furthermore, by construction, $(y - \xi_2)^\infty$ is in D^∞. It follows as desired that $x - \xi_1 - \xi_2$ is in D.

2. By definition we have $D = \mathbb{A}_k^{S_\infty} \times D^\infty \subseteq \mathbb{A}_{k,S_\infty}$. Since the adèlic measure coincides with the product measure on subsets of the form $(\mathbb{A}_k)_S$, the adèlic measure of D is

$$\mu_{\mathbb{A}_k}(D) = \left(\int_{\mathbb{A}_k^{S_\infty}} d\mu_{\mathbb{A}_k^{S_\infty}}\right) \left(\int_{D^\infty} \prod_{v \in S^\infty} d\mu_{x_v}\right).$$

13.1 The Additive Group: The Adèles

The second factor is equal to $\sqrt{|d_k|}$ by Lemma 13.4. As for the first, we have

$$\int_{\mathbb{A}_k^{S_\infty}} d\mu_{\mathbb{A}_k^{S_\infty}} = \prod_{v \notin S_\infty} \int_{\mathcal{O}_v} d\mu_v = \prod_{v \notin S_\infty} N(\mathfrak{d}_v)^{-1/2} = \prod_{v \notin S_\infty} |d_{k_v}|^{-1/2},$$

where we used our normalisation for the additive Haar measures on the local fields k_v (Definition 12.2) and Theorem 10.5. Finally, using Theorem 10.6 we conclude that $\prod_{v \notin S_\infty} |d_{k_v}|^{-1/2} = |d_k|^{-1/2}$, which simplifies $\int_{D^\infty} \prod_{v \in S^\infty} d\mu_{x_v} = |d_k|^{1/2}$, giving the result. □

Corollary 13.1 (Position of k Inside $\mathbb{A}_k$) *The field k is a discrete subgroup of $\mathbb{A}_k$ and the quotient $\mathbb{A}_k/k$ is compact.*

Proof It is clear that Theorem 13.2 remains true if we replace our choice of D^∞ with the set $\tilde{D}^\infty = \{\sum_{i=1}^n x_i \omega_i : x_i \in [-1/2, 1/2)\}$. Define $\tilde{D}$ as the corresponding additive fundamental domain. Since $\tilde{D}$ contains a neighbourhood of 0, the decomposition $\mathbb{A}_k = \bigsqcup_{\xi \in k}(\xi + \tilde{D})$ of Theorem 13.2(1) shows that each point of ξ has a neighbourhood that is disjoint from a neighbourhood of any other point. Thus, k is discrete in $\mathbb{A}_k$. The quotient $\mathbb{A}_k/k$ is compact since there is a continuous surjection $\overline{D} \twoheadrightarrow \mathbb{A}_k/k$ with $\overline{D}$ compact. □

Lemma 13.5 *The character Λ of Eq. (13.1) vanishes on k.*

Proof For each place v of k, let $\mathbb{Q}_v$ be the completion of $\mathbb{Q}$ at its unique place lying under v (equivalently: the closure of $\mathbb{Q}$ in k_v). By definition,

$$\Lambda(\xi) = \sum_v \Lambda_v(\xi) = \sum_v \lambda_v(\operatorname{tr}_{k_v/\mathbb{Q}_v} \xi)$$

$$= \sum_{w \in \Omega_\mathbb{Q}} \lambda_w \left(\sum_{v|w} \operatorname{tr}_{k_v/\mathbb{Q}_w}(\xi) \right) = \sum_{w \in \Omega_\mathbb{Q}} \lambda_w \left(\operatorname{tr}_{k/\mathbb{Q}}(\xi) \right),$$

where we used Theorem 10.7. Setting $x = \operatorname{tr}_{k/\mathbb{Q}}(\xi)$ we are then reduced to showing

$$\sum_{w \in \Omega_\mathbb{Q}} \lambda_w(x) \equiv 0 \pmod 1 :$$

we have reduced the lemma to the case $k = \mathbb{Q}$. To treat this, we need to show that $\sum_v \lambda_v(x)$ is an integer for every $x \in \mathbb{Q}$. Clearly it suffices to show that it is

q-integral at each (finite) prime q. This is achieved by looking at the decomposition

$$\sum_v \lambda_v(x) = \left(\sum_{p \neq q, \infty} \lambda_p(x)\right) + \lambda_q(x) + \lambda_\infty(x)$$

$$= \left(\sum_{p \neq q, \infty} \lambda_p(x)\right) + \left(\lambda_q(x) - x\right) \bmod \mathbb{Z} :$$

each $\lambda_p(x)$ is a rational number with denominator a power of p, hence it is q-integral, while $\lambda_q(x) - x$ is q-integral by definition. □

Theorem 13.3 (Dual of $\mathbb{A}_k/k$) *The map*

$$\beta : k \to \widehat{\mathbb{A}_k/k}$$
$$\eta \mapsto \exp(2\pi i \Lambda(\eta \cdot))$$

is an isomorphism of topological groups.

Proof By Proposition 9.1, we have $\widehat{\mathbb{A}_k/k} \cong k^\perp$, where $k^\perp$ is the closed subgroup of $\widehat{\mathbb{A}_k}$ given by those characters that vanish on k. By Lemma 13.5 we have $k \subseteq k^\perp$. We now show that they are equal, by combining the following three observations:

1. By Proposition 9.1 there is an isomorphism of $\widehat{\mathbb{A}_k/k}$ with $k^\perp$. Note that $\widehat{\mathbb{A}_k/k}$ is discrete, because $\mathbb{A}_k/k$ is compact (apply Theorem 9.1 and Corollary 13.1). We have already observed that $k \subseteq k^\perp$, so we can consider the quotient $k^\perp/k$, which is therefore discrete.[1]
2. On the other hand, via the self-duality $\mathbb{A}_k \cong \widehat{\mathbb{A}_k}$, the quotient $k^\perp/k$ can be considered as a subgroup $\mathbb{A}_k/k$, which is compact by Corollary 13.1. Combined with (1), this shows that $k^\perp/k$ is both discrete and compact, hence finite.
3. Finally, $k^\perp$ has a natural structure of k-vector space (for $\psi \in k^\perp, \xi \in k$ we set $(\xi \cdot \psi)(\eta) := \psi(\xi\eta))$, and k is a k-vector subspace. Thus, $k^\perp/k$ is a k-vector space of finite cardinality. Since k is infinite, this implies that $k^\perp/k$ is trivial, hence $k^\perp = k$, as desired.

□

13.1.2 The Poisson Formula

In this section we work with continuous functions $\tilde{\varphi} : \mathbb{A}_k/k \to \mathbb{C}$. It is technically simpler to consider them as functions $\mathbb{A}_k \to \mathbb{C}$ that are invariant under translation by any $\xi \in k$. We give this as a formal definition:

---

[1] Note that if X is a discrete topological group and Y is any subgroup, then X/Y is discrete: each point is open.

13.1 The Additive Group: The Adèles

Definition 13.3 (Periodic Function) Let $\pi : \mathbb{A}_k \to \mathbb{A}_k/k$ be the canonical projection. A function $\varphi : \mathbb{A}_k \to \mathbb{C}$ is called *periodic* if $\varphi(x + \xi) = \varphi(x)$ for all $x \in \mathbb{A}_k$ and all $\xi \in k \subseteq \mathbb{A}_k$. Such a function induces, by passing to the quotient, a function $\tilde{\varphi} : \mathbb{A}_k/k \to \mathbb{C}$. Conversely, given $\tilde{\varphi} : \mathbb{A}_k/k \to \mathbb{C}$, we denote by $\varphi = \tilde{\varphi} \circ \pi$ the corresponding function on $\mathbb{A}_k$.

Remark 13.1 Notice that $\tilde{\varphi} : \mathbb{A}_k/k \to \mathbb{C}$ is continuous if and only if the corresponding function $\varphi : \mathbb{A}_k \to \mathbb{C}$ is continuous and periodic.

Lemma 13.6 *Let $\varphi(x) : \mathbb{A}_k \to \mathbb{C}$ be continuous and periodic, and let $\tilde{\varphi} : \mathbb{A}_k/k \to \mathbb{C}$ be the corresponding continuous function on $\mathbb{A}_k/k$. We have*

$$\int_D \varphi(x)\, d\mu_{\mathbb{A}_k}(x) = \int_{\mathbb{A}_k/k} \tilde{\varphi}(x)\, d\mu(x),$$

where the measure $d\mu$ on $\mathbb{A}_k/k$ is the unique Haar measure on this (compact, by Corollary 13.1) group such that $\mu(\mathbb{A}_k/k) = 1$.

Proof Follows from Theorem 13.2(2). More precisely, denote by π the canonical projection $\mathbb{A}_k \to \mathbb{A}_k/k$ and introduce the function

$$\begin{aligned} I : L^1(\mathbb{A}_k/k) &\to \mathbb{C} \\ \tilde{\varphi}(x) &\mapsto \int_D \tilde{\varphi} \circ \pi(x)\, d\mu_{\mathbb{A}_k}(x). \end{aligned}$$

It is easy to check that I has the properties required to be a Haar integral.[2] Moreover, it gives measure 1 to $\mathbb{A}_k/k$ by Theorem 13.2(2).

We check in greater detail the invariance of I under translation. Let $y \in \mathbb{A}_k$ and let $\psi(x) = \varphi(x+y)$. Denote by $\tilde{\psi}$ the corresponding function on the quotient $\mathbb{A}_k/k$. We need to check that

$$\int_D \tilde{\varphi} \circ \pi(x)\, d\mu_{\mathbb{A}_k}(x) = \int_D \tilde{\psi} \circ \pi(x)\, d\mu_{\mathbb{A}_k}(x),$$

or equivalently

$$\int_D \varphi(x)\, d\mu_{\mathbb{A}_k}(x) = \int_D \psi(x)\, d\mu_{\mathbb{A}_k}(x),$$

---

[2] To be precise: if we only consider functions $\tilde{\varphi}$ that are the characteristic functions of subsets of $\mathbb{A}_k/k$, the functional I clearly gives a measure on $\mathbb{A}_k/k$. We will see below that this measure is translation-invariant and gives mass 1 to $\mathbb{A}_k/k$, so it is the unique normalised Haar measure. A posteriori, this implies that I is the integration against this Haar measure, hence it is well-defined on all of $L^1(\mathbb{A}_k/k)$.

which can further be rewritten as

$$\int_D \varphi(x)\,d\mu_{\mathbb{A}_k}(x) = \int_D \varphi(x+y)\,d\mu_{\mathbb{A}_k}(x).$$

Since $d\mu_{\mathbb{A}_k}$ is translation-invariant, we are reduced to showing

$$\int_D \varphi(x)\,d\mu_{\mathbb{A}_k}(x) = \int_{y+D} \varphi(x)\,d\mu_{\mathbb{A}_k}(x).$$

Now, since $\mathbb{A}_k = \bigsqcup_{\xi \in k}(\xi + D)$, we have

$$y + D = \bigsqcup_{\xi \in k} \left((\xi + D) \cap (y + D) \right),$$

where only finitely many sets in the union are non-empty.

We rewrite the integral $\int_{y+D} \varphi(x)\,d\mu_{\mathbb{A}_k}(x)$ as the series (really a finite sum)

$$\sum_{\xi \in k} \int_{(\xi+D) \cap (y+D)} \varphi(x)\,d\mu_{\mathbb{A}_k}(x) = \sum_{\xi \in k} \int_{(\xi+D) \cap (y+D)} \varphi(x+\xi)\,d\mu_{\mathbb{A}_k}(x)$$

$$= \sum_{\xi \in k} \int_{D \cap (y-\xi+D)} \varphi(x)\,d\mu_{\mathbb{A}_k}(x), \tag{13.2}$$

where we have used the translation-invariance of both the measure $d\mu_{\mathbb{A}_k}$ and the function φ. Now, it is immediate to check that $y + D$ is another additive fundamental domain for $\mathbb{A}_k$, hence the sets $\{-\xi + (y+D)\}_{\xi \in k}$ are disjoint and cover $\mathbb{A}_k$. It follows that the sets $\{D \cap (y - \xi + D)\}_{\xi \in k}$ are disjoint and cover D, and therefore the last integral in (13.2) is also equal to $\int_D \varphi(x)\,d\mu_{\mathbb{A}_k}(x)$, as desired. $\square$

Recall from Theorem 13.3 that k is the character group of $\mathbb{A}_k/k$. It follows that the Fourier transform of a function on $\mathbb{A}_k/k$ can be identified with a function on k, and precisely we have:

Definition 13.4 (Fourier Transform on $\mathbb{A}_k/k$) Let φ be a complex-valued, periodic, continuous function on $\mathbb{A}_k$. Its Fourier transform is the function

$$\hat{\varphi} : k \to \mathbb{C}$$
$$\xi \mapsto \int_D \varphi(x) e^{-2\pi i \Lambda(\xi x)}\,d\mu_{\mathbb{A}_k}(x).$$

Note that, by compactness of $\mathbb{A}_k/k$ (Corollary 13.1), any continuous function on this quotient is automatically L^1. We exploit this in the next lemma:

13.1 The Additive Group: The Adèles

Lemma 13.7 *Let $\varphi(x) : \mathbb{A}_k \to \mathbb{C}$ be continuous and periodic with $\sum_{\xi \in k} |\hat{\varphi}(\xi)| < \infty$. We have the Fourier inversion formula*

$$\varphi(x) = \sum_{\xi \in k} \hat{\varphi}(\xi) e^{2\pi i \Lambda(x\xi)}.$$

Proof As already observed, $\varphi(x)$ induces a continuous function $\tilde{\varphi}(x)$ in $L^1(\mathbb{A}_k/k)$. The hypothesis of the lemma means that the Fourier transform of $\tilde{\varphi}(x)$ is in $L^1(k)$. Thus, $\tilde{\varphi}(x)$ satisfies the assumptions of the abstract Fourier inversion theorem (Theorem 9.2). The conclusion of the lemma is then simply the inversion formula, once we check that the measure μ_k on k dual to the Haar measure we fixed on $\mathbb{A}_k/k$ is the counting measure (and not a nontrivial multiple thereof).

To see that this holds, we apply part (2) of Lemma 11.6 to $H_i = \mathbb{A}_k/k$. The group $H_i^\perp$ is clearly trivial (hence can be identified to the singleton $\{0\}$ of $k \cong \widehat{\mathbb{A}_k/k}$), so we obtain $1 = \mu_{\mathbb{A}_k/k}(\mathbb{A}_k/k) \cdot \mu_k(\{0\}) = 1 \cdot \mu_k(\{0\})$, which shows that μ_k is the counting measure, as desired. $\square$

It is now natural to ask where we can find periodic functions on the adèles to which the above theory applies. The simplest and by far most common way to construct a periodic function is to take an arbitrary function $f(x)$ and consider the sum of all its translates $f(x + \xi)$. The next lemma describes what assumptions are necessary to obtain a well-behaved function in this way.

In order to state it formally, we need a notion of *uniform convergence* for sums indexed by elements of k. Since k is a number field (which is a discrete object, without any natural topology), the only possible definition is the following:

Definition 13.5 (Uniform Convergence of a Series of Functions) Let $a_\xi(x) : \mathbb{A}_k \to \mathbb{C}$ be a set of complex-valued functions indexed by elements $\xi \in k$ and let X be a subset of $\mathbb{A}_k$. We say that the series $\sum_{\xi \in k} a_\xi(x)$ **converges uniformly** for $x \in X$ if the following holds: for every $\varepsilon > 0$ there exists a finite set $F \subset k$ such that

$$\sum_{\xi \notin F} |a_\xi(x)| < \varepsilon$$

for all $x \in X$.

Note that, by standard arguments, the sum of a uniformly convergent series of continuous functions is itself a continuous function.

Lemma 13.8 *Let $f(x)$ be a continuous function in $L^1(\mathbb{A}_k)$ and suppose that $\sum_{\eta \in k} f(x+\eta)$ is uniformly convergent for $x \in D$. The continuous periodic function $\varphi(x) = \sum_{\eta \in k} f(x + \eta)$ satisfies $\hat{\varphi}(\xi) = \hat{f}(\xi)$ for all $\xi \in k$.*

Remark 13.2 Note that the equality $\hat{\varphi}(\xi) = \hat{f}(\xi)$ only makes sense for $\xi \in k$: we consider φ as a function on $\mathbb{A}_k/k$, so its Fourier transform is defined on k.

Proof This is essentially a direct calculation. We have

$$\hat{\varphi}(\xi) = \int_D \varphi(x) e^{-2\pi i \Lambda(\xi x)} \, d\mu_{\mathbb{A}_k}(x)$$

$$= \int_D \left(\sum_{\eta \in k} f(x+\eta) e^{-2\pi i \Lambda(x\xi)} \right) d\mu_{\mathbb{A}_k}(x)$$

$$\stackrel{(1)}{=} \sum_{\eta \in k} \int_D f(x+\eta) e^{-2\pi i \Lambda(x\xi)} \, d\mu_{\mathbb{A}_k}(x)$$

$$\stackrel{(2)}{=} \sum_{\eta \in k} \int_{\eta+D} f(x) e^{-2\pi i \Lambda((x-\eta)\xi)} \, d\mu_{\mathbb{A}_k}(x)$$

$$\stackrel{(3)}{=} \int_{\mathbb{A}_k} f(x) e^{-2\pi i \Lambda(x\xi)} \, d\mu_{\mathbb{A}_k}(x)$$

$$= \hat{f}(\xi),$$

where

- in (1) we have used the fact that the series converges uniformly in D, which is of finite measure;
- in (2) we have used the translation-invariance of the Haar measure;
- in (3) we have applied the relation $\Lambda(\eta\xi) = 0$ for all $\eta, \xi \in k$, which follows from Lemma 13.5.

□

We now have all the ingredients to prove the two main results of this section:

Proposition 13.1 (Poisson Formula) *Let* $f(x) : \mathbb{A}_k \to \mathbb{C}$ *satisfy the following three conditions:*

1. $f(x)$ *is continuous and in* $L^1(\mathbb{A}_k)$;
2. $\sum_{\xi \in k} f(x+\xi)$ *is uniformly convergent for* $x \in D$;
3. $\sum_{\xi \in k} |\hat{f}(\xi)|$ *converges.*

The following equality holds:

$$\sum_{\xi \in k} \hat{f}(\xi) = \sum_{\xi \in k} f(\xi).$$

13.1 The Additive Group: The Adèles

Proof Let $\varphi(x) = \sum_{\eta \in k} f(x + \eta)$. By assumption, we can apply Lemma 13.8 to obtain

$$\hat{\varphi}(\xi) = \hat{f}(\xi),$$

where $\varphi(x)$ is continuous and periodic. Taking absolute values and summing over $\xi \in k$ we obtain

$$\sum_{\xi \in k} |\hat{\varphi}(\xi)| = \sum_{\xi \in k} |\hat{f}(\xi)| < \infty,$$

so the assumption of Lemma 13.7 is satisfied. That lemma then implies

$$\varphi(x) = \sum_{\xi \in k} \hat{\varphi}(\xi) e^{2\pi i \Lambda(x\xi)}.$$

Replacing $\varphi(x)$ with its definition and $\hat{\varphi}(\xi)$ with $\hat{f}(\xi)$ (Lemma 13.8 again) we arrive at

$$\sum_{\eta \in k} f(x + \eta) = \sum_{\xi \in k} \hat{f}(\xi) e^{2\pi i \Lambda(x\xi)}.$$

Setting $x = 0$ gives the result. □

In turn, the Poisson formula leads immediately to the most important result of this section, which Tate calls *an arithmetic analogue of the Riemann-Roch theorem*. We will explain in the next subsection where the name comes from.

Theorem 13.4 (Riemann-Roch) *Let $f(x) : \mathbb{A}_k \to \mathbb{C}$ satisfy the following three conditions:*

1. *$f(x)$ is continuous and in $L^1(\mathbb{A}_k)$;*
2. *$\sum_{\xi \in k} f(a(x + \xi))$ is convergent for all idèles a and all adèles x, uniformly for $x \in D$;*
3. *$\sum_{\xi \in k} |\hat{f}(a\xi)|$ is convergent for all $a \in I_k$.*

The following holds for every idèle $a \in I_k$:

$$\frac{1}{\|a\|} \sum_{\xi \in k} \hat{f}(\xi/a) = \sum_{\xi \in k} f(a\xi).$$

Proof Fix an idèle a and define $g(x) = f(ax)$. We check that $g(x)$ satisfies the hypotheses of Proposition 13.1. It is clear that $g(x)$ is continuous and L^1, and

$\sum_{\xi \in k} g(x + \xi)$ is uniformly convergent for $x \in D$ by assumption. As for the third condition, we compute the Fourier transform of $g(x)$ as follows:

$$\hat{g}(x) = \int g(\eta) e^{-2\pi i \Lambda(x\eta)} d\mu_{\mathbb{A}_k}(\eta)$$

$$= \int f(a\eta) e^{-2\pi i \Lambda(x\eta)} d\mu_{\mathbb{A}_k}(\eta)$$

$$= \frac{1}{\|a\|} \int f(\eta) e^{-2\pi i \Lambda(x\eta/a)} d\mu_{\mathbb{A}_k}(\eta)$$

$$= \frac{1}{\|a\|} \hat{f}(x/a),$$

where in the only non-trivial equality we used Lemma 13.2. We now have $\sum_{\xi \in k} |\hat{g}(\xi)| = \frac{1}{\|a\|} \sum_{\xi \in k} |\hat{f}(a^{-1}\xi)| < \infty$ by assumption. Thus, the Poisson formula holds and yields

$$\sum_{\xi \in k} g(\xi) = \sum_{\xi \in k} \hat{g}(\xi),$$

that is,

$$\sum_{\xi \in k} f(a\xi) = \frac{1}{\|a\|} \sum_{\xi \in k} \hat{f}(\xi/a).$$

$\square$

13.1.3 Analogy with the Geometric Riemann-Roch Theorem

This section should be taken as an extended aside. In it, we connect Theorem 13.4 with the classical statement known as the Riemann-Roch theorem in the geometry of curves (over finite fields). The presentation is heavily inspired by [1, §7.2], to which the reader is referred for further details. Of course, this section will make more sense to readers who already have some basic knowledge of algebraic geometry.

Let k be the function field of a curve C over a finite field $\mathbb{F}_q$. In this setting, the *places* of k (that are trivial on $\mathbb{F}_q$) are in bijection with the Galois orbits of points of $C(\overline{\mathbb{F}_q})$ (for the reader familiar with scheme theory, these are simply the closed points of the scheme corresponding to C). Given a point $P \in C(\overline{\mathbb{F}_q})$, let $\mathbb{F}_{q^{\deg P}}$ be the minimal field extension of $\mathbb{F}_q$ over which P is defined (in the language of scheme theory, this is the residue field of P). The valuation v on k corresponding to P can be interpreted geometrically as the order of vanishing of $f \in k^\times$ at the point P. We identify places and valuations.

13.1 The Additive Group: The Adèles

With this terminology in place, one can define the adèles $\mathbb{A}_k$ and idèles I_k of k in complete analogy with the case of number fields. We stress an important difference, however: the function field $k = \mathbb{F}_q(C)$ does not have any Archimedean places. There is a suitable notion of *global field* that encompasses both the case of number fields we discussed so far and the case of function fields of curves over finite fields. All the main theorems discussed in this book extend to the setting of global fields: in the discussion below we will therefore use some of the theorems we stated and proved for number fields also for the function field k.

Each place v has a **degree** $\deg v$, defined as the degree over $\mathbb{F}_q$ of the residue field of the point corresponding to v. The **local ring at** v, or **ring of integers at** v, is the subring O_v of k consisting of those elements for which $v(f) \geq 0$. This ring is local, with unique maximal ideal $\mathfrak{p}_v = \{0\} \cup \{f \in k^\times : v(f) > 0\}$.

We define a **divisor** to be a formal linear combination $\sum_v n_v v$, where each n_v is an integer and all but finitely many of them are zero. We denote by $\mathrm{Div}(k)$ the set of all divisors. There is a notion of **degree**, namely $\deg(\sum n_v v) = \sum n_v \deg(v)$, and a notion of **principal divisor**: given $f \in k^\times$, the principal divisor corresponding to f is

$$\mathrm{div}(f) = \sum_v v(f) v.$$

The (analogue of the) product formula, Theorem 13.6 below, holds in this context, and gives

$$1 = \|f\|_{\mathbb{A}_k} = \prod_v \|f\|_v = \prod_v (q^{\deg v})^{v(f)} = q^{\sum_v v(f) \deg(v)},$$

which shows that $\deg(\mathrm{div}\, f) = 0$. Finally, it is not hard to see that an element $f \in k^\times$ satisfies $\mathrm{div}\, f = 0$ if and only if $f \in \mathbb{F}_q^\times$.

We define a partial ordering on the set of divisors by

$$\sum_v n_v v \geq \sum_v n'_v v \iff n_v \geq n'_v \ \forall v.$$

With each divisor D we can then associate the **linear system**

$$L(D) = \{0\} \cup \{f \in k^\times : \mathrm{div}(f) \geq -D\}.$$

It is immediate to see that $L(D)$ is an $\mathbb{F}_q$-vector space. We write $l(D)$ for its dimension. It is not hard to show that $l(D)$ is finite for every D, but we will simply admit this.

Before comparing Theorem 13.4 with the classical statement of Riemann-Roch, we make two further remarks. One is that, given a divisor $D = \sum n_v v$, we can always find an idèle $x(D) = (x(D)_v)$ such that $v(x(D)_v) = n_v$. If we further define

$$f = \prod_v 1_{O_v}, \tag{13.3}$$

the product of the characteristic functions of the local rings at all valuations, for every $\xi \in k$ we have

$$f(\xi x(D)) = \prod_v 1_{O_v}(\xi x(D)_v)$$

$$= \prod_v \begin{cases} 1, & \text{if } v(x(D)_v) + v(\xi) \geq 0 \\ 0, & \text{otherwise} \end{cases}$$

$$= \begin{cases} 1, & \text{if } v(\xi) \geq -n_v \; \forall v \\ 0, & \text{otherwise} \end{cases},$$

and therefore, by definition, this number is 1 if ξ is in $L(D)$ and is 0 otherwise. Thus,

$$\sum_{\xi \in k} f(\xi x(D)) = \#L(D) = q^{l(D)}.$$

Theorem 13.5 (Riemann-Roch, Geometric Form) *There exists an integer $g \geq 0$ and a divisor K_C of degree $2g - 2$ such that*

$$l(D) - l(K_C - D) = \deg(D) - g + 1.$$

Proof Fix a character $\Lambda : \mathbb{A}_k \to \mathbb{S}^1$ which gives rise to a self-duality, as in Theorem 13.1. For each place v, let $\mathfrak{p}_v^{m_v}$ be the minimal power of $\mathfrak{p}_v$ on which Λ_v is trivial (note that m_v can easily be negative, see Lemma 12.1). We define

$$K_C = -\sum_v m_v v;$$

it is a divisor, because $m_v = 0$ for almost all v.

We now compute the Fourier transform of the function f defined in (13.3). By Lemma 11.7, this is given by the product of the Fourier transforms of the functions 1_{O_v}. By exactly the same calculation as in the proof of Theorem 12.2, we obtain that

$$\widehat{1_{O_v}}(x) = N(\mathfrak{p}_v^{m_v})^{1/2} \cdot 1_{\mathfrak{p}_v^{m_v}}.$$

13.1 The Additive Group: The Adèles

Multiplying over all v, we obtain the Fourier transform of f as

$$\prod_v N(\mathfrak{p}_v^{m_v})^{1/2} \cdot \prod_v \mathbf{1}_{\mathfrak{p}_v^{m_v}} = q^{\frac{1}{2}\sum m_v \deg(v)} \cdot \prod_v \mathbf{1}_{\mathfrak{p}_v^{m_v}} = q^{-\frac{1}{2}\deg K_C} \cdot \prod_v \mathbf{1}_{\mathfrak{p}_v^{m_v}}.$$

Applying this formula to $\xi x(D)^{-1}$, where $x(D)$ is as above, we obtain

$$\hat{f}\left(\xi x(D)^{-1}\right) = \begin{cases} q^{-\frac{1}{2}\deg K_C}, & \text{if } v(\xi) \geq m_v + n_v \ \forall v \\ 0, & \text{otherwise.} \end{cases}$$

Notice that this is nonzero precisely if $v(\xi) \geq -v(D) + v(K_C) = -v(D - K_C)$, so

$$\sum_{\xi \in k} \hat{f}\left(\xi x(D)^{-1}\right) = q^{-\frac{1}{2}\deg K_C} \#L(K_C - D) = q^{-\frac{1}{2}\deg K_C + l(K_C - D)}.$$

Finally, we apply Theorem 13.4 to the function f and the idèle $x(D)$. It yields

$$\sum_{\xi \in k} f(\xi x(D)) = \|x(D)\|^{-1} \sum_{\xi \in k} \hat{f}(\xi x(D)^{-1}),$$

that is,

$$q^{l(D)} = \|x(D)\|^{-1} q^{-\frac{1}{2}\deg K_C + l(K_C - D)}.$$

Since $\|x(D)\|^{-1} = \prod_v (q^{\deg v})^{n_v} = q^{\deg D}$, we have obtained

$$q^{l(D)} = q^{\deg D - \frac{1}{2}\deg(K_C) + l(K_C - D)},$$

that is,

$$l(D) - l(K_C - D) = \deg D - \frac{1}{2}(\deg K_C).$$

We conclude by setting $\deg(K_C) = 2g - 2$: it is clear from the statement of the theorem that g is automatically an integer (since every other term in the formula is), and non-negativity follows from the fact that replacing $D = K_C$ in the statement we obtain $l(K_C) - l(0) = 2g - 2 - g + 1$, that is, $g = l(K_C) - l(0) + 1 = l(K_C) \geq 0$. □

13.2 The Multiplicative Group: The Idèles

Recall from Definition 11.3 the group of idèles I_k of k. We shall regard I_k as a topological group, with the topology coming from its structure as a restricted product.

Definition 13.6 (Map to Ideals) For every idèle $a = (a_v)_v$ we define $\varphi(a)$ as the fractional ideal of O_k given by $\varphi(a) = \prod_{\mathfrak{p} \notin S_\infty} \mathfrak{p}_v^{v_\mathfrak{p}(a_\mathfrak{p})}$, where we identify the set of finite places of k to the set of (non-zero) prime ideals of O_k.

Remark 13.3 The map $a \mapsto \varphi(a)$ is a homomorphism with kernel I_{S_∞} (the idèles that are units at all finite places).

Following the general structure of our analysis, we now fix a measure on I_k and describe its quasi-characters.

Definition 13.7 (Idèlic Haar Measure) We take as Haar measure on the idèles the product $\prod_v d\alpha_v$ of the local multiplicative measures $d\alpha_v$ of Definition 12.5.

By Theorem 11.1, the quasi-characters of $I_k = \prod'_v (k_v^\times, \mathfrak{u}_v)$ are of the form $c(a) = \prod_v c_v(a_v)$, where—for almost all v—the character c_v belongs to $\mathfrak{u}_v^\perp$, that is to say, $c_v(\mathfrak{u}_v) = \{1\}$. Such a character, as already mentioned, is said to be *unramified*.

We embed $k^\times$ into I_k by sending $\alpha \in k^\times$ to the idèle $\alpha = (\alpha, \alpha, \ldots, \alpha, \ldots)$. As was the case with the additive group, the most interesting questions about I_k concern the 'position' of $k^\times$ inside it. The next result is well-known—in fact, we have already met it as Theorem 10.2—but we give a proof in our language.

Theorem 13.6 (Product Formula) *The following hold.*

1. *Let $\alpha \in k^\times$. The ideal $\varphi(\alpha)$ is the principal ideal (α).*
2. *For every $\alpha \in k^\times$ we have*

$$\|\alpha\| = \prod_v \|\alpha\|_v = 1.$$

Proof

1. This is almost tautological: for any finite place $\mathfrak{p}$ of k we have $v_\mathfrak{p}(\varphi(\alpha)) = v_\mathfrak{p}(\alpha_\mathfrak{p}) = v_\mathfrak{p}(\alpha)$, so the fractional ideals $\varphi(\alpha)$ and (α) have the same prime factorisation, hence they are equal.
2. Let D be the additive fundamental domain of Definition 13.2. The point is that αD is another additive fundamental domain, hence one expects D and αD to have the same volume. Since $\mu(\alpha D) = \|\alpha\| \mu(D)$ by Lemma 13.2, this should imply $\|\alpha\| = 1$. We now make this precise.

13.2 The Multiplicative Group: The Idèles

Notice first that $\bigsqcup_{\xi \in k}(\xi + \alpha D) = \bigsqcup_{\xi \in k}(\alpha \xi + \alpha D) = \bigsqcup_{\xi \in k}\alpha(\xi + D)$ is still a disjoint union, and that this union covers $\bigsqcup_{\xi \in k}\alpha(\xi + D) = \alpha\left(\bigsqcup_{\xi \in k}(\xi + D)\right) = \alpha \mathbb{A}_k = \mathbb{A}_k$.

Now

$$\mu_{\mathbb{A}_k}(\alpha D) = \sum_{\xi \in k} \int_{\alpha D \cap (\xi + D)} d\mu_{\mathbb{A}_k}(x) = \sum_{\xi \in k} \int_{(-\xi + \alpha D) \cap D} d\mu_{\mathbb{A}_k}(x)$$

$$\stackrel{\xi \mapsto -\alpha\xi}{=} \sum_{\xi \in k} \int_{(\alpha\xi + \alpha D) \cap D} d\mu_{\mathbb{A}_k}(x) = \int_{\bigsqcup_\xi((\alpha\xi + \alpha D) \cap D)} d\mu_{\mathbb{A}_k}(x)$$

$$= \int_D d\mu_{\mathbb{A}_k}(x) = \mu_{\mathbb{A}_k}(D).$$

Using $\mu_{\mathbb{A}_k}(\alpha D) = \|\alpha\| \mu_{\mathbb{A}_k}(D)$ (Lemma 13.2) and the fact that $\mu_{\mathbb{A}_k}(D) \neq 0$ (Theorem 13.2(2)), the proof is complete. $\square$

Definition 13.8 (Idèles of Norm 1) We let J denote the kernel of the map

$$I_k \to \mathbb{R}^+$$
$$a \mapsto \|a\|.$$

Let v_0 be an (arbitrarily chosen) Archimedean place of k. Denote by T the subgroup of idèles that are trivial away from v_0, and that are positive real numbers in the component $k_{v_0}^\times$ (which can be either $\mathbb{R}^\times$ or $\mathbb{C}^\times$). For a positive real number t, we also denote by t the unique idèle in T with absolute value t (notice that, when $k_{v_0} \cong \mathbb{C}$, the identification makes t correspond to the idèle $(1, 1, \ldots, 1, \sqrt{t}, 1, \ldots)$).

Remark 13.4 Tate writes that it is *aesthetically disturbing and not really necessary* to make this arbitrary choice of T. To see that the choice really is unnecessary, one can note that we could take as T the subgroup of I_k (isomorphic to $\mathbb{R}^+$) given by the image of the map that sends $t \in \mathbb{R}^+$ to the idèle that is 1 at each finite place and is equal to t (respectively, $\sqrt{t}$) at each real (respectively, complex) place. However, Tate's choice of T *is* more convenient for certain calculations, so we will use it despite its aesthetic drawbacks.

The following lemma is obvious:

Lemma 13.9 *We have* $I = J \times T$.

We will write idèles as $a = \|a\| \cdot b$, where $\|a\| \in T$ and $b = a/\|a\| \in J$. Since T is isomorphic to a copy of $(\mathbb{R}_{>0}, \cdot)$, its Haar measure is (proportional to) $\frac{dt}{t}$. Moreover, the Haar measure(s) on a product are products of Haar measures on the factors, so, if we choose $\frac{dt}{t}$ as our Haar measure on T, this determines a unique measure db on J such that our already-defined measure da is equal to $db\frac{dt}{t}$. Fubini's theorem then yields the equalities

$$\int_I f(a)\,da = \int_0^\infty \left(\int_J f(tb)db\right)\frac{dt}{t} = \int_J \left(\int_0^\infty f(tb)\frac{dt}{t}\right)db \quad (13.4)$$

for any $f \in L^1(I_k)$.

13.2.1 Multiplicative Fundamental Domain

The product formula (Theorem 13.6) implies that $k^\times$ is contained in J. Our next objective is to describe the relative position of $k^\times$ inside of J, and in particular to find a fundamental domain for $J/k^\times$.

Definition 13.9 We set $J_{S_\infty} = J \cap I_{S_\infty}$. Thus, J_{S_∞} is the group of idèles of norm 1 that are units at all finite places. We further introduce the set $S'_\infty = S_\infty \setminus \{v_0\}$ of the infinite places of k different from our chosen place v_0, and the function

$$\begin{aligned} l : J_{S_\infty} &\to \prod_{v \in S'_\infty} \mathbb{R} \\ b &\mapsto (\log \|b_v\|_v)_{v \in S'_\infty}. \end{aligned}$$

Lemma 13.10 *l is a continuous, surjective homomorphism.*

Proof That l is a homomorphism is obvious from the elementary properties of the logarithm. Continuity is equally easy. To see that it is surjective, simply notice that, given any point $(t_v)_{v \in S'_\infty} \in \mathbb{R}^{S'_\infty}$, we can easily find $(b_v)_{v \in S'_\infty}$ such that $\log \|b_v\|_v = t_v$ for all $v \in S'_\infty$. We extend this to an idèle in J_{S_∞} by choosing $b_{v_0} \in k_{v_0}^\times$ in such a way that $\prod_{v \in S_\infty} \|b_v\|_v = 1$, at which point the idèle with finite components equal to 1 and infinite components equal to the b_v maps to $(t_v)_{v \in S'_\infty} \in \mathbb{R}^{S'_\infty}$ under l, as desired. □

We now describe the intersection $k^\times \cap J_{S_\infty}$.

Lemma 13.11 *The following hold*

1. $k^\times \cap J_{S_\infty} = O_k^\times$;
2. $k^\times \cap \ker l = \mu(k)$, *the (finite cyclic) group of roots of unity in k.*

Proof

1. By definition, an idèle in J_{S_∞} is a unit at all finite places. If it also an element of k, then it is a unit of O_k.
2. It is a well-known theorem of Kronecker that the units of O_k that are of absolute value 1 under all complex embeddings are precisely the roots of unity in k, see Problem 13.2. Notice that, given $\alpha \in k^\times \cap \ker l$, we know that $\|\alpha\|_v = 1$ for all

13.2 The Multiplicative Group: The Idèles

finite v and for all $v \in S'_\infty$. The product formula (Theorem 13.6) then implies that also $\|\alpha\|_{v_0}$ is equal to 1. □

Recall now Theorem 3.9. Together with its proof, it shows the following. Let $r = r_1 + r_2 - 1$ and $\varepsilon_1, \ldots, \varepsilon_r$ be a system of generators[3] for the free part of $O_k^\times$. The images $l(\varepsilon_1), \ldots, l(\varepsilon_r)$ form a full-dimensional lattice inside $\mathbb{R}^r$. In particular, the $l(\varepsilon_i)$ form an $\mathbb{R}$-basis of $\mathbb{R}^r$. Thus, for every $b \in J_{S_\infty}$, one can write uniquely

$$l(b) = \sum_{i=1}^{r} x_i l(\varepsilon_i) \tag{13.5}$$

for some $x_i \in \mathbb{R}$. By analogy with the additive case (see Definition 13.2), it is natural to consider the 'fundamental parallelotope'

$$P = \left\{ \sum_{i=1}^{r} x_i l(\varepsilon_i) \mid x_i \in [0, 1) \right\}. \tag{13.6}$$

This is not yet the good fundamental domain for the multiplicative action, for reasons that we will explain below, but is closely related to it. For this reason, it is useful to know the measure of P, which we compute in the next lemma.

Lemma 13.12 *We have*

$$\int_{l^{-1}(P)} db = \frac{2^{r_1}(2\pi)^{r_2}}{\sqrt{|d_k|}} R,$$

where

$$R = \left| \det (\log \|\varepsilon_i\|_v)_{\substack{1 \leq i \leq r \\ v \in S'_\infty}} \right|$$

is the regulator of k (see Definition 3.6).

Proof Let Q be the unit cube $Q = \{(x_v)_{v \in S'_\infty} : 0 \leq x_v < 1 \ \forall v \in S'_\infty\}$ and let $X = \prod_{v \in S'_\infty} \mathbb{R}$ be the real vector space in which P, Q live. Since l is a surjective homomorphism, denoting by μ_X any Haar measure on X (for example, the standard Lebesgue measure), by Problem 13.3 we have

$$\frac{\mu_{I_k}(l^{-1}(P))}{\mu_{I_k}(l^{-1}(Q))} = \frac{\mu_X(P)}{\mu_X(Q)}.$$

---

[3] This means that the classes of $\varepsilon_1, \ldots, \varepsilon_r$ in $O_k^\times / \{\text{roots of unity}\} \cong \mathbb{Z}^r$ form a basis of the quotient.

By definition, there is a linear map taking Q to P whose matrix has the $l(\varepsilon_i)$ as columns, so the above ratio is equal to the absolute value of the determinant of this matrix, that is, R. It remains to show that

$$\int_{l^{-1}(Q)} db = \frac{2^{r_1}(2\pi)^{r_2}}{\sqrt{|d_k|}}.$$

By definition, $l^{-1}(Q)$ is the set of $b \in J_{S_\infty}$ that satisfy $1 \leq |b|_v < e$ for $v \in S'_\infty$. Consider the more canonical set Q^* given by

$$Q^* = \{a \in I_{S_\infty} : 1 \leq \|a\|_v < e \quad \forall v \in S_\infty\}$$

(notice the change from S'_∞ to S_∞). Fubini's theorem gives

$$\int_{Q^*} da = \int_J \left(\int_{t:tb\in Q^*} \frac{dt}{t}\right) db = \int_{l^{-1}(Q)} \left(\int_{t:1\leq\|tb\|_{v_0}<e} \frac{dt}{t}\right) db$$

$$= \int_{l^{-1}(Q)} \left(\int_{\|b\|_{v_0}^{-1}}^{e\|b\|_{v_0}^{-1}} \frac{dt}{t}\right) db = \int_{l^{-1}(Q)} db,$$

where we have used:

1. $tb \in l^{-1}(Q)$ if and only if $b \in l^{-1}(Q)$ and $1 \leq \|tb\|_{v_0} < e$ (recall that t is an idèle whose only non-unitary component is along v_0);
2. the integral $\int_{\|b\|_{v_0}^{-1}}^{e\|b\|_{v_0}^{-1}} \frac{dt}{t}$ can be evaluated exactly: it gives

$$\log(e\|b\|_{v_0}^{-1}) - \log(\|b\|_{v_0}^{-1}) = \log(e) = 1.$$

Thus, the claim of the lemma is equivalent to the fact that

$$\int_{Q^*} da = \frac{2^{r_1}(2\pi)^{r_2}}{\sqrt{|d_k|}}.$$

Now, Q^* is a product set: it can be written as $I^{S_\infty} \times \prod_{v\in S_\infty}\{a_v : 1 \leq \|a_v\|_v < e\}$. By (11.1), its measure is therefore given by

$$\prod_{v\in S_\infty} \int_{\{a_v:1\leq\|a_v\|<e\}} d\mu_{k_v^\times} \times \prod_{v \text{ finite}} \int_{u_v} d\mu_{k_v^\times}.$$

13.2 The Multiplicative Group: The Idèles

We compute the factors in this product:

1. if v is real,

$$\int_{\{a_v : 1 \leq \|a_v\|_v < e\}} d\mu_{k_v^\times} = \int_{(-e,-1] \cup [1,e)} \frac{dt}{t} = 2 \int_1^e \frac{dt}{t} = 2.$$

2. if v is complex,

$$\int_{\{a_v : 1 \leq \|a_v\|_v < e\}} d\mu_{k_v^\times} = \int_{\{z \in \mathbb{C} : 1 \leq |z| < \sqrt{e}\}} \frac{dz}{|z|^2} = \int_1^{\sqrt{e}} \frac{2r\, dr}{r^2} \int_0^{2\pi} d\vartheta = 2\pi.$$

For this calculation, one needs to pay attention to the fact that $\|z\| = |z|^2$, where $|\cdot|$ is the usual complex norm. This justifies both the appearance of $\sqrt{e}$ and the denominator r^2.

3. for v finite, by Lemma 12.4 we have $\mu_{k_v^\times}(\mathfrak{u}_v) = N(\mathfrak{o}_{k_v})^{-1/2}$.

Putting everything together we obtain that the measure of Q^* (hence of $l^{-1}(Q)$) is

$$2^{r_1}(2\pi)^{r_2} \prod_{v \text{ finite}} N(\mathfrak{o}_{k_v})^{-1/2} = 2^{r_1}(2\pi)^{r_2}|d_k|^{-1/2},$$

where the last equality is obtained using Theorem 10.6. □

Definition 13.10 (Multiplicative Fundamental Domain) Let h be the class number of k and let $b_1, \ldots, b_h \in J$ be chosen so that the corresponding ideals $\varphi(b_1), \ldots, \varphi(b_h)$ represent all the ideal classes. Let w be the number of roots of unity in k. Let l be the logarithmic map of Eq. (13.5) and P be the fundamental parallelotope of Eq. (13.6). Define

$$E_0 = \{b \in l^{-1}(P) : 0 \leq \arg(b_{v_0}) < \frac{2\pi}{w}\} \subseteq J_{S_\infty}$$

and

$$E := E_0 b_1 \cup E_0 b_2 \cup \cdots \cup E_0 b_h.$$

We call E the **multiplicative fundamental domain** for $J \bmod k^\times$.

Theorem 13.7 (Properties of the Multiplicative Fundamental Domain)

1. The union

$$E_0 b_1 \cup E_0 b_2 \cup \cdots \cup E_0 b_h$$

appearing in the definition is disjoint.

2. $J = \bigsqcup_{\alpha \in k^\times} \alpha E$ (so that E deserves its name: it is a set of representatives for $J \mod k^\times$).
3.
$$\int_E db = \frac{2^{r_1}(2\pi)^{r_2} h R}{\sqrt{|d_k|} w}.$$

Proof

1. It suffices to observe that $\varphi(E_0 b_i) = \varphi(b_i)$ since E_0 is by definition a subset of J_{S_∞}, which is in the kernel of φ by Lemma 13.11.
2. To show that the union is disjoint, it suffices to prove that if we have a solution to

$$\alpha e_1 = e_2$$

with $e_1, e_2 \in E$ and $\alpha \in k^\times$, then $\alpha = 1$. Suppose that e_i is in $E_0 b_i$ for $i = 1, 2$: applying φ we then obtain $(\alpha)\varphi(b_1) = \varphi(b_2)$, hence (since the ideal classes $\varphi(b_i)$ are distinct) $b_1 = b_2$. Write $e_i = c_i b_i$ with $c_i \in E_0$. After simplifying a factor $b_1 = b_2$, we are left with showing that the equation $\alpha c_1 = c_2$ with $c_1, c_2 \in E_0$ and $\alpha \in k^\times$ has solutions only for $\alpha = 1$.

Applying again φ shows $\varphi(\alpha)\varphi(c_1) = \varphi(c_2)$, where the principal ideals $\varphi(c_1), \varphi(c_2)$ are trivial, since $c_i \in E_0 \subseteq J_{S_\infty}$. Hence $\varphi(\alpha)$ is not just principal, but also trivial, and α is a unit of $O_k^\times$. Writing $\alpha = \zeta \prod_{j=1}^r \varepsilon_j^{n_j}$ and applying the homomorphism l we then obtain

$$\sum_j n_j l(\varepsilon_j) = l(\alpha) = l(c_2) - l(c_1).$$

Now observe that the n_j are integers, while the coefficients of $l(c_2) - l(c_1)$ in the basis $l(\varepsilon_1), \ldots, l(\varepsilon_r)$ lie in the open interval $(-1, 1)$. Hence, the n_j are all 0, and $\alpha = \zeta$ is a root of unity in E_0. It is also a root of unity in $k^\times$, hence its order divides w. On the other hand, by definition, $0 \leq \arg \alpha_{v_0} < \frac{2\pi}{w}$, which implies $\alpha_{v_0} = \zeta_{v_0} = 1$. Thus, the Archimedean embedding v_0 sends α to 1, and therefore $\alpha = 1$ as desired.

To show that the union is all of J we reason essentially in the same way. Start with any idèle $b \in J$. There is a unique $i \in \{1, \ldots, r\}$ such that bb_i^{-1} represents a principal ideal, say αO for some $\alpha \in k^\times$. The idèle $bb_i^{-1}\alpha^{-1}$ is then an element of J representing the trivial ideal, so it is in J_{S_∞}. Apply l and write

$$l(bb_i^{-1}\alpha^{-1}) = \sum_{j=1}^r (n_j + x_j) l(\varepsilon_j),$$

13.2 The Multiplicative Group: The Idèles

with $n_j \in \mathbb{Z}$ and $x_j \in [0, 1)$. The idèle

$$bb_i^{-1}\alpha^{-1}\prod_{j=1}^{r}\varepsilon_j^{-n_j}$$

is in $l^{-1}(P)$. To land in E_0 we need to adjust the argument of the v_0-component so that it lies in the interval $[0, \frac{2\pi}{w})$. There is a (unique) choice of root of unity ζ such that the v_0-component of $\zeta^{-1}\left(bb_i^{-1}\alpha^{-1}\prod_{j=1}^{r}\varepsilon_j^{-n_j}\right)$ satisfies the desired inequality on the argument. The idèle $bb_i^{-1}\zeta^{-1}\alpha^{-1}\prod_{j=1}^{r}\varepsilon_j^{-n_j}$ is then in E_0, which means that we have $b \in \left(\zeta\alpha\prod_{j=1}^{r}\varepsilon_j^{n_j}\right)b_i E_0 \subseteq k^\times b_i E_0 \subseteq k^\times E$, as desired.
3. By definition we have the equalities

$$E = \bigsqcup_{j=1}^{h} E_0 b_j, \quad l^{-1}(P) = \bigsqcup_{\zeta \in \mu_w} E_0 \zeta,$$

which (together with Lemma 13.12) imply

$$\int_E db = h\int_{E_0} db, \quad \frac{2^{r_1}(2\pi)^{r_2}}{\sqrt{|d_k|}}R = \int_{l^{-1}(P)} db = w\int_{E_0} db.$$

Combining these equations gives the claim. $\square$

Corollary 13.2 (Position of $k^\times$ Inside J) *The subgroup $k^\times$ is discrete in J (hence in I_k). The quotient J mod $k^\times$ is compact.*

Proof As in Corollary 13.1: E has non-empty interior (so, up to translation, contains 1 in its interior) and is contained in a compact set. $\square$

Remark 13.5 Let $\tilde{E}$ be another fundamental domain for the multiplicative action (for example, $\tilde{E} = E^{-1}$). Arguing as in the proof of Theorem 13.6(2), one sees that if $f(x) : \mathbb{A}_k \to \mathbb{C}$ satisfies $f(\xi x) = f(x)$ for all $\xi \in k^\times$, then $\int_{\tilde{E}} f(a)\,da = \int_E f(a)\,da$.

13.2.2 The Quasi-Characters of $I_k/k^\times$

Just like in Sect. 13.1.2 we worked with functions on $\mathbb{A}_k$ invariant under translation by k (because we were secretly interested in functions on $\mathbb{A}_k/k$), we now consider functions (in fact, quasi-characters) of I_k that are trivial on $k^\times$. We shall work exclusively with such quasi-characters: we will call them **(quasi-)characters of** $I_k/k^\times$, but we will always treat them as functions defined on I_k and $k^\times$-periodic.

Remark 13.6 Let c be a quasi-character of $I_k/k^\times$. The restriction of c to J is a character (that is, it takes values in $\mathbb{S}^1$), because $\|c(J)\| = \|c(J \bmod k^\times)\|$ is a continuous image of a compact set, but it is also a group, so it is a compact subgroup of $\mathbb{R}_{>0}$. The only such subgroup is $\{1\}$.

Suppose moreover that c is a quasi-character of $I_k/k^\times$ that is trivial on J. We claim that c is of the form $c(a) = \|a\|^s$ for some complex number s uniquely determined by c. Indeed, c factors via $I_k/J = T \cong \mathbb{R}^\times$, and the quotient map is precisely $a \mapsto \|a\|$. We are thus reduced to describing the characters of $\mathbb{R}^\times$, which we have already done in Problem 12.4.

Definition 13.11 (Exponent of a Quasi-Character of $I_k/k^\times$) Let c be a quasi-character of $I_k/k^\times$. By the previous remark, $\|c(a)\| = \|a\|^s$ for some $s \in \mathbb{C}$. Since $\|c(a)\|$ is real and positive, s is a real number σ. We call this σ the **exponent** of c.

Note that a quasi-character of $I_k/k^\times$ is a character if and only if its exponent is 0.

13.3 Global Zeta Functions

We now develop the global analogue of the local zeta functions of Sect. 12.3. We begin by introducing the analogue of the class of functions of Definition 12.6 in the global setting.

Definition 13.12 (Class of $\mathfrak{z}$-Functions, Global Case) We denote by $\mathfrak{z}$ be the class of all functions $f : \mathbb{A}_k \to \mathbb{C}$ that satisfy

1. $f(x) \in \mathfrak{V}^1(\mathbb{A}_k)$ (that is, $f(x)$ and its Fourier transform are in $L^1(\mathbb{A}_k)$ and $f(x)$ is continuous) and $\hat{f}(x)$ is continuous.
2. the series $\sum_{\xi \in k} f(a(x+\xi))$ and $\sum_{\xi \in k} \hat{f}(a(x+\xi))$ are convergent for every idèle a and adèle x. The convergence is uniform in (a, x) ranging over any (fixed) subset of the form $K \times D$, where D is the additive fundamental domain and K is a compact subset of I_k.
3. $f(a)\|a\|^\sigma$ and $\hat{f}(a)\|a\|^\sigma$ are in $L^1(I_k)$ for $\sigma > 1$.

Remark 13.7 We have not asked that f be continuous on I_k. However, Problem 13.1 shows that the topology on I_k is finer than the subspace topology, hence $f|_{I_k}$ is automatically continuous (both for the subspace topology, which is obvious, and for the restricted product topology).

It should be clear that properties (1) and (2) in Definition 13.12 are precisely the hypotheses needed to apply the Riemann-Roch theorem 13.4. On the other hand, property (3) is what is needed to mimic the definition of the local zeta functions:

13.3 Global Zeta Functions

Definition 13.13 (ζ-function, Global Case) Let $f \in \mathfrak{z}$. We introduce the function $\zeta(f, c)$ of quasi-characters c of $I_k/k^\times$, defined for all quasi-characters of exponent greater than 1, by

$$\zeta(f, c) = \int_{I_k} f(a)c(a)\,da.$$

We call such a function a ζ-function of the global field k.

Notice the complete analogy with the local case: these global ζ functions are essentially Fourier transforms on the multiplicative group of idèles, just like the local ones were defined as Fourier transforms on $k^\times$. As was the case in Sect. 12.3, we are interested in working with equivalence classes of quasi-characters. The *local* definition of equivalence is that two characters are equivalent if the coincide on the units; its *global* counterpart is the following:

Definition 13.14 (Equivalence Class of Quasi-Character) Let c_1, c_2 be two quasi-characters of $I_k/k^\times$. We say that they are **equivalent** if they coincide on J. The equivalence class of the quasi-character c is the set of all quasi-characters of the form $c(a)\|a\|^s$ for $s \in \mathbb{C}$.

All the considerations of Sect. 12.3 now apply: $\zeta(f, c)$ can be considered locally (that is, on each equivalence class of quasi-characters) as a function of a complex variable s, and—when regarded as such—it is a holomorphic function in the domain of quasi-characters of exponent greater than 1 (see Lemma 12.5).

The next, and most important, step is now to establish the functional equation and analytic continuation of these global ζ functions.

Theorem 13.8 (Analytic Continuation and Functional Equation of the Global ζ-functions) *Let k be a number field with standard invariants (r_1, r_2) (signature), h_k, R_k (class number and regulator), d_k (discriminant) and w_k (number of roots of unity). Let f be a function of class $\mathfrak{z}$ and define the constant*

$$\kappa := \frac{2^{r_1}(2\pi)^{r_2} h_k R_k}{\sqrt{|d_k|} w_k},$$

which by Theorem 13.7(3) gives the volume of the multiplicative fundamental domain E. The ζ-function $\zeta(f, c)$ may be extended by analytic continuation to the domain of all quasi-characters. The extended function is meromorphic and has poles only at the quasi-characters $c(a) = 1$ and $c(a) = \|a\|$, where it has simple poles with residues $-\kappa f(0)$ and $+\kappa \hat{f}(0)$. Moreover, $\zeta(f, c)$ satisfies the functional equation

$$\zeta(f, c) = \zeta(\hat{f}, \hat{c}),$$

where $\hat{c}(a) = \|a\| c(a)^{-1}$, as in the local theory.

Remark 13.8 The reader might find it strange that these ζ functions also have a pole at 0, whereas the global zeta functions we are used to from Chap. 1 (say, the Dedekind zeta functions) only have a pole at $s = 1$. Two remarks are in order: the first and most important one is that Tate's global ζ functions are analogues of the 'completed' ζ functions (see e.g. Remark 1.2 or the definition of Λ_K in Theorem 3.13), and not of the Dedekind zeta functions themselves. The second is that a functional equation of the form $\Lambda(s) = \Lambda(1-s)$ as in Theorem 3.13 certainly implies that any pole at $s = 1$ should also show up at $s = 0$.

Before showing the theorem we state and prove some auxiliary lemmas. We begin with the following simple observation, which can be shown much in the spirit of Lemma 13.6:

Lemma 13.13 *Let* $f : J \to \mathbb{C}$ *be a continuous function such that* $f(x) = f(\alpha x)$ *for every* $x \in J, \alpha \in k^\times$. *Then, for any two fundamental domains* E, E' *for* $J/k^\times$ *having the same measure*[4] *we have* $\int_E f(b)db = \int_{E'} f(b)db$.

Proof The function f induces a function on the quotient $J/k^\times$, and both integrals equal $\int_{J/k^\times} f(b)db$, where the Haar measure on the quotient is normalised so that $\mathrm{vol}(J/k^\times) = \mathrm{vol}(E) = \mathrm{vol}(E')$. □

Remark 13.9 In fact, the lemma is true (with the same proof) under the slightly weaker assumptions that f is measurable, satisfies $f(x) = f(\alpha x)$, and that the induced function $\overline{f} : J/k^\times \to \mathbb{C}$ is in $L^1(J/k^\times)$. Note that a continuous function on the compact set $J/k^\times$ is automatically bounded and hence in L^1.

Our next lemma is a consequence of the Riemann-Roch theorem (Theorem 13.4) and will be the crucial ingredient in the proof of Theorem 13.8.

Lemma 13.14 *For a fixed* $t \in T$ *define*

$$\zeta_t(f, c) = \int_J f(tb)c(tb)\,db.$$

For all quasi-characters c *of* $I_k/k^\times$ *we have*

$$\zeta_t(f, c) + f(0)\int_E c(tb)\,db = \zeta_{1/t}(\hat{f}, \hat{c}) + \hat{f}(0)\int_E \hat{c}\left(\frac{1}{t}b\right)db.$$

Proof Recall from Theorem 13.7 that $J = \bigsqcup_{\alpha \in k^\times} \alpha E$. We start by writing

$$\zeta_t(f, c) + f(0)\int_E c(tb)\,db = \sum_{\alpha \in k^\times} \int_{\alpha E} f(tb)c(tb)\,db + f(0)\int_E c(tb)\,db.$$

---

[4] This property is in fact automatic: one can show it as in the proof of Theorem 13.6.

13.3 Global Zeta Functions

Using the translation-invariance of the Haar measure and writing $b = \alpha b$, we rewrite this as

$$\sum_{\alpha \in k^\times} \int_E f(\alpha t b) c(\alpha t b)\, db + f(0) \int_E c(tb)\, db.$$

Now use the fact that $c(\alpha) = 1$ since c is trivial on $k^\times$ and the uniform convergence of the sum $\sum_{\alpha \in k^\times} f(\alpha t b)$ (property (2) in Definition 13.12, applied to the relatively compact subset E) to further rewrite the above as

$$\zeta_t(f, c) + f(0) \int_E c(tb)\, db = \int_E \left(\sum_{\alpha \in k^\times} f(\alpha t b) \right) c(tb)\, db + \int_E f(0) c(tb)\, db$$

$$= \int_E \left(\sum_{\xi \in k} f(\xi t b) \right) c(tb)\, db. \tag{13.7}$$

We are now in a position to apply Theorem 13.4 to the inner sum, which yields

$$\zeta_t(f, c) + f(0) \int_E c(tb)\, db = \int_E \left(\sum_{\xi \in k} \hat{f}\left(\frac{\xi}{tb}\right) \right) \frac{1}{\|tb\|} c(tb)\, db.$$

We now observe that $b \mapsto 1/b$ is an involution of the abelian group J, hence sends the Haar measure to itself (indeed, let g be this automorphism. Then $g^* db = \gamma\, db$ for some constant $\gamma > 0$, hence $db = g^* g^* db = \gamma^2 db$ and $\gamma = 1$). We also recall that E^{-1} is another multiplicative fundamental domain (Remark 13.5), and observe that the function $g(tb) = \left(\sum_{\xi \in k} \hat{f}\left(\frac{\xi}{tb}\right) \right) \frac{1}{\|tb\|} c(tb)$ satisfies $g(\eta tb) = g(tb)$ for all $\eta \in k^\times$, so that $\int_E f(a)\, da = \int_{E^{-1}} f(a)\, da$ (see Lemma 13.13). Combining these observations, we arrive at the representation

$$\zeta_t(f, c) + f(0) \int_E c(tb)\, db = \int_E \left(\sum_{\xi \in k} \hat{f}\left(\frac{\xi b}{t}\right) \right) \frac{\|b\|}{\|t\|} c(t/b)\, db$$

$$= \int_E \left(\sum_{\xi \in k} \hat{f}\left(\xi \frac{1}{t} b\right) \right) \hat{c}(b/t)\, db.$$

On the other hand, we can restart from Eq. (13.7) and replace $f \to \hat{f}, t \to 1/t, c \to \hat{c}$ to obtain

$$\zeta_{1/t}(\hat{f}, \hat{c}) + \hat{f}(0) \int_E \hat{c}(b/t) \, db = \int_E \left(\sum_{\xi \in k} \hat{f}\left(\xi \frac{1}{t} b\right) \right) \hat{c}(b/t) \, db.$$

Comparing the last two equations yields the lemma. □

Lemma 13.15 *Let c be a quasi-character of $I_k/k^\times$ and let $t \in T$. We have*

$$\int_E c(tb) \, db = \begin{cases} \kappa t^s, & \text{if } c(a) = \|a\|^s \\ 0, & \text{if } c \text{ is non-trivial on } J \end{cases}$$

Proof This is similar to several other results we proved (see Propositions 6.3 and 9.2). The condition $c(a) = \|a\|^s$ is equivalent to c being trivial on J, see Remark 13.6. Since E is a fundamental domain for J mod $k^\times$ (Theorem 13.7), the integral in the statement is simply

$$\int_{J \bmod k^\times} c(tb) \, db = c(t) \int_{J \bmod k^\times} c(b) \, db.$$

Furthermore, $c(b)$ is a character on J mod $k^\times$ (Remark 13.6 again), so we are integrating a character on a compact group: the result is either the measure of that compact group, if the character is trivial, or 0, if it is not. In the former case, we also need to apply Theorem 13.7(3) and observe that $\|t\| = t$ since t is essentially a positive real number (hence $c(t) = \|t\|^s = t^s$). □

Proof of Theorem 13.8 Using Fubini's theorem, for c of exponent greater than 1 we can write the integral over $I_k = J \times T$ that defines $\zeta(f, c)$ as

$$\zeta(f, c) = \int_{I_k} f(a) c(a) \, da = \int_0^\infty \left(\int_J f(tb) c(tb) \, db \right) \frac{dt}{t} = \int_0^\infty \zeta_t(f, c) \frac{dt}{t},$$

where

$$\zeta_t(f, c) = \int_J f(tb) c(tb) \, db$$

as in Lemma 13.14. We split the integral from 0 to ∞ as

$$\zeta(f, c) = \int_0^1 \zeta_t(f, c) \frac{dt}{t} + \int_1^\infty \zeta_t(f, c) \frac{dt}{t}.$$

13.3 Global Zeta Functions

Notice that (using Fubini in reverse) we have

$$\int_1^\infty \zeta_t(f, c) \frac{dt}{t} = \int_{\substack{a \in I \\ \|a\| \geq 1}} f(a)c(a) \, da.$$

By assumption (property (3) in Definition 13.12), the integral over all of I converges absolutely for c of exponent greater than 1. But if the exponent of c' is smaller than the exponent of c we have $|f(a)c'(a)| \leq |f(a)c(a)|$ for $\|a\| \geq 1$, so convergence of the integral $\int_{\substack{a \in I \\ \|a\| \geq 1}} f(a)c(a) \, da$ for c implies convergence for c'. Since the integral converges for c of exponent greater than 1, it converges for all c. Consider then $\int_0^1 \zeta_t(f, c) \frac{dt}{t}$. We rewrite this integral using Lemmas 13.14 and 13.15. Consider first the case when c is non-trivial on J. Then, Lemmas 13.14 and 13.15 together imply $\zeta_t(f, c) = \zeta_{1/t}(\hat{f}, \hat{c})$, hence

$$\int_0^1 \zeta_t(f, c) \frac{dt}{t} = \int_0^1 \zeta_{1/t}(\hat{f}, \hat{c}) \frac{dt}{t}.$$

On the other hand, if c is trivial on J then it is of the form $c(a) = \|a\|^s$ for some $s \in \mathbb{C}$, and Lemmas 13.14 and 13.15 yield

$$\int_0^1 \zeta_t(f, c) \frac{dt}{t} = \int_0^1 \left(\zeta_{1/t}(\hat{f}, \hat{c}) + \hat{f}(0) \int_E \hat{c}\left(\frac{1}{t}b\right) db - f(0) \int_E c(tb) \, db \right) \frac{dt}{t}$$

$$= \int_0^1 \left(\zeta_{1/t}(\hat{f}, \hat{c}) + \hat{f}(0) \int_E \|\tfrac{1}{t}b\|^{1-s} db - f(0)\kappa t^s \right) \frac{dt}{t}$$

$$= \int_0^1 \left(\zeta_{1/t}(\hat{f}, \hat{c}) + \hat{f}(0)\kappa t^{s-1} - f(0)\kappa t^s \right) \frac{dt}{t}.$$

Now observe that $t \mapsto t^{-1}$ preserves the Haar measure on $\mathbb{R}_{>0}$, yielding

$$\int_0^1 \zeta_{1/t}(\hat{f}, \hat{c}) \frac{dt}{t} = \int_1^\infty \zeta_t(\hat{f}, \hat{c}) \frac{dt}{t}.$$

By the same argument used above, this integral is analytic for all c.

Carrying out the (trivial) integration of t^{s-1}, t^s (assuming $\Re s > 1$), we have thus obtained

$$\zeta(f, c) = \int_1^\infty \zeta_t(f, c) \frac{dt}{t} + \int_1^\infty \zeta_t(\hat{f}, \hat{c}) \frac{dt}{t} + \mathbf{1}_{c = \|\cdot\|^s} \cdot \kappa \left(\frac{\hat{f}(0)}{s-1} - \frac{f(0)}{s} \right)$$

for all quasi-characters of exponent greater than 1. The first two summands in this expression are analytic for all c, while the last two terms (when $c = \|\cdot\|^s$) clearly have meromorphic continuation to $\mathbb{C}$. They also let us read off the poles and residues of $\zeta(f, c)$ directly.

Finally, the functional equation follows trivially: if c is not of the form $c(a) = \|a\|^s$, the expression above is unchanged under the substitution $(f, c) \leftrightarrow (\hat{f}, \hat{c})$. When $c(a) = \|a\|^s$, the dual character is $\hat{c}(a) = \|a\|^{1-s}$, so the substitution $(f, c) \leftrightarrow (\hat{f}, \hat{c})$ also replaces $s \leftrightarrow 1 - s$, which shows the desired invariance. □

Problems

13.1 (Two Topologies on I_k) Check that the restricted product topology of I_k is **strictly** finer than the subspace topology that I_k inherits from $\mathbb{A}_k$.

13.2 (Kronecker's Theorem) Let $\alpha \in O_k^\times$ have absolute value 1 under all complex embeddings of k. Prove that α is a root of unity.

Sketch of Solution. Let α^n be any power of α. The coefficients of the minimal polynomial of α^n over $\mathbb{Q}$ are bounded (since they are combinations with constant coefficients of $\sigma_i(\alpha^n) = \sigma_i(\alpha)^n$, which are of absolute value 1). The degrees of these characteristic polynomials are also bounded. Thus, the numbers $\{\alpha^n\}_{n \in \mathbb{N}}$ are roots of finitely many polynomials; in particular, the set $\{\alpha^n\}$ is finite. Hence, there exist m, n such that $\alpha^m = \alpha^n$.

13.3 Let $f : G \to H$ be a surjective homomorphism of locally compact abelian groups and let X_1, X_2 be measurable subsets of H, with $0 < \mu_H(X_2) < \infty$. Prove that

$$\frac{\mu_G(f^{-1}(X_1))}{\mu_G(f^{-1}(X_2))} = \frac{\mu_H(X_1)}{\mu_H(X_2)}.$$

Hint. Prove that $X \mapsto \mu_G(f^{-1}(X))$ is a Haar measure on H, and rescale by the correct factor.

Reference

1. Ramakrishnan, D., Valenza, R.J.: Fourier Analysis on Number Fields, Graduate Texts in Mathematics, vol. 186. Springer, New York (1999). https://doi.org/10.1007/978-1-4757-3085-2

Chapter 14
Hecke L-functions

Abstract In this chapter we give a complete definition of Hecke L-functions, we explain the relation between Hecke's point of view in terms of ideals and the modern language of idèles, and we investigate the relations of these notions with global class field theory.

Following Tate, in the development of the global theory we have worked with (quasi-)characters of I_k that are trivial on $k^\times \subset I_k$. We have not fully motivated this choice yet. In this section we try to give some background for why this choice is natural, and how characters of $I_k/k^\times$ relate to our previous definition of Hecke L-functions (Definition 5.2). This is a good time for the following definition:

Definition 14.1 (Hecke Character, Idèlic Version) A **Hecke (quasi-)character** of k is a quasi-character of I_k that is trivial on $k^\times$.

Remark 14.1 Following the general use, in this section we shall simply write *Hecke characters* even when we mean *Hecke quasi-characters*.

Hecke had an ideal-theoretic definition of his characters that is however substantially less easy to work with than the previous one. We shall essentially re-derive Hecke's formulas for his characters below, where however they will not be taken as the *definition*, but purely as a *consequence* of the general theory, see Remark 14.4.

To each Hecke character we can attach an L-function in a way that looks different from Definition 5.2. We will discuss below the relationship between the two.

Definition 14.2 (Hecke L-function) Given an (idèlic) Hecke character χ, let S be the set of finite places of k at which χ is ramified (=not unramified). We define the corresponding L-function as

$$L(s, \chi) = \prod_{\mathfrak{p} \notin S} \left(1 - \frac{\chi(\mathfrak{p})}{(N\mathfrak{p})^s}\right)^{-1},$$

for $\Re s$ sufficiently large to make the product converge (when χ is of exponent 0—that is, when χ is a character in the strict sense of the word—the product converges at least over $\{\Re s > 1\}$).

In the previous formula, the symbol $\chi(\mathfrak{p})$ is taken to mean the character χ evaluated at any idèle $b(\mathfrak{p})$ that is trivial at all $v \neq \mathfrak{p}$ and such that $b(\mathfrak{p})_\mathfrak{p}$ is a uniformiser at $\mathfrak{p}$.

Remark 14.2 By assumption, χ is unramified at all places $v \notin S$. Any two choices of uniformisers at $\mathfrak{p}$ differ by a unit in $u_\mathfrak{p}$, and since χ is unramified at $\mathfrak{p}$ (=trivial on $u_\mathfrak{p}$) we obtain that $\chi(\mathfrak{p})$ is well-defined.

Connecting Definitions 5.2 and 14.2 is not at all trivial, and largely depends on class field theory. Developing class field theory would require a substantial effort, so we keep this discussion to a minimum. A beautiful modern exposition of class field theory is given in Harari's book [1]. The following theorem condenses many results in class field theory in a form that is suitable for our application:

Theorem 14.1 (Class Field Theory for Number Fields) *Let k be a number field and denote by k^{ab} the maximal abelian extension of k (inside a fixed algebraic closure). There is a canonical surjective map $\vartheta : I_k \to \mathrm{Gal}(k^{\mathrm{ab}}/k)$ that satisfies:*

1. *ϑ is trivial on $k^\times$.*
2. *let L/k be a finite Galois extension of k with abelian Galois group, so that there is a canonical surjection $I_k \xrightarrow{\vartheta} \mathrm{Gal}(k^{\mathrm{ab}}/k) \xrightarrow{\pi} \mathrm{Gal}(L/k)$. Let $\mathfrak{p}$ be a place of k that is unramified in L, and define a corresponding idèle[1] $b(\mathfrak{p})$ as in Definition 14.2. Then, the Artin symbol $\left(\frac{L/k}{\mathfrak{p}}\right)$—which is a well-defined element of $\mathrm{Gal}(L/k)$, because the extension is abelian— coincides with $\pi(\vartheta(b(\mathfrak{p})))$.*
3. *with the same notation as in the previous part, for every prime $\mathfrak{q}$ of k, $\pi(\vartheta(u_\mathfrak{q}))$ is the inertia group at (any place lying over) $\mathfrak{q}$ for the extension L/k.*

The first statement in the theorem explains the relevance of the quotient $I_k/k^\times$, which serves as a handy proxy for the Galois group $\mathrm{Gal}(k^{\mathrm{ab}}/k)$, while actually containing more information.

With this (hard) theorem in hand, it is not too difficult to reconnect Definitions 5.2 and 14.2. Fix a finite Galois extension L/k with group G and a character $\chi : G \to \mathbb{C}^\times$. The character χ factors via the abelianisation of G, hence it factors via $\tilde{\chi} : \mathrm{Gal}(F/k) \to \mathbb{C}^\times$, where F/k is the maximal abelian sub-extension of L/k. By Theorem 5.2, the L-functions of χ and $\tilde{\chi}$ coincide, hence we can and do assume that $L = F$ is abelian over k and that $\tilde{\chi} = \chi$. The character χ now gives an idèlic Hecke character χ_{Hecke}, defined simply as the composition

$$\chi_{\mathrm{Hecke}} = \chi \circ \pi \circ \vartheta,$$

---

[1] By abuse of notation, the idèle $b(\mathfrak{p})$ is usually denoted simply by $\mathfrak{p}$. From now on, we adopt this convention.

14.1 Characters of the Idèles

where ϑ is the map of Theorem 14.1 and π is the canonical projection $\mathrm{Gal}(k^{\mathrm{ab}}/k) \to \mathrm{Gal}(L/k)$. The two Hecke L-functions are now easily seen to have the same local factors:

1. both L-functions have trivial local factors at primes $v \in S$. This is true by definition for the L-function of Definition 14.2, and is easy to check for the (Artin) L-function of Definition 5.2: we simply need to show that $V^{I_\mathfrak{p}}$ is trivial, and this follows immediately from part (3) of Theorem 14.1 together with the construction of the set S in Definition 14.2.
2. both L-functions have the same local factor at primes $\mathfrak{p} \notin S$. We have to check that

$$\chi_{\mathrm{Hecke}}(\mathfrak{p}) = \chi\left(\left(\frac{L/k}{\mathfrak{p}}\right)\right),$$

and this is a consequence of Theorem 14.1(2) (simply use the definition $\chi_{\mathrm{Hecke}}(\mathfrak{p}) = \chi(\pi(\vartheta(\mathfrak{p})))$).

Thus, we see that Definition 5.2 captures a large part of the class of Hecke L-functions, but not all of them: essentially, in Definition 5.2 we were only considering those characters that factor via the Galois group $\mathrm{Gal}(k^{\mathrm{ab}}/k)$, and in particular, only characters with finite image (so that they further factor via the Galois group of some *finite* abelian extension L/k). Hecke's original definition is more general. However, we will see below (Example 14.1) that in the case $k = \mathbb{Q}$ both definitions are essentially the same, and reduce to Dirichlet's L-functions.

14.1 Characters of the Idèles

We now describe the equivalence classes of (quasi-)characters of I_k that are trivial on $k^\times$, where k is a number field. We begin by noticing that each equivalence class of quasi-characters of I_k contains a character. If c is a quasi-character of I_k, then $c|_J$ is a quasi-character of a compact group, hence a character. In particular, the restriction of $|c|$ to J is trivial. This implies that $|c|$ is a quasi-character of the group $I_k/J \cong \mathbb{R}_{>0}$ (recall that J is the kernel of the norm map $I_k \to \mathbb{R}_{>0}$ sending a to $\|a\|$), and as such, $|c(a)| = \|a\|^s$ for some $s \in \mathbb{C}$. Since $|c(a)|$ is a real number, s is also a real number. The quasi-character c lies in the same equivalence class as the character $c(a) \cdot \|a\|^{-s}$. From now on, we shall therefore only work with *characters*.

By Theorem 11.1, any character of I_k is a product

$$c(a) = \prod_v c_v(a_v)$$

of local characters c_v, where all but finitely many c_v are trivial on $\mathfrak{u}_v$. We say that c is **unramified** at v if and only if $c_v|_{\mathfrak{u}_v}$ is trivial.

We fix a finite set S of places, containing all the Archimedean ones, such that c_v is trivial on $\mathfrak{u}_v$ for every $v \notin S$ (which we express concisely as 'c is unramified outside S'). We start by observing that, for $v \notin S$, the characters c_v factor via the group of ideals prime to S, in the following precise sense.

Definition 14.3 We let φ_S be the map

$$\begin{aligned} \varphi_S: \quad I_k &\to \mathcal{F}(S) := \{\text{fractional ideals } I \mid v(I) = 0 \quad \forall v \in S, v \text{ finite}\} \\ a = (a_v)_v &\mapsto \prod_{v \notin S} \mathfrak{p}_v^{v(a_v)}. \end{aligned}$$

It is a homomorphism with respect to the natural product structure on the set on the right and its kernel is I_S.

Define now $c^*(a) = \prod_{v \notin S} c_v(a_v)$. Since $c_v(a_v)$ only depends on the v-adic valuation of a_v, the character c^* factors via φ_S. We write

$$c^*(a) = \chi(\varphi_S(a))$$

for some character χ of $\mathcal{F}(S)$. On the other hand, by Theorem 12.3, each character c_v for $v \in S$ can be written as

$$c_v(a_v) = \tilde{c}_v(\tilde{a}_v) \|a_v\|_v^{it_v},$$

where for each $a_v \in K_v^\times$ we have written $a_v = \rho \cdot \tilde{a}_v$ for some $\tilde{a}_v$ of norm 1 and ρ either real (if v is Archimedean) or a power of a fixed uniformiser at v (if v is finite). Here $\tilde{c}_v$ is a character of $\mathfrak{u}_v$. We have thus expressed our character c in the form

$$c(a) = \prod_{v \in S} \tilde{c}_v(\tilde{a}_v) \cdot \prod_{v \in S} \|a_v\|_v^{it_v} \cdot \chi(\varphi_S(a)).$$

Remark 14.3 The previous considerations show that c is described by the following data:

1. the characters $\tilde{c}_v$ of $\mathfrak{u}_v$, for $v \in S$;
2. the real numbers t_v, for $v \in S$;
3. a character χ of the group $\mathcal{F}(S)$,

subject to the condition that $c(\alpha) = 1$ for all $\alpha \in k^\times$.

We now make the description of Remark 14.3 more explicit. Suppose that $|S| = m + 1$ and fix a system of generators[2] $\varepsilon_1, \ldots, \varepsilon_m$ for the group of S-units of k (see Theorem 3.10). Also denote by ε_0 a generator of the finite group of roots of unity in $k^\times$, so that $\langle \varepsilon_0, \ldots, \varepsilon_m \rangle = O_{k,S}^\times$. Notice that every $\varepsilon \in O_{k,S}^\times = k^\times \cap I_S$ satisfies

---

[2] A **system of generators** is a set of elements in $O_{k,S}^\times$ whose images in $O_{k,S}^\times/\text{torsion}$ form a basis of this free abelian group.

14.1 Characters of the Idèles

$\varphi_S(\varepsilon) = (1)$, and so $\chi(\varphi_S(\varepsilon)) = 1$. The condition that c be trivial on $O_{k,S}^\times$ implies in particular $c(\varepsilon_0) = 1$, hence

$$\prod_{v \in S} \tilde{c}_v(\varepsilon_0) = 1. \tag{14.1}$$

Notice that $\|\varepsilon_0\|_v = 1$ for all $v \in \Omega_k$.

Suppose now that we have fixed a family $\tilde{c}_v$ for $v \in S$ that satisfies condition (14.1). The equality $c(\varepsilon_i) = 1$ for $i = 1, \ldots, m$ is satisfied if and only if (for *some* determination of the logarithms) we have

$$\prod_{v \in S} \tilde{c}_v(\tilde{\varepsilon}_{i,v}) \|\varepsilon_{i,v}\|_v^{it_v} = 1 \Leftrightarrow \sum_{v \in S} t_v \log \|\varepsilon_{i,v}\|_v = i \log \left(\prod_{v \in S} \tilde{c}_v(\tilde{\varepsilon}_v) \right), \tag{14.2}$$

which is a linear system of m equations in the $m + 1$ unknowns t_v. One can show (it is part of Theorem 3.10) that the matrix of the linear system (14.2) has maximal rank. We can be even more explicit: since for every S-unit ε and every place $v \notin S$ we have $\|\varepsilon\|_v = 1$, it follows from Theorem 10.2 (which we have now proved as Theorem 13.6) that

$$0 = \log 1 = \log \prod_v \|\varepsilon\|_v = \sum_v \log \|\varepsilon\|_v = \sum_{v \in S} \log \|\varepsilon\|_v.$$

In particular, $\sum_{v \in S} \log \|\varepsilon_{i,v}\|_v = 0$ for all $i = 1, \ldots, m$, so a generator for the kernel of the matrix corresponding to the linear system (14.2) is the vector all of whose coordinates are equal to 1. Thus, given a solution $\{t_v\}_{v \in S}$, all other solutions are of the form $\{t_v + t\}_{v \in S}$ for some $t \in \mathbb{R}$.

Finally, for a given choice of $(\tilde{c}_v)_{v \in S}$ satisfying (14.1) and a choice of $(t_v)_{v \in S}$ satisfying (14.2), what conditions should χ satisfy? The choice is very constrained (often unique): for all $\alpha \in k^\times$ we must have

$$1 = c(\alpha) = \prod_{v \in S} \tilde{c}_v(\tilde{\alpha}_v) \|\alpha\|_v^{it_v} \cdot \chi(\varphi_S(\alpha)),$$

which means that $\chi(\varphi_S(\alpha))$ is uniquely determined by the formula

$$\chi(\varphi_S(\alpha)) = \prod_{v \in S} \tilde{c}_v(\tilde{\alpha}_v)^{-1} \|\alpha\|_v^{-it_v}. \tag{14.3}$$

Remark 14.4 Equations (14.1), (14.2), and (14.3) essentially give Hecke's original description of the Hecke characters.

Let $f : k^\times \to \mathcal{F}(S)$ be the map sending each element $x \in k^\times$ to the ideal obtained from (x) by deleting all primes in S from its factorisation. Formally, if $(x) = \prod_{\mathfrak{p} \in S} \mathfrak{p}^{v_\mathfrak{p}(x)} \cdot \prod_{\mathfrak{p} \notin S} \mathfrak{p}^{v_\mathfrak{p}(x)}$, then

$$f(x) = \prod_{\mathfrak{p} \notin S} \mathfrak{p}^{v_\mathfrak{p}(x)}. \tag{14.4}$$

Equation (14.3) shows that χ is uniquely determined on the subgroup $f(k^\times)$. In order to determine χ, we must choose one of the (finitely many) extensions of $\chi|_{f(k^\times)}$ to $\mathcal{F}(S)$. Problem 14.1 shows that $f(k^\times)$ is indeed a subgroup and bounds its index inside $\mathcal{F}(S)$; as an easy special case, when k has class number 1, every ideal in $\mathcal{F}(S)$ is principal, hence of the form $\varphi_S(\alpha)$ for some α, and so the character χ is uniquely determined by the other data.

Example 14.1 (Dirichlet Characters as Idèlic Hecke Characters of $\mathbb{Q}$) Take $k = \mathbb{Q}$. We claim that the idèles I_k decompose as the direct product

$$\widehat{\mathbb{Z}}^\times \times \mathbb{Q}^\times \times \mathbb{R}^\times_{>0}; \tag{14.5}$$

this is essentially a reflection of the fact that $\mathbb{Z}$ has unique factorisation. More precisely, given any idèle $(a_v)_{v \in \Omega_\mathbb{Q}} = ((\tilde{a}_p p^{e_p})_{p \text{ prime}}, t)$ with $t \in \mathbb{R}^\times$, $\tilde{a}_p \in \mathbb{Z}_p^\times$, and $e_p = 0$ for almost all p, introduce the rational number $r := \text{sgn}(t) \prod_p p^{-e_p}$. It is then clear that ra is an idèle that lies in $\prod_p \mathbb{Z}_p^\times \times \mathbb{R}_{>0}$; from this, the decomposition (14.5) follows easily. Moreover, under this isomorphism, $\mathbb{Q}^\times \subset I_\mathbb{Q}$ corresponds to the direct factor $\{1\} \times \mathbb{Q}^\times \times \{1\}$. Thus, a quasi-character of $I_\mathbb{Q}$ trivial on $\mathbb{Q}^\times$ is simply a quasi-character of $\widehat{\mathbb{Z}}^\times \times \mathbb{R}^\times_{>0}$, hence is the product of a quasi-character of $\widehat{\mathbb{Z}}^\times$ and a quasi-character of $\mathbb{R}^\times_{>0}$.

1. We claim that every continuous quasi-character χ of $\widehat{\mathbb{Z}}^\times$ factors via a finite quotient. This follows from Problem 12.6: if U is a sufficiently small neighbourhood of 1 in $\mathbb{C}^\times$ (containing no non-trivial subgroups), $\chi^{-1}(U)$ is open. On the other hand, $\chi(\chi^{-1}(U))$ is a subgroup contained in U, so it is trivial. It follows that $\ker \chi$ contains the open neighbourhood of the identity $\chi^{-1}(U)$, and therefore is open in $\widehat{\mathbb{Z}}^\times$. Since every open subgroup of $\widehat{\mathbb{Z}}^\times$ has finite index, the claim follows. In particular, χ is a character (and not just a quasi-character), because it factors via a quasi-character of a finite group, and every quasi-character of a finite group is a character.

   Finally, it is easy to see that the subgroups of the form $\{x \in \widehat{\mathbb{Z}}^\times : x \equiv 1 \pmod{m}\}$ give a fundamental system of neighbourhoods of the identity in $\widehat{\mathbb{Z}}^\times$, as m varies in $\mathbb{N}$. We obtain that every quasi-character of $\widehat{\mathbb{Z}}^\times$ factors via a quotient $\frac{\widehat{\mathbb{Z}}^\times}{\{x \in \widehat{\mathbb{Z}}^\times : x \equiv 1 \pmod{m}\}} \cong (\mathbb{Z}/m\mathbb{Z})^\times$. Thus, there is a bijection between the characters of $\widehat{\mathbb{Z}}^\times$ and the pairs $(m, \tilde{\chi})$ where m is a positive integer and $\tilde{\chi}$ is a primitive character modulo m.

14.1 Characters of the Idèles

2. The quasi-characters of $\mathbb{R}_{>0}$ are easy to describe: we have seen in Problem 12.4 that they are all of the form $t \mapsto t^s$ with $s \in \mathbb{C}$.

We can then write every Hecke character of $\mathbb{Q}$ as

$$\chi((a_v)_{v \in \Omega_{\mathbb{Q}}}) = \tilde{\chi}(\pi_m(a_v)) \cdot |a_\infty|^s,$$

where $\tilde{\chi}$ is a primitive character modulo m and π_m is the composition of the isomorphism $I_{\mathbb{Q}} \cong \hat{\mathbb{Z}}^\times \times \mathbb{Q}^\times \times \mathbb{R}_{>0}^\times$ with the canonical projection $\hat{\mathbb{Z}}^\times \to (\mathbb{Z}/m\mathbb{Z})^\times$. It is easy to see from the definition that χ is ramified precisely at the primes dividing m. Moreover, as $b(p)$ (see Definition 14.2) we can take the idèle $(1, 1, \ldots, p, 1, \ldots)$ with p in the position corresponding to the factor $\mathbb{Q}_p$. The L-function of the character χ is therefore

$$\prod_{p \nmid m} (1 - \chi((b_p)) p^{-s})^{-1} = \prod_{p \nmid m} (1 - \tilde{\chi}(p \bmod m)) p^{-s})^{-1} = L(s, \tilde{\chi}),$$

where $L(s, \tilde{\chi})$ is the Dirichlet L-function of the character $\tilde{\chi}$ from Definition 1.9. Since we have already checked that Dirichlet L-functions are Artin L-functions (Proposition 5.3), we see that for $k = \mathbb{Q}$ all our many definitions of (abelian) L-functions coincide.

Problems

14.1 Let f be the function of Eq. (14.4).

1. Prove that f is a homomorphism, and therefore its image is a subgroup of $\mathcal{F}(S)$.
2. Show that the image of f is a subgroup of finite index of $\mathcal{F}(S)$. Prove that this index is at most h, the class number of k.

14.2

1. Given a Gaussian integer $\alpha \in \mathbb{Z}[i]$ with $\alpha \not\equiv 0 \pmod{(1+i)}$, show that there exists $r(\alpha) \in \mathbb{Z}$, unique modulo 4, such that $\alpha \cdot i^{r(\alpha)} \equiv 1 \pmod{(1+i)^3}$.
2. Define a function $\chi : \mathbb{Z}[i] \to \mathbb{C}$ by setting

$$\chi(\alpha) = \begin{cases} \alpha \cdot i^{r(\alpha)}, & \text{if } \alpha \not\equiv 0 \pmod{(1+i)} \\ 0, & \text{otherwise} \end{cases}$$

3. Explain in what sense χ is a Hecke character of the field $\mathbb{Q}(i)$.
   Hint. This requires an analysis similar to that of Example 14.1. In particular, you will need to use that $\mathbb{Z}[i]$ has unique factorisation.
4. Show that $\chi \neq \overline{\chi}$. Prove on the other hand that the corresponding Hecke L-functions $L(s, \chi)$ and $L(s, \overline{\chi})$ coincide.

5. Describe the coefficients a_p, b_p in the Euler product

$$L(s, \chi) = \prod_p \frac{1}{1 - a_p p^{-s} + b_p p^{-2s}}.$$

Hint. Group primes of $\mathbb{Z}[i]$ according to the prime of $\mathbb{Z}$ over which they lie.

Reference

1. Harari, D.: Galois Cohomology and Class Field Theory. Universitext. Springer, Cham (2020). https://doi.org/10.1007/978-3-030-43901-9. Translated from the 2017 French original by Andrei Yafaev

Chapter 15
Recovering the Classical Theory

Abstract In this final chapter we apply the methods developed so far to prove all the analytic statements that we have assumed in Part I: the functional equation and meromorphic continuation of Dedekind zeta functions and Hecke L-functions and the analytic class number formula.

Our purpose in this final chapter is to prove the main analytic results we have assumed in Part I, namely the functional equation and analytic continuation for Dedekind ζ-functions (Theorem 3.13) and Hecke L-functions (Theorem 5.1), and the analytic class number formula (Theorem 6.4). We will of course do this by showing that the completed ζ functions of number fields, and suitable completed Hecke L-functions, are global ζ functions in the sense of Tate.

Before treating the general case in detail, we sketch the special case of the Dedekind zeta functions, starting with the case $k = \mathbb{Q}$ of the Riemann zeta function. Below we will check—in much greater generality—that all the results we invoke can indeed be applied, in the sense that the necessary technical assumptions (e.g., convergence of certain integrals) are all satisfied. We will therefore be somewhat brief in our treatment of the Riemann and Dedekind zeta functions, in the hope of showing how the deep results of Chap. 1 now follow easily from Tate's approach.

15.1 The Riemann Zeta Function

The ground field is $k = \mathbb{Q}$. We take c to be the trivial character of I_k, and define a function f on the adèles by setting

$$f(x) = \prod_p f_p(x_p) \cdot f_\infty(x_\infty),$$

where f_p is the characteristic function of $\mathbb{Z}_p \subset \mathbb{Q}_p$ and f_∞ is $t \mapsto e^{-\pi t^2}$ (cf. Sect. 12.4). One can show that f is of class ʒ. We compute the Fourier transform of f as

$$\hat{f}(y) = \prod_v \hat{f}_v(y) = f_\infty(y) \cdot \prod_p f_p(y) = f(y),$$

where we have used the results of Sect. 12.4, that show $\hat{f}_p = f_p$ for each prime p and $\widehat{f_\infty} = f_\infty$. Theorem 13.8 now implies

$$\zeta(f, \|\cdot\|^s) = \zeta(f, \|\cdot\|^{1-s}).$$

Finally, we express $\zeta(f, \|\cdot\|^s)$ as a product of local factors as

$$\prod_v \zeta(f_v, \|\cdot\|^s) = \pi^{-s/2}\Gamma\left(\frac{s}{2}\right)\prod_p (1-p^{-s})^{-1} = \pi^{-s/2}\Gamma\left(\frac{s}{2}\right)\zeta(s).$$

Thus, Theorem 13.8 gives us the functional equation and analytic continuation for the ζ function, together with the information that the residue at $s = 1$ is $\kappa \hat{f}(0) = 1$. It also finally justifies Remark 1.2, in the sense that the function $\pi^{-s/2}\Gamma\left(\frac{s}{2}\right)\zeta(s)$ naturally appears as a global ζ function of the field $\mathbb{Q}$.

15.2 Dedekind ζ Functions

The case of Dedekind zeta functions is not much harder. We take $c = 1$ and f_v to be the standard function considered in Sect. 12.4 for the trivial character. Explicitly,

$$f_v(\xi) = \begin{cases} e^{-\pi\|\xi\|_v^2}, & v \text{ real} \\ e^{-2\pi\|\xi\|_v}, & v \text{ complex} \\ 1_{\mathcal{O}_v}, & v \text{ finite and unramified in } k \\ 1_{\mathfrak{d}_v^{-1}}(\xi), & v \text{ finite and ramified.} \end{cases}$$

Note that for all but finitely many v we have $f_v = 1_{\mathfrak{u}_v}$. The Fourier transform is $\prod_v \hat{f}_v$, where

$$\hat{f}_v(\xi) = \begin{cases} f_v(\xi), & v \text{ real} \\ f_v(\xi), & v \text{ complex} \\ f_v(\xi), & v \text{ finite and unramified in } k \\ (N\mathfrak{d}_v)^{1/2} 1_{\mathcal{O}_v}(\xi), & v \text{ finite and ramified in } k. \end{cases}$$

15.2 Dedekind ζ Functions

Write
$$\Gamma_{\mathbb{R}}(s) = \pi^{-s/2}\Gamma(s/2), \quad \Gamma_{\mathbb{C}}(s) = (2\pi)^{1-s}\Gamma(s) \tag{15.1}$$

for the local ζ functions $\zeta(f_v, \|\cdot\|^s)$ at the Archimedean places (we get different functions according to whether v is real or complex).

The ζ function $\zeta(f, \|\cdot\|^s)$ corresponding to f and the trivial equivalence class of characters is then

$$\prod_v \zeta(f_v, \|\cdot\|^s) = \Gamma_{\mathbb{R}}(s)^{r_1}\Gamma_{\mathbb{C}}^{r_2} \prod_{v \text{ finite}} \zeta(f_v, \|\cdot\|^s)$$

$$= \Gamma_{\mathbb{R}}(s)^{r_1}\Gamma_{\mathbb{C}}^{r_2} \prod_{v \text{ finite}} \frac{(N\mathfrak{d}_v)^{s-1/2}}{1 - (N\mathfrak{p}_v)^{-s}}$$

$$= \Gamma_{\mathbb{R}}(s)^{r_1}\Gamma_{\mathbb{C}}(s)^{r_2}|d_k|^{s-1/2}\zeta_k(s),$$

where we used Eq. (12.10) and Theorem 10.6.

On the other hand, we now determine the ζ function of $\hat{f}$, $\widehat{\|\cdot\|^s} = \|\cdot\|^{1-s}$. Let S_0 be the set of finite places of k that are ramified over $\mathbb{Q}$. The calculations in Sect. 12.4 show that for every finite v—ramified or unramified—we have

$$\zeta(\hat{f}_v, \|\cdot\|^{1-s}) = \left(1 - (N\mathfrak{p}_v)^{-(1-s)}\right)^{-1}.$$

The desired ζ function is then given by

$$\zeta(\hat{f}, \|\cdot\|^{1-s}) = \prod_v \zeta(\hat{f}_v, \|\cdot\|^{1-s})$$

$$= \Gamma_{\mathbb{R}}(1-s)^{r_1}\Gamma_{\mathbb{C}}(1-s)^{r_2} \prod_{v \text{ finite}} (1 - (N\mathfrak{p}_v)^{-(1-s)})^{-1}$$

$$= |d_k|^{s-1/2}\zeta(f, \|\cdot\|^{1-s}),$$

where we used Theorem 12.5. (Notice that the product $\prod_{v \text{ finite}}(1-(N\mathfrak{p}_v)^{-(1-s)})^{-1}$ does not converge for $\Re s > 1$; what we mean is that we already know analytic continuation of both sides of the equation, and this equality is true wherever both sides are defined. Also pay attention to the change of variables $s \to 1-s$, which changes $|d_k|^{s-1/2}$ to $|d_k|^{1/2-s}$.)

Using Theorem 13.8, which gives $\zeta(f, \|\cdot\|^s) = \zeta(\hat{f}, \|\cdot\|^{1-s})$, we obtain

$$\zeta(f, \|\cdot\|^s) = \zeta(\hat{f}, \|\cdot\|^{1-s}) = |d_k|^{s-1/2}\zeta(f, \|\cdot\|^{1-s}).$$

Multiplying by $|d_k|^{1/2-s/2}$ on both sides we obtain

$$|d_k|^{(1-s)/2}\zeta(f,\|\cdot\|^s) = |d_k|^{s/2}\zeta(f,\|\cdot\|^{1-s}),$$

that is, the function

$$\tilde{\Lambda}_k(s) := |d_k|^{(1-s)/2}\zeta(f,\|\cdot\|^s) = |d_k|^{s/2}\Gamma_{\mathbb{R}}(s)^{r_1}\Gamma_{\mathbb{C}}(s)^{r_2}\zeta_k(s)$$

satisfies analytic continuation and the functional equation $\tilde{\Lambda}_k(s) = \tilde{\Lambda}_k(1-s)$.

It is easy to check that $\tilde{\Lambda}_k(s) = (2\pi)^{r_2}\Lambda_k(s)$, where $\Lambda_k(s)$ is the completed ζ function appearing in Theorem 3.13: indeed, the *fudge factors*

$$|d_k|^{s/2}\Gamma_{\mathbb{R}}(s)^{r_1}\Gamma_{\mathbb{C}}(s)^{r_2}$$

multiply out to give

$$|d_k|^{s/2}\Gamma_{\mathbb{R}}(s)^{r_1}\Gamma_{\mathbb{C}}^{r_2} = |d_k|^{s/2}\pi^{-r_1 s/2}\Gamma(s/2)^{r_1}(2\pi)^{r_2-r_2 s}\Gamma(s)^{r_2}$$

$$= |d_k|^{s/2}(2\pi)^{r_2}\frac{1}{\pi^{s/2(r_1+2r_2)}}\Gamma(s/2)^{r_1}\Gamma(s)^{r_2}$$

$$= |d_k|^{s/2}(2\pi)^{r_2}\frac{1}{\pi^{ns/2}2^{r_2 s}}\Gamma(s/2)^{r_1}\Gamma(s)^{r_2}$$

$$= (2\pi)^{r_2}\left(\frac{|d_k|}{4^{r_2}\pi^n}\right)^{s/2}\Gamma(s/2)^{r_1}\Gamma(s)^{r_2},$$

where we used $r_1 + 2r_2 = n$. Thus, we also get the functional equation of Theorem 3.13.

Theorem 13.8 tells us that the residue at 1 of $\zeta(f,\|\cdot\|^s)$ is

$$\kappa\hat{f}(0) = \kappa\prod_{v\in S_0}(N\partial_v)^{1/2} = \kappa|d_k|^{1/2}.$$

From the above we have

$$\zeta_k(s) = \left(|d_k|^{s-1/2}\Gamma_{\mathbb{R}}(s)^{r_1}\Gamma_{\mathbb{C}}(s)^{r_2}\right)^{-1}\zeta(f,\|\cdot\|^s). \tag{15.2}$$

The factor $\left(|d_k|^{s-1/2}\Gamma_{\mathbb{R}}(s)^{r_1}\Gamma_{\mathbb{C}}(s)^{r_2}\right)^{-1}$ is regular at $s = 1$, and its value is

$$(|d_k|^{1/2}\pi^{-r_1/2}\Gamma(1/2)^{r_1})^{-1} = |d_k|^{-1/2}.$$

15.3 The General Case: L-functions of Characters

Thus, the residue of $\zeta_k(s)$ at 1 is

$$|d_k|^{-1/2} \operatorname{Res}_{s=1} \zeta(f, \|\cdot\|^s) = |d_k|^{-1/2} \cdot \kappa |d_k|^{1/2} = \kappa = \frac{2^{r_1}(2\pi)^{r_2} h_k R_k}{\sqrt{|d_k|} w_k}.$$

This proves Theorem 6.4. As for Theorem 1.4, we know that $\zeta(f, \|\cdot\|^s)$ has a simple pole at $s = 0$ and $s = 1$. Since $\Gamma_{\mathbb{R}}(s)^{-1}$ and $\Gamma_{\mathbb{C}}(s)^{-1}$ are everywhere holomorphic and both vanish at $s = 0$, Eq. (15.2) shows that $\zeta_k(s)$ is everywhere meromorphic with a unique pole at $s = 1$, because the pole at $s = 0$ cancels out with the zero of the Gamma factors. This completes the proof of Theorem 1.4.

15.3 The General Case: L-functions of Characters

Our final task is to discuss the analytic properties of Dirichlet L-functions, and more generally Hecke L-functions (taken as in Definition 14.2—we have already checked that these contain all the Hecke L-functions of Definition 5.2).

15.3.1 Idèlic Character

We have discussed the characters of I_k in Sect. 14.1. With notation as in that section, fix a character

$$c(a) = \prod_{v \in S} \tilde{c}_v(\tilde{a}_v) \cdot \prod_{v \in S} \|a_v\|_v^{it_v} \cdot \chi(\varphi_S(a)) \tag{15.3}$$

where $\tilde{c}_v$, t_v and χ satisfy conditions (14.1), (14.2) and (14.3).

15.3.2 Choice of the Adèlic Function

For each $v \in S$, let $f_v(x_v)$ be the function used in Sect. 12.4 to compute the function $\rho(c_v \| \cdot \|_v^s)$ corresponding to the character c_v. For $v \notin S$, let $f_v(x_v)$ be the characteristic function of O_v. We set

$$f(x) = \prod_{v \in S} f_v(x_v).$$

We claim, and we will show below, that $f(x)$ is a function of class 3. We begin by noticing that each function $f_v(x_v)$ is in $\mathfrak{V}^1(k_v)$ by the direct computations of Sect. 12.4, and that $f_v = \mathbf{1}_{O_v}$ for almost all v. Lemma 11.7 yields that f is in $\mathfrak{V}^1(\mathbb{A}_k)$, and that the Fourier transform of $f(x)$ is $\hat{f}(x) = \prod_v \hat{f}_v(x_v)$. We have thus shown that $f(x)$ satisfies condition 1 in Definition 13.12.

15.3.3 Fourier Transform

We have already seen that the Fourier transform of $f(x)$ is

$$\hat{f}(x) = \prod_v \hat{f}_v(x_v).$$

For $v \notin S$ and v unramified in k, the Fourier transform of $f_v = \mathbf{1}_{O_v}$ is given by f_v itself, as shown by Eq. (12.9). For $v \notin S$ but ramified in k, we have

$$\hat{f}_v = (N\mathfrak{d}_v)^{-1/2} \mathbf{1}_{\mathfrak{d}_v^{-1}}$$

by a simple direct calculation. Moreover, since each local factor f_v is one of the standard functions of Sect. 12.4, the Fourier transforms $\hat{f}_v$ are in $\mathfrak{V}^1(k_v^\times)$, and in fact, $\hat{f}_v \| \cdot \|^\sigma$ is in L^1 for every $\sigma > 0$ (for $v \notin S$, this follows from the explicit expression above; for $v \in S$, notice that all the standard functions f_v considered in Sect. 12.4 are of class 3 for the local field k_v).

We will only need to know this: the Fourier transform $\hat{f}$ is a product of local functions $\hat{f}_v$, almost all of which agree with the corresponding $f_v(x)$, and all of which satisfy $\widehat{f_v}\| \cdot \|^\sigma \in L^1(k_v^\times)$ for $\sigma > 1$.

15.3.4 The ζ-Function

The function $|f(a)| \cdot \|a\|^\sigma = \prod_v \left(|f_v(a_v)| \cdot \|a_v\|_v^\sigma\right)$ is a product of local functions, and all but finitely many of these are equal to 1 on $\mathfrak{u}_v$ by definition. We can then use Theorem 11.2 to check that $|f(a)| \cdot \|a\|^\sigma$ is in $L^1(I_k)$. It suffices to verify that the infinite product

$$\prod_{v \in \Omega_k} \int_{k_v^\times} |f_v(a_v)| \|a_v\|_v^\sigma \, d\mu_{k_v^\times}(a_v)$$

converges. By the choice of standard functions in Sect. 12.4, each single function $|f_v(a_v)| \|a_v\|_v^\sigma$ is in $L^1(k_v^\times)$ for all $\sigma > 1$, so it suffices to check that the product of $\int_{k_v^\times} |f_v(a_v)| \|a_v\|^\sigma \, d\mu_{k_v^\times}(a_v)$ over all but finitely many v converges. In particular,

15.3 The General Case: L-functions of Characters

we can consider the product

$$\prod_{\substack{v \notin S \\ v \text{ unramified in } k}} \int_{k_v^\times} |f_v(a_v)| \|a_v\|_v^\sigma \, d\mu_{k_v^\times}(x_v) = \prod_{\substack{v \notin S \\ v \text{ unramified in } k}} \int_{O_v \setminus \{0\}} \|a_v\|_v^\sigma \, d\mu_{k_v^\times}(a_v)$$

$$= \prod_{\substack{v \notin S \\ v \text{ unramified in } k}} \frac{1}{1 - (N\mathfrak{p}_v)^{-\sigma}},$$

where the last equality uses Eq. (12.10) and the fact that $N\mathfrak{d}_v = 1$ for v unramified in k. This infinite product converges for $\sigma > 1$ by the classical calculation in Proposition 5.2. We already observed that the Fourier transform of f is a product of local functions, and all but finitely many factors coincide with those of f. From this, it follows easily that $\hat{f}$ is continuous and that $\hat{f}(a)\|a\|^\sigma$ is also in L^1 for every $\sigma > 1$. We have thus checked that f satisfies the third condition in Definition 13.12.

We now verify condition 2 in Definition 13.12, that is, we show that the sum

$$\sum_{\xi \in k} f(a(x + \xi))$$

is uniformly convergent for $(a, x) \in C \times D$, where C is a compact subset of I_k. As before, the argument for $\sum_{\xi \in k} \hat{f}(a(x + \xi))$ is completely analogous, so we only treat the case of f.

Lemma 15.1 *Let f and C be fixed. There is a fractional ideal A of O_k such that, for all $a \in C$ and $x \in D$, we have $f(a(x + \xi)) = 0$ unless ξ is in A.*

Proof Let v be a finite place. By definition, f_v vanishes outside of a compact subset B_v of k_v, and we can take $B_v = O_v$ for all but finitely many v. Thus, $f(a(x + \xi))$ can only be non-zero if $a(x + \xi)$ belongs to $\prod_v B_v$. Equivalently, $x + \xi$ has to belong to $\prod_v a_v^{-1} B_v$ (which is still a compact set), and ξ has to belong to $k \cap \prod_v (-x_v + a_v^{-1} B_v)$. We claim that the intersection $k \cap \prod_v (-x_v + a_v^{-1} B_v)$ is contained in a fractional ideal A of O_k independent of x and a (provided that these elements lie in D and C, respectively). To prove this, we notice that:

1. for each v, the v-valuation of elements in $-x_v + a_v^{-1} B_v$ is bounded below, uniformly as x varies in D and a in C. This follows from the compactness of B_v, of C and of $\overline{D}$, which implies the compactness of $-\overline{D} + C^{-1} B_v$, combined with the fact that v (being continuous) is bounded on any compact subset of $k_v^\times$.
2. for all but finitely many v, all elements in $-x_v + a_v^{-1} B_v$ are v-integral (for any choice of $x \in D$ and $a \in C$). To see this, recall from Lemma 11.1 that a compact subset of $I_k = \prod_v' (k_v^\times, \mathfrak{u}_v)$ is contained in a product of compact subsets C_v, where all but finitely many C_v are equal to $\mathfrak{u}_v$. Moreover, by definition, the elements in D are v-integral for all finite v. Thus, if v is any place for which $B_v = O_v$ (all but finitely many) and $C_v = \mathfrak{u}_v$ (all but finitely many), then $-x_v + a_v^{-1} B_v \subset O_v + \mathfrak{u}_v O_v \subseteq O_v$.

Combining the previous two statements, we obtain that $e_v = \min\{0, v(-x_v + a_v^{-1}b_v) : x \in D, a \in C, b_v \in B_v\}$ is a well-defined integer, equal to 0 for almost all v. Thus, $A := \prod_{v \text{ finite}} \mathfrak{p}_v^{e_v}$ is a well-defined fractional ideal of k, which by construction contains $k \cap \prod_v (-x_v + a_v^{-1} B_v)$. This concludes the proof of the lemma. □

Let A be the ideal given by the lemma. Like all fractional ideals of a number field, it is a free $\mathbb{Z}$-module of finite rank $n = [k : \mathbb{Q}]$. Let $\omega_1, \ldots, \omega_n$ be a $\mathbb{Z}$-basis of A. The series $\sum_{\xi \in k} f(a(x + \xi))$ can then be rewritten as a sum over $\xi \in A$, hence as

$$\sum_{(c_1,\ldots,c_n) \in \mathbb{Z}^n} f\left(a\left(x + \sum_{i=1}^n c_i \omega_i\right)\right).$$

By definition, $f(x) = \prod_v f_v(x_v)$, and all the non-Archimedean factors are bounded by 1 in absolute value (see Eq. (12.8)). Each Archimedean factor is also bounded (though not necessarily by 1), see Equations (12.5) and (12.6). Thus, it suffices to show that

$$\sum_{(c_1,\ldots,c_n) \in \mathbb{Z}^n} \prod_{v \text{ Archimedean}} f_v\left(a_v\left(x_v + \sum_{i=1}^n c_i \omega_i\right)\right) \tag{15.4}$$

converges uniformly. The product $\prod_{v \text{ Archimedean}} f_v(a_v(x_v + \sum_{i=1}^n c_i \omega_i))$ is bounded above by a function on $\mathbb{R}^n$ of the form $w \mapsto p(w)\exp(c_{a,x} + d_{a,x}\|w\| - B_{a,x}\|w\|^2)$, where $p(w)$ is a polynomial, $c_{a,x}, d_{a,x}$ are numbers depending continuously on a, x, and $B_{a,x}$ is a non-singular matrix depending continuously on a, x. Since a_v, x_v are bounded (lying respectively in the compact set C and in the relatively compact set D) and a_v is also bounded away from zero (for the same reason), it is easy to see that $w \mapsto p(w)\exp(c_{a,x} - B_{a,x}\|w\|^2)$ is bounded above by a function $g(w)$, independent of a, x, of the same form, for which the sum converges (see Problem 15.1; notice that we are summing over a full-rank lattice of $\mathbb{R}^n$, so as $\max |c_i| \to \infty$, also $\|\sum c_i \omega_i\|$ tends to infinity). This completes the verification that our function f lies in class $\mathfrak{z}$.

15.3.5 Conclusion: Analytic Continuation and Functional Equation

Let $\tilde{c}$ be a quasi-character of exponent greater than 1. Applying Theorem 11.2, we obtain that the ζ function $\zeta(f, \tilde{c}) = \int_{I_k} f(a)\tilde{c}(a)\, d\mu_{I_k}(a)$ decomposes as the product

$$\prod_v \left(\int_{k_v^\times} f_v(a_v)\tilde{c}_v(a_v)\, d\mu_{k_v^\times}(a_v)\right) = \prod_v \zeta(f_v, \tilde{c}_v) \tag{15.5}$$

15.3 The General Case: L-functions of Characters

of the local ζ functions of the quasi-characters $\tilde{c}_v$. In particular, if c is our fixed character (of exponent 0)

$$c(a) = \prod_v c_v(a_v) = \prod_{v \in S} c_v(a_v) \cdot \chi(\varphi_S(a)),$$

we can apply the decomposition (15.5) to the quasi-character $\tilde{c} = c\|\cdot\|^s$ for every s with $\Re s > 1$.

Next, we show that for every $v \notin S$ and $\Re s > 1$ we can explicitly compute the integral defining $\zeta(f_v, c_v\|\cdot\|_v^s)$. To do this, notice first that every $a_v \in O_v$ can be written uniquely as $\pi_v^i \tilde{a}_v$ with $i \in \mathbb{N}$ and $\tilde{a}_v \in \mathfrak{u}_v$. By definition, c_v is unramified, so $c_v(a_v) = c_v(\pi_v^i) = c_v(\pi_v)^i$ only depends on i, the valuation of a_v. Furthermore, $c_v(\pi_v)$ is by definition $\chi(\mathfrak{p}_v)$, where $\mathfrak{p}_v$ is the prime ideal corresponding to the valuation v. We may now compute

$$\begin{aligned}
\zeta(f_v, c_v\|\cdot\|_v^s) &= \int_{O_v \setminus \{0\}} c_v(a_v) \|a_v\|_v^s \, d\mu_{k_v^\times}(a_v) \\
&= \sum_{i=0}^\infty \mu_{k_v^\times}(\pi_v^i \mathfrak{u}_k)(N\mathfrak{p}_v)^{-is} \chi(\mathfrak{p}_v)^i \\
&= \sum_{i=0}^\infty \mu_{k_v^\times}(\mathfrak{u}_k)(N\mathfrak{p}_v)^{-is} \chi(\mathfrak{p}_v)^i \\
&= (N\mathfrak{d}_v)^{-1/2} \left(1 - \frac{\chi(\mathfrak{p}_v)}{(N\mathfrak{p}_v)^s}\right)^{-1}.
\end{aligned} \tag{15.6}$$

(We take this opportunity to mention that this is basically the same computation necessary to check Theorem 12.5.) Thus,

$$\begin{aligned}
\zeta(f, c\|\cdot\|^s) &= \prod_{v \in S} \zeta(f_v, c_v\|\cdot\|_v^s) \cdot \prod_{v \notin S} \left((N\mathfrak{d}_v)^{-1/2} \left(1 - \frac{\chi(\mathfrak{p}_v)}{(N\mathfrak{p}_v)^s}\right)^{-1}\right) \\
&= \prod_{v \in S} \zeta(f_v, c_v\|\cdot\|_v^s) \cdot \prod_{v \notin S} (N\mathfrak{d}_v)^{-1/2} \cdot L(s, c),
\end{aligned}$$

where $L(s, c)$ is the Hecke L-function of the idèlic character c in the sense of Definition 14.2.

To justify the appearance of $L(s, c)$, notice that

$$L(s, c) = \prod_{v \notin S} \left(1 - \frac{c(\mathfrak{p}_v)}{(N\mathfrak{p}_v)^s}\right)^{-1} = \prod_{v \notin S} \left(1 - \frac{\chi(\mathfrak{p}_v)}{(N\mathfrak{p}_v)^s}\right)^{-1} = L(s, \chi),$$

because by definition the Hecke L-function $L(s, c)$ only depends on $c(b(\mathfrak{p}))$, where $\mathfrak{p}$ ranges over the primes at which c is unramified. The explicit description (15.3) shows immediately that $c(b(\mathfrak{p})) = \chi(b(\mathfrak{p}))$. We have thus shown that $L(s, c)$ can be represented as

$$L(s, c) = \prod_{v \in S} \zeta(f_v, c_v \| \cdot \|_v^s)^{-1} \prod_{v \notin S} (N\mathfrak{d}_v)^{-1/2} \cdot \zeta(f, c \| \cdot \|^s).$$

Theorem 13.8 shows that $\zeta(f, c\|\cdot\|^s)$ has meromorphic continuation to $\mathbb{C}$, with poles at $s = 0, 1$ if c is trivial. Proposition 12.5 shows that the factor $\prod_{v \in S} \zeta(f_v, c_v \| \cdot \|_v^s)^{-1}$ is everywhere analytic, and $(N\mathfrak{d}_v)^{-1/2}$ is clearly a constant. Thus, we obtain analytic continuation to $\mathbb{C}$ as soon as c is not the trivial character; when c is the trivial character, we still need to check that the pole at $s = 1$ is not cancelled by the local zeta factors, while the pole at $s = 0$ *does* cancel out. This follows easily from the explicit expressions for the local zeta functions. In particular, the local ζ function at any Archimedean place is of one of the two forms given in Eq. (15.1): both these functions are regular and non-vanishing at $s = 1$, while they have a pole at $s = 0$. Their inverses thus vanish at $s = 0$, and cancel out the pole of $\zeta(f, \|\cdot\|^s)$. As for the situation at $s = 1$, it suffices to check that none of the local zeta functions computed in Sect. 12.4 has a pole at $s = 1$, which is easily seen to be the case. This completes the proof of Theorem 5.1. In fact, we have established the following stronger result:

Theorem 15.1 (Analytic Continuation for the Hecke L-functions) *Let $c : I_k/k \to \mathbb{C}^\times$ be a non-trivial Hecke character. The Hecke L-function $L(s, c)$ admits analytic continuation to the full complex plane.*

Finally, for the sake of completeness we also briefly discuss the functional equation satisfied by the Hecke L-functions. We can write the ζ function for the dual pair $(\hat{f}, \widehat{c\|\cdot\|^s}) = (\hat{f}, c^{-1}\|\cdot\|^{1-s})$ as

$$\zeta(\hat{f}, \widehat{c\|\cdot\|^s}) = \prod_{v \in S} \zeta(\hat{f}_v, \widehat{c_v\|\cdot\|_v^s}) \prod_{v \notin S} \chi(\mathfrak{d}_v)(N\mathfrak{d}_v)^{-s} \cdot L(1-s, \chi^{-1})$$

for $\Re s < 0$. The global functional equation of Theorem 13.8 then yields

$$\zeta(f, c\|\cdot\|^s) = \zeta(\hat{f}, \widehat{c\|\cdot\|^s}) \iff$$

$$\prod_{v \in S} \zeta(f_v, c_v\|\cdot\|_v^s) \cdot \prod_{v \notin S} (N\mathfrak{d}_v)^{-1/2} \cdot L(s, \chi)$$

$$= \prod_{v \in S} \zeta(\hat{f}_v, \widehat{c_v\|\cdot\|_v^s}) \prod_{v \notin S} \chi(\mathfrak{d}_v)(N\mathfrak{d}_v)^{-s} \cdot L(1-s, \chi^{-1}).$$

15.3 The General Case: *L*-functions of Characters

Dividing both sides by $\prod_{v \in S} \zeta(\hat{f}_v, \widehat{c_v \| \cdot \|_v^s}) \prod_{v \notin S} \chi(\mathfrak{d}_v)(N\mathfrak{d}_v)^{-s}$ we finally obtain the functional equation

$$\prod_{v \in S} \rho(c_v \| \cdot \|_v^s) \cdot \prod_{v \notin S} \left(\chi(\mathfrak{d}_v)(N\mathfrak{d}_v)^{s-1/2} \right) \cdot L(s, \chi) = L(1-s, \chi^{-1}),$$

where—using the local functional equations $\zeta(f_v, c_v \| \cdot \|_v^s) = \rho(c_v)\zeta(\hat{f}_v, \widehat{c_v \| \cdot \|_v^s})$ of Theorem 12.4—we have rewritten the ratios

$$\frac{\zeta(f_v, c_v \| \cdot \|_v^s)}{\zeta(\hat{f}_v, \widehat{c_v \| \cdot \|_v^s})}$$

in terms of ρ-factors. Recall that these functions have been explicitly computed in Sect. 12.4 and only depend on the character c_v, not on the choice of f_v. Replacing each ρ-factor with its explicit expression yields the classical functional equation for Hecke *L*-functions.

Problems

15.1 Complete the argument for the uniform convergence of (15.4).

15.2

1. Describe a cubic Dirichlet character χ (that is, a non-trivial Dirichlet character whose non-zero values are third roots of unity) in three ways: as a map $\mathbb{Z} \to \mathbb{C}$, as a 1-dimensional representation of a suitable Galois group, and as a character of the idèle group of $\mathbb{Q}$.
2. What is the general shape of the functional equation for $L(s, \chi)$? How would you determine the exact form of this functional equation?

Chapter 16
An Extended Example: The L-function of a CM Elliptic Curve

Abstract We briefly consider L-functions of a different nature from those covered in the previous parts of the book. In particular, we give an example of an L-function of geometric origin and use the theory developed in the previous chapters to prove that it admits analytic continuation.

In this chapter we look at a specific L-function whose origin is different from anything we have considered so far: the L-function of an elliptic curve over the rational numbers. Our example enjoys some special properties (specifically, it is an elliptic curve with complex multiplication) which enable us to prove analytic continuation for its L-function by reducing to the Hecke case. The reader should be aware that, in general, it is extremely difficult to prove analytic continuation of L-functions of elliptic curves over $\mathbb{Q}$—it is a problem that has only been solved in general thanks to the full modularity theorem. Some of the material in this chapter relies on tools that are beyond the scope of this book: nevertheless, we hope that the reader can get a glimpse of the beautiful mathematics behind this example, and perhaps be motivated to investigate it further.

16.1 The L-function of an Elliptic Curve

Consider the elliptic curve $E/\mathbb{Q}$ defined by the homogeneous cubic equation $Y^2Z + YZ^2 = X^3$. We will work with the corresponding Weierstrass equation $y^2 + y = x^3$. We note that—up to isomorphism—E can also be represented by the Weierstrass model $v^2 = u^3 + 16$: the isomorphism is given by $y = v/8 - 1/2$, $x = u/4$. Note that the same isomorphism makes sense over $\mathbb{F}_p$ for any prime $p \neq 2$.

Remark 16.1 It is easy to show that E has good reduction at all primes except 3. This means that if we reduce the equation $Y^2Z + YZ^2 = X^3$ modulo any prime $p \neq 3$, we get a *smooth* projective curve of genus 1.

We start by studying the elliptic curve defined by $Y^2Z + YZ^2 = X^3$ over $\mathbb{F}_p$, where p is any prime other than 3. We define the zeta function of a general elliptic curve $E/\mathbb{F}_p$ as follows:

Definition 16.1 Let E be an elliptic curve over the finite field $\mathbb{F}_p$. For every $k \geq 1$, let $N_k := \#E(\mathbb{F}_{p^k})$ be the number of projective solutions to the equation $Y^2Z + YZ^2 = X^3$ in $\mathbb{F}_{p^k}$. We set

$$Z(E/\mathbb{F}_p, t) = \exp\left(\sum_{k \geq 1} \frac{1}{k} N_k t^k\right).$$

Remark 16.2 It may not be clear why this is called a zeta function. The language of schemes provides a unified interpretation for the ζ functions of number fields and the geometric zeta function above. To see the connection, let X be a scheme of finite type over $\mathbb{Z}$ and define

$$\zeta(X, s) = \prod_{x \in X_{(0)}} \frac{1}{1 - N(x)^{-s}},$$

where the product is indexed by the closed points of X and $N(x)$ is the size of the residue field at x. When $X = \operatorname{Spec} \mathbb{Z}$, the closed points correspond bijectively to the non-zero primes of $\mathbb{Z}$, and we get back Riemann's zeta function:

$$\zeta(\operatorname{Spec} \mathbb{Z}, s) = \prod_p \frac{1}{1 - p^{-s}}.$$

More generally, when K is a number field and $X = \operatorname{Spec} O_K$, we have $\zeta(X, s) = \zeta_K(s)$, where $\zeta_K(s)$ is the Dedekind ζ function of Definition 1.7. On the other hand, it is an exercise in the manipulation of formal power series (Problem 16.1) to show that $\zeta(E, s) = Z(E/\mathbb{F}_p, p^{-s})$ when E is an elliptic curve over the finite field $\mathbb{F}_p$.

We now wish to compute $Z(E/\mathbb{F}_p, t)$ as explicitly as possible for our specific example. It is a classical theorem in the theory of elliptic curves [2, p. 136] that there exist two complex numbers α, β (unique up to exchanging α and β), of absolute value $p^{1/2}$, with the property that $\#E(\mathbb{F}_{p^k}) = p^k + 1 - \alpha^k - \beta^k$ for all $k \geq 1$. These numbers are in fact uniquely identified by the conditions

$$\alpha\beta = p, \quad \alpha + \beta = p + 1 - \#E(\mathbb{F}_p). \tag{16.1}$$

Lemma 16.1 *Let α, β be as above. We have*

$$Z(E/\mathbb{F}_p, t) = \frac{1 - (\alpha + \beta)t + pt^2}{(1 - t)(1 - pt)}.$$

16.1 The L-function of an Elliptic Curve

Proof Consider the power series $\exp(t) = \sum_{n\geq 0} \frac{t^n}{n!}$ and $\log(1-t) = -\sum_{n\geq 1} \frac{t^n}{n}$. For any complex number γ we have the elementary identities of formal power series

$$L_\gamma(t) := -\log(1-\gamma t) = \sum_{k\geq 1} \frac{(\gamma t)^k}{k}$$

and

$$\exp(L_\gamma(t)) = \frac{1}{\exp(\log(1-\gamma t))} = \frac{1}{1-\gamma t}.$$

Writing $N_k = p^k + 1 - \alpha^k - \beta^k$ and replacing in the definition of the zeta function we find

$$Z(E/\mathbb{F}_p, t) = \exp\left(\sum_{k\geq 1} \frac{(pt)^k}{k} + \frac{t^k}{k} - \frac{(\alpha t)^k}{k} - \frac{(\beta t)^k}{k}\right)$$

$$= \exp\left(L_p(t) + L_1(t) - L_\alpha(t) - L_\beta(t)\right)$$

$$= \frac{\exp(L_p(t))\exp(L_1(t))}{\exp(L_\alpha(t))\exp(L_\beta(t))}$$

$$= \frac{(1-\alpha t)(1-\beta t)}{(1-t)(1-pt)}$$

$$= \frac{1 - (\alpha+\beta)t + \alpha\beta t^2}{(1-t)(1-pt)}$$

$$= \frac{1 - (\alpha+\beta)t + pt^2}{(1-t)(1-pt)}.$$

$\square$

Our next objective is to understand the nature of the numbers α, β in our particular example. The case $p \equiv 2 \pmod{3}$ is easy:

Lemma 16.2 *Let $p \equiv 2 \pmod{3}$. We have $\{\alpha, \beta\} = \{\pm i\sqrt{p}\}$.*

Proof As $p \equiv 2 \pmod{3}$, the map $x \mapsto x^3$ induces a bijection of $\mathbb{F}_p$ with itself. Writing the equation of the elliptic curve in Weierstrass form, $y^2 + y = x^3$, this implies that for each value of $y \in \mathbb{F}_p$ there is exactly one solution $x \in \mathbb{F}_p$. This shows that E has exactly p points in the affine plane (x, y). Taking into account the point at infinity, we have $\#E(\mathbb{F}_p) = p+1$, so $\alpha + \beta = 0$ and $\alpha\beta = p$. Solving these equations gives $\{\alpha, \beta\} = \{\pm i\sqrt{p}\}$. $\square$

To deal with the case $p \equiv 1 \pmod{3}$, we rely on the classical theory of *Gauss and Jacobi sums*:

Definition 16.2 (Gauss and Jacobi Sums) Let $\chi : \mathbb{F}_p^\times \to \mathbb{C}^\times, \lambda : \mathbb{F}_p^\times \to \mathbb{C}^\times$ be characters of the finite group $\mathbb{F}_p^\times$. We set conventionally $\chi(0) = 0, \lambda(0) = 0$, so that we can consider χ, λ as functions from $\mathbb{F}_p$ to $\mathbb{C}$. We define the **Gauss sum**

$$g(\chi) := \sum_{t \in \mathbb{F}_p} \chi(t) \zeta_p^t,$$

where $\zeta_p := \exp(2\pi i/p)$, and the **Jacobi sum**

$$J(\chi, \lambda) = \sum_{\substack{a,b \in \mathbb{F}_p \\ a+b=1}} \chi(a)\lambda(b).$$

Note that ζ_p^t is well defined for $t \in \mathbb{F}_p$: for any two lifts $t_1, t_2 \in \mathbb{Z}$ of t we have $\zeta_p^{t_1} = \zeta_p^{t_2}$.

As shown in Remark 12.14, we have $|g(\chi)| = \sqrt{p}$ for any non-trivial character χ. We now prove some more relations satisfied by Gauss and Jacobi sums, which we will later use to determine exactly the values of the invariants α, β associated with our elliptic curve $E/\mathbb{F}_p$. The fact that this calculation can be done explicitly in relatively elementary terms is related to the fact that E has the non-trivial automorphism $x \mapsto \zeta_3 x$ (in particular, it is a CM curve).

Lemma 16.3 *Let χ, λ be non-trivial characters of the group $\mathbb{F}_p^\times$ such that $\chi\lambda$ is also non-trivial. We have*

$$J(\chi, \lambda) = \frac{g(\chi)g(\lambda)}{g(\chi\lambda)}.$$

In particular, $|J(\chi, \lambda)| = p^{1/2}$.

Proof We have

$$g(\chi)g(\lambda) = \sum_{t_1, t_2 \in \mathbb{F}_p} \chi(t_1)\lambda(t_2)\zeta_p^{t_1}\zeta_p^{t_2}$$

$$= \sum_{s=0}^{p-1} \zeta_p^s \sum_{t_1+t_2=s} \chi(t_1)\lambda(t_2)$$

$$= \sum_{t_1+t_2=0} \chi(t_1)\lambda(t_2) + \sum_{s=1}^{p-1} \zeta_p^s \sum_{t_1+t_2=s} \chi(t_1)\lambda(t_2).$$

16.1 The L-function of an Elliptic Curve

We claim that the first sum vanishes. Indeed,

$$\sum_{t_1+t_2=0} \chi(t_1)\lambda(t_2) = \sum_{t\in\mathbb{F}_p} \chi(t)\lambda(-t) = \lambda(-1)\sum_{t\in\mathbb{F}_p}(\chi\lambda)(t) = 0,$$

where the last equality follows from the fact that $\chi\lambda$ is not the trivial character together with Proposition 6.3.

We now consider the second sum. Since every $s \in \{1,\ldots,p-1\}$ is invertible modulo p, the relation $t_1 + t_2 = s$ is equivalent to $t_1 = st_1', t_2 = st_2'$ for certain (unique) $t_1', t_2' \in \mathbb{F}_p$ with $t_1' + t_2' = 1$. In particular, for every $s \in \{1,\ldots,p-1\}$ we have

$$\sum_{t_1+t_2=s} \chi(t_1)\lambda(t_2) = \sum_{t_1'+t_2'=1} \chi(st_1')\lambda(st_2')$$

$$= \chi(s)\lambda(s) \sum_{t_1'+t_2'=1} \chi(t_1')\lambda(t_2')$$

$$= (\chi\lambda)(s)J(\chi,\lambda),$$

and therefore

$$g(\chi)g(\lambda) = \sum_{s=1}^{p-1} \zeta_p^s \sum_{t_1+t_2=s} \chi(t_1)\lambda(t_2)$$

$$= J(\chi,\lambda) \sum_{s=1}^{p-1} \zeta_p^s \cdot (\chi\lambda)(s) = J(\chi,\lambda)g(\chi\lambda).$$

Note that in the last equality we have used $(\chi\lambda)(0) = 0$.

The last statement follows from the fact that each Gauss sum has absolute value $\sqrt{p}$. $\square$

Lemma 16.4 *Let χ be a non-trivial character of $\mathbb{F}_p^\times$. We have*

$$\overline{g(\chi)} = \chi(-1)g(\overline{\chi}).$$

Proof We compute

$$\overline{g(\chi)} = \overline{\sum_{t\in\mathbb{F}_p}\chi(t)\zeta_p^t} = \sum_{t\in\mathbb{F}_p}\overline{\chi(t)}\zeta_p^{-t} = \sum_{u\in\mathbb{F}_p}\overline{\chi(-u)}\zeta_p^u$$

$$= \overline{\chi(-1)} \sum_{u\in\mathbb{F}_p}\overline{\chi(u)}\zeta_p^u = \overline{\chi(-1)}g(\overline{\chi}) = \chi(-1)g(\overline{\chi}),$$

where the last equality uses that $\chi(-1) \in \{\pm 1\}$ is a real number. $\square$

Starting with the next lemma, it will be useful to denote by $N(x^n = a)$ the number of solutions $x \in \mathbb{F}_p$ of the equation $x^n = a$ (for a fixed $a \in \mathbb{F}_p$ and $n \geq 1$). The following is a simple exercise in character theory (see Problem 16.2):

Lemma 16.5 *Let $\chi : \mathbb{F}_p^\times \to \mathbb{C}^\times$ be a character of exact order n. For every $a \in \mathbb{F}_p$ we have*

$$N(x^n = a) = \sum_{i=0}^{n-1} \chi^i(a),$$

where by convention $\chi^0(0) = 1$.

Lemma 16.6 *Let p be an odd prime, ρ the unique character of $\mathbb{F}_p^\times$ of order 2, and χ a non-trivial character of $\mathbb{F}_p^\times$. We have*

$$J(\rho, \chi) = \chi(4) J(\chi, \chi).$$

Proof Using the equality $\sum_{t_2 \in \mathbb{F}_p} \chi(t_2) = 0$ (which follows from Proposition 6.3) and Lemma 16.5 we can write

$$\sum_{t_1+t_2=1} \rho(t_1)\chi(t_2) = \sum_{t_1+t_2=1} (1 + \rho(t_1))\chi(t_2)$$

$$= \sum_{t_1+t_2=1} N(t^2 = t_1)\chi(t_2) = \sum_{t_1 \in \mathbb{F}_p} N(t^2 = t_1)\chi(1 - t_1)$$

$$= \sum_{t \in \mathbb{F}_p} \chi(1 - t^2) = \chi(4) \sum_{t \in \mathbb{F}_p} \chi\left(\frac{1-t}{2}\right) \chi\left(\frac{1+t}{2}\right)$$

$$= \chi(4) \sum_{x+y=1} \chi(x)\chi(y) = \chi(4) J(\chi, \chi).$$

□

The invariants α, β of $E/\mathbb{F}_p$ for $p \equiv 1 \pmod{3}$ are closely related to $J(\chi, \chi)$, where χ is a character of $\mathbb{F}_p^\times$ of order 3. To show this, we need the following lemma:

Lemma 16.7 *Let p be a prime congruent to 1 modulo 3 and $\chi : \mathbb{F}_p^\times \to \mathbb{C}^\times$ be a character of order 3. The following hold:*

1. $J(\chi, \chi) = \frac{g(\chi)^2}{g(\chi^2)}$ *and* $p J(\chi, \chi) = g(\chi)^3$.
2. $J(\chi, \chi) J(\overline{\chi}, \overline{\chi}) = J(\chi, \chi) \overline{J(\chi, \chi)} = p$.
3. $J(\chi, \chi)$ *belongs to $\mathbb{Z}[\zeta_3]$ and we have $J(\chi, \chi) \equiv -1 \pmod{3\mathbb{Z}[\zeta_3]}$.*

Proof The equality $J(\chi, \chi) = \frac{g(\chi)^2}{g(\chi^2)}$ is a special case of Lemma 16.3. Since $\chi^2 = \chi^{-1} = \overline{\chi}$, we also have

$$g(\chi)^3 = J(\chi, \chi) g(\chi^2) g(\chi) = J(\chi, \chi) g(\overline{\chi}) g(\chi).$$

16.1 The L-function of an Elliptic Curve

Observe that $\chi(-1) = 1$, since $\chi(-1)$ has order dividing both 2 (because -1 does) and 3 (because χ does). Applying Lemma 16.4 we then obtain

$$g(\chi)^3 = J(\chi, \chi)\overline{g(\chi)}g(\chi) = J(\chi, \chi)|g(\chi)|^2 = pJ(\chi, \chi),$$

where we further used the fact that non-trivial Gauss sums have absolute value $\sqrt{p}$. This proves (1).

Repeating the same argument with $\overline{\chi}$ in place of χ we find $g(\overline{\chi})^3 = pJ(\overline{\chi}, \overline{\chi})$. Multiplying together this relation with the analogous one for χ we obtain

$$g(\chi)^3 g(\overline{\chi})^3 = p^2 J(\chi, \chi)J(\overline{\chi}, \overline{\chi}).$$

We have already proved that $g(\overline{\chi}) = \overline{g(\chi)}$, so (using again that Gauss sums have absolute value $\sqrt{p}$) this relation yields

$$p^3 = p^2 J(\chi, \chi)J(\overline{\chi}, \overline{\chi}),$$

which proves (2); note that by definition $\overline{J(\chi, \chi)} = J(\overline{\chi}, \overline{\chi})$.

Finally, cubic characters take values in the finite group $\langle \zeta_3 \rangle$, which is contained in $\mathbb{Z}[\zeta_3]$, so clearly $J(\chi, \chi)$ belongs to this ring. Working modulo the ideal (3) of $\mathbb{Z}[\zeta_3]$ we have

$$J(\chi, \chi) \equiv J(\chi, \chi)^3 \equiv \sum_{t_1+t_2=1} \chi(t_1)^3 \chi(t_2)^3 \equiv p - 2 \equiv -1 \pmod{3},$$

where the sum evaluates to $p - 2$ since all p summands are equal to 1, except those corresponding to $(t_1, t_2) = (1, 0), (0, 1)$. □

Corollary 16.1 *Let $p \equiv 1 \pmod{3}$ be a prime. There are unique (up to order) algebraic integers $\pi, \overline{\pi} \in \mathbb{Z}[\zeta_3]$ such that $p = \pi\overline{\pi}$ and $\pi \equiv 1 \pmod{3\mathbb{Z}[\zeta_3]}$. Up to exchanging π and $\overline{\pi}$, they are given by $-J(\chi, \chi)$ and $-\overline{J(\chi, \chi)}$.*

Proof Existence is obvious from Lemma 16.7. Uniqueness follows from unique factorisation in $\mathbb{Z}[\zeta_3]$ (which is a PID), together with the fact that the only units in $\mathbb{Z}[\zeta_3]^\times$ are the powers of $-\zeta_3$. More precisely: suppose given another such factorisation $p = \pi'\overline{\pi}'$. Unique factorisation yields that π' is associate to π or $\overline{\pi}$; without loss of generality, assume that π' and π generate the same ideal. We then have $\pi' = \pi u$ for some $u \in \mathbb{Z}[\zeta_3]^\times = \langle -\zeta_3 \rangle$. Write $u = (-\zeta_3)^i$ for some integer i. Since both π and π' are congruent to 1 modulo 3, we obtain $u \equiv 1 \pmod{3\mathbb{Z}[\zeta_3]}$. It is easy to check that the only power of $-\zeta_3$ congruent to 1 modulo $3\mathbb{Z}[\zeta_3]$ is 1. □

Lemma 16.8 *Let $p \equiv 1 \pmod{3}$. Up to reordering, the numbers α, β corresponding to the elliptic curve $E/\mathbb{F}_p$ as in (16.1) are given by $-J(\chi, \chi)$ and $-\overline{J(\chi, \chi)}$, where χ is a character of $\mathbb{F}_p^\times$ of order 3.*

Proof We work with the alternative Weierstrass model $y^2 = x^3 + 16$. Let $\rho : \mathbb{F}_p^\times \to \{\pm 1\}$ be the unique non-trivial character of order 2. Recalling that $\chi(-1) = 1$ and using Lemma 16.5, we have

$$\#\{(x,y) \in \mathbb{F}_p^2 : y^2 = x^3 + 16\} = \sum_{s+t=16} N(y^2 = s) N(x^3 = -t)$$

$$= \sum_{s+t=16} (1 + \rho(s))(1 + \chi(-t) + \chi^2(-t))$$

$$= \sum_{s+t=16} 1 + \sum_{t \in \mathbb{F}_p} (\chi(t) + \chi^2(t)) + \sum_{s \in \mathbb{F}_p} \rho(s)$$

$$+ \sum_{s+t=16} \rho(s)(\chi(t) + \chi^2(t)).$$

The first sum clearly evaluates to p. The second and third sum vanish by orthogonality of characters (Proposition 6.3). We are left with the last sum, which we rewrite as

$$\sum_{s+t=16} \rho(s)\chi(t) + \overline{\sum_{s+t=16} \rho(s)\chi(t)} :$$

to justify the equality $\sum_{s+t=16} \rho(s)\chi^2(t) = \overline{\sum_{s+t=16} \rho(s)\chi(t)}$, note that $\rho(s) \in \{\pm 1\}$ is a real number, while $\chi(t)$ is a third root of unity (or zero), hence $\chi^2(t) = \chi^{-1}(t) = \overline{\chi(t)}$. Setting $s = 16t_1$, $b = 16t_2$ (note that 16 is invertible in $\mathbb{F}_p$, since $p \neq 2$), the condition $s + t = 16$ becomes $t_1 + t_2 = 1$, and we can rewrite the above as

$$\sum_{s+t=16} \rho(s)\chi(t) + \overline{\sum_{s+t=16} \rho(s)\chi(t)}$$

$$= \sum_{t_1+t_2=1} \rho(16t_1)\chi(16t_2) + \overline{\sum_{t_1+t_2=1} \rho(16t_1)\chi(16t_2)}$$

$$= \rho(16)\chi(16) \sum_{t_1+t_2=1} \rho(t_1)\chi(t_2) + \overline{\rho(16)\chi(16) \sum_{t_1+t_2=1} \rho(t_1)\chi(t_2)}$$

$$= \rho(16)\chi(16) J(\rho, \chi) + \overline{\rho(16)\chi(16) J(\rho, \chi)}.$$

Since ρ has order 2 we have $\rho(16) = \rho(4)^2 = 1$. Using Lemma 16.6 and the fact that $\chi(t)^3 = 1$ for every $t \in \mathbb{F}_p^\times$ we then obtain $\rho(16)\chi(16)J(\rho, \chi) = \chi(16)\chi(4)J(\chi, \chi) = \chi(4)^3 J(\chi, \chi) = J(\chi, \chi)$. Thus,

$$\#\{(x,y) \in \mathbb{F}_p^2 : y^2 = x^3 + 16\} = p + J(\chi, \chi) + \overline{J(\chi, \chi)}$$

16.1 The L-function of an Elliptic Curve

and
$$\#E(\mathbb{F}_p) = p + 1 + J(\chi, \chi) + \overline{J(\chi, \chi)}.$$

Set $\alpha := -J(\chi, \chi)$ and $\beta = -\overline{J(\chi, \chi)}$. We have just proved $\alpha + \beta = p + 1 - \#E(\mathbb{F}_p)$, and on the other hand $\alpha\beta = p$ by Lemma 16.7 (2). These properties uniquely determine the invariants α, β of the elliptic curve $E/\mathbb{F}_p$ (see Eq. (16.1)), so we are done. □

The combination of Lemma 16.8, Corollary 16.1 and Lemma 16.2 shows the following:

Corollary 16.2 *Let $p \equiv 1 \pmod{3}$ be a prime. Factor $p = \pi\overline{\pi}$ with $\pi, \overline{\pi} \in \mathbb{Z}[\zeta_3]$ and $\pi \equiv 1 \pmod{3\mathbb{Z}[\zeta_3]}$. We have*

$$\#E(\mathbb{F}_p) = p + 1 - \pi - \overline{\pi},$$

hence $a_p = \pi + \overline{\pi}$. If instead $p \equiv 2 \pmod{3}$, we have $\#E(\mathbb{F}_p) = p+1$ and $a_p = 0$.

Example 16.1 Take $p = 7$ in the above corollary and let χ be a character of order 3 of $\mathbb{F}_7^\times$. One computes that $\{J(\chi, \chi), \overline{J(\chi, \chi)}\} = \{2+3\zeta_3, 2+3\zeta_3^2\}$. The ideal (7) then factors as $(2+3\zeta_3)(2+3\zeta_3^2)$ in $\mathbb{Z}[\zeta_3]$. It is apparent that $2+3\zeta_3 \equiv 2+3\zeta_3^2 \equiv -1 \pmod{3\mathbb{Z}[\zeta_3]}$. We set $\alpha := -2 - 3\zeta_3$ and $\beta := -2 - 3\zeta_3^2$; they satisfy $\alpha\beta = 7$, $\alpha + \beta = -1$ and $\alpha \equiv \beta \equiv 1 \pmod{3\mathbb{Z}[\zeta_3]}$. We should then have $\#E(\mathbb{F}_7) = 7 + 1 - (\alpha + \beta) = 7 + 1 + 1 = 9$. A quick calculation confirms this: working with the model $y^2 = x^3 + 16$, the 9 points are ∞ and

$$(0, \pm 3), (3, \pm 1), (5, \pm 1), (6, \pm 1).$$

From now on, we denote by α_p, β_p the invariants attached to $E/\mathbb{F}_p$ as in Eq. (16.1) and set $a_p := \alpha_p + \beta_p$. Recall that we have proved $a_p = 0$ for $p \equiv 2 \pmod{3}$, and $\alpha_p = \pi, \beta_p = \overline{\pi}$ if $p \equiv 1 \pmod{3}$ factors in $\mathbb{Z}[\zeta_3]$ as $\pi\overline{\pi}$ with $\pi \equiv 1 \pmod{3\mathbb{Z}[\zeta_3]}$. We now turn to the *global* setting, considering E as an elliptic curve over $\mathbb{Q}$. We define the L-function of E as

$$L(E, s) := \prod_{\substack{p \text{ prime} \\ p \neq 3}} \frac{1}{(1 - \alpha_p p^{-s})(1 - \beta_p p^{-s})}$$

$$= \prod_{\substack{p \text{ prime} \\ p \neq 3}} \frac{1}{1 - a_p p^{-s} + p \cdot p^{-2s}} \quad (16.2)$$

$$= \prod_{p \equiv 1 \pmod 3} \frac{1}{(1 - \pi p^{-s})(1 - \overline{\pi} p^{-s})} \prod_{p \equiv 2 \pmod 3} \frac{1}{1 + p^{1-2s}},$$

where in the case $p \equiv 1 \pmod{3}$ we have again adopted the notation $p = \pi\overline{\pi}$ with $\pi, \overline{\pi} \equiv 1 \pmod{3\mathbb{Z}[\zeta_3]}$.

Remark 16.3 The reader who is not familiar with the theory of elliptic curves should simply take this definition at face value. We will show below that this is a Hecke L-function, and also the Artin L-function of a Galois representation (or rather a compatible system of Galois representations) naturally associated with E.

16.2 Interpretation as a Hecke L-function

In this section we define a Hecke character of the field $k := \mathbb{Q}(\zeta_3)$ and check that its Hecke L-function coincides with (16.2). We note at the outset that the ring of integers of k has unique factorisation, and in particular, the class number of k is 1.

The field k has a single place at infinity, with completion $\mathbb{C}$. We denote it by ∞. It will also be important to consider the unique place v_3 of k above 3; the completion of k at v_3 is $\mathbb{Q}_3(\zeta_3)$, with ring of integers $\mathbb{Z}_3[\zeta_3]$. We take $\sqrt{-3} := 2\zeta_3 + 1$ as uniformiser at v_3 (of course, we could have taken $1 - \zeta_3$ instead, but this choice would have led to slightly more complicated calculations with logarithms below). We identify the completion k_∞ with $\mathbb{C}$ in such a way that $\zeta_3 \in k \subset k_\infty$ corresponds to $\frac{-1+\sqrt{3}i}{2}$ in $\mathbb{C}$; this implies that $\sqrt{-3} \in k$ corresponds to $\sqrt{3}i$ in $\mathbb{C}$.

We describe a quasi-character $\psi : I_k/k^\times \to \mathbb{C}^\times$ (a Hecke character) using the presentation of Sect. 14.1, see in particular Remark 14.3. First of all, we write our Hecke (quasi-)character as

$$\psi((a_v)) = c((a_v)) \cdot \|(a_v)\|^{-1/2}$$

for a character $c : I_k \to \mathbb{S}^1$ that is trivial on $k^\times$.

We describe c as in Remark 14.3. We need to specify a set of places S, local characters $\tilde{c}_v : \mathfrak{u}_v \to \mathbb{S}^1$ for $v \in S$, real numbers t_v for $v \in S$, and a character χ of the group $\mathcal{F}(S)$. We take $S = \{\infty, v_3\}$. The group of S-units of O_k has rank 1; a system of generators is given by $\varepsilon_1 := \sqrt{-3}$ and the torsion part is generated by $\varepsilon_0 := \zeta_6 = -\zeta_3^2$. We define

$$\tilde{c}_\infty : \mathfrak{u}_\infty = \mathbb{S}^1 \to \mathbb{S}^1$$
$$u \mapsto u^{-1}$$

and

$$\tilde{c}_{v_3} : \mathbb{Z}_3[\zeta_3]^\times \to \left(\frac{\mathbb{Z}_3[\zeta_3]}{(1-\zeta_3)^2}\right)^\times \to \mathbb{S}^1$$
$$\zeta_6 \mapsto \exp\left(\frac{2\pi i}{6}\right),$$

16.2 Interpretation as a Hecke L-function

where the first map is the natural projection. It is easy to check that the residue class of ζ_6 in the quotient $\frac{\mathbb{Z}_3[\zeta_3]}{(1-\zeta_3)^2}$ generates the unit group of this finite ring, so the above formula uniquely determines $\tilde{c}_{v_3}$. By a slight abuse of notation, we will denote by ζ_6 the complex number $\exp\left(\frac{2\pi i}{6}\right)$.

We also set $t_\infty = 0$ and $t_{v_3} = -\frac{\pi}{2\log 3}$. We check that this data satisfies the conditions in Eqs. (14.1) and (14.2). For the former we have $\varepsilon_0 = \zeta_6$ and $\tilde{c}_{v_3}(\zeta_6) = \zeta_6$, $\tilde{c}_\infty(\zeta_6) = 1/\zeta_6$, so (14.1) is clearly satisfied. For the latter, we need to check that for $\varepsilon_1 = \sqrt{-3}$ we have

$$\sum_{v \in S} t_v \log \|\varepsilon_{1,v}\|_v = i \log \left(\prod_{v \in S} \tilde{c}_v(\tilde{\varepsilon}_{1,v})\right),$$

where for each place v we write $\varepsilon_{1,v} = \tilde{\varepsilon}_{1,v} \cdot \rho$, with ρ either a positive real number (if v is infinite) or a power of the chosen uniformiser (if it is finite).

For the place at infinity we have $\varepsilon_{1,\infty} = \sqrt{3} \cdot i$, so $\tilde{\varepsilon}_{1,\infty} = i$. For the 3-adic place, $\sqrt{-3}$ is our chosen uniformiser, so $\tilde{\varepsilon}_{1,v_3} = 1$ by definition. The equation above then reads

$$t_{v_3} \log \|\sqrt{-3}\|_{v_3} = i \log(\tilde{c}_\infty(i)) = i \log(1/i).$$

Choosing $-\frac{\pi}{2}i$ as determination for $\log(1/i)$, using $\|\sqrt{-3}\|_{v_3} = 1/3$, and taking $\log(1/3)$ to be its real determination, we have

$$-t_{v_3} \cdot \log 3 = i \cdot \left(-\frac{\pi}{2}i\right) \Leftrightarrow t_{v_3} = -\frac{\pi}{2\log 3},$$

so Eq. (14.2) is satisfied as well.

Remark 16.4 The condition $t_{v_3} = -\frac{\pi}{2\log 3}$, together with our choice of determinations of the logarithms, simply amounts to $c(\sqrt{-3}) = i$.

By the discussion after Eq. (14.4), the character χ on $\mathcal{F}(S)$ is uniquely determined by the other data. Specifically, Eq. (14.3) gives

$$\chi(\varphi_S(\alpha)) = \prod_{v \in S} \tilde{c}_v(\tilde{\alpha}_v)^{-1} \|\alpha\|_v^{-it_v},$$

and since every ideal of O_k is principal, in particular every ideal in $\mathcal{F}(S)$ is of the form $\varphi_S(\alpha)$, so this formula determines χ on all of $\mathcal{F}(S)$ (see also Problem 14.1). Putting everything together, we have defined the Hecke character

$$\psi((a_v)) = \tilde{c}_{v_3}(\tilde{a}_{v_3}) \cdot \|a_{v_3}\|_{v_3}^{-i\pi/2\log 3} \cdot \frac{1}{a_\infty} \cdot \chi(\varphi_S((a_v))) \cdot \|(a_v)\|^{-1/2}$$

$$= \tilde{c}_{v_3}(\tilde{a}_{v_3}) \cdot i^j \cdot \frac{1}{a_\infty} \cdot \chi(\varphi_S((a_v))) \cdot \|(a_v)\|^{-1/2}$$

(16.3)

where $\tilde{a}_\infty = \frac{a_\infty}{\sqrt{\|a_\infty\|_\infty}}$ and we have written $a_{v_3} = \tilde{a}_{v_3} \cdot \sqrt{-3}^j$ for some $j \in \mathbb{Z}$ and $\tilde{a}_{v_3} \in \mathbb{Z}_3[\zeta_3]^\times$.

By construction, the only finite place at which the Hecke character ψ is ramified is v_3. Let now v be a place of k different from ∞ and v_3. As v is finite, it corresponds to a certain prime $\mathfrak{p}_v$ of the ring of integers, and we write Nv for $N\mathfrak{p}_v = \#\frac{O_k}{\mathfrak{p}_v}$. We choose a uniformiser π_v at v in the following way:

1. if v corresponds to a prime $p \in \mathbb{Z}$ with $p \equiv 2 \pmod{3}$, we take $\pi_v = p$. Note that in this case $Nv = p^2$.
2. if v corresponds to a prime $\pi \in \mathbb{Z}[\zeta_3]$ with $\pi\bar{\pi} = p \equiv 1 \pmod{3}$, we choose π_v to be the unique generator of the ideal (π) that is congruent to 1 modulo $3\mathbb{Z}[\zeta_3]$. In particular, we have $\pi_v\bar{\pi_v} = p$ and $\pi_v \equiv 1 \pmod{3\mathbb{Z}[\zeta_3]}$, so π_v satisfies the conditions in Corollary 16.2. In this case, $Nv = p$.

We define $b(v)$ as the idèle that is trivial at all $w \neq v$ and is π_v at v. Consider the Hecke L-function attached to ψ. By definition, it is the product

$$\prod_{\substack{v \text{ finite} \\ v \neq v_3}} \left(1 - \frac{\psi(b(v))}{(Nv)^s}\right)^{-1}.$$

We can compute the local factors as follows.

1. Suppose first that v corresponds to $p \equiv 2 \pmod{3}$. Then—using the fact that ψ is trivial on $k^\times$—we can write $\psi(b(v)) = \psi(1/p \cdot b(v))$, where $\beta := 1/p \cdot b(v)$ is the idèle that is trivial at v and equal to $1/p$ at all other places. In particular, for every finite place w of k the element $\beta_w \in k_w^\times$ is a unit, and so $\varphi_S(\beta) = (1)$, and a fortiori $\chi(\varphi_S(\beta)) = 1$. The adèlic norm of β is $\|\beta_\infty\|_\infty = \|1/p\|_\infty = 1/p^2$. The class of $1/p \equiv -1 \pmod{3}$ in $\mathbb{Z}_3[\zeta_3]/(1-\zeta_3)^2$ is -1, so $\tilde{c}_{v_3}(1/p) = -1$. Finally, $\tilde{\beta}_\infty = 1$ since $1/p$ is real.

   The value of ψ at β is then given by $\tilde{c}_3(1/p)\|\beta\|^{-1/2} = \tilde{c}_3(1/p)(1/p^2)^{-1/2} = -p$. The local factor is therefore

   $$\left(1 - \frac{\psi(b(v))}{(Nv)^s}\right)^{-1} = \left(1 + \frac{p}{p^{2s}}\right)^{-1} = \left(1 + p^{1-2s}\right)^{-1} = (1 - a_p p^{-s} + p \cdot p^{-2s})^{-1},$$

   since $a_p = 0$ in this case (see Lemma 16.2).

2. Suppose instead that v corresponds to a prime $\pi \in \mathbb{Z}[\zeta_3]$ with $\pi\bar{\pi} = p \equiv 1 \pmod{3}$. As before, we write $\psi(b(v)) = \psi(1/\pi_v \cdot b(v))$ and let β be the idèle $1/\pi_v \cdot b(v)$, which is trivial at v and equal to the w-unit $1/\pi_v$ at all finite places $w \neq v$. We again have $\chi(\varphi_S(\beta)) = 1$. The adèlic norm of β is $\|\beta_\infty\|_\infty = \|1/\pi_v\|_\infty = 1/|\pi_v|^2 = 1/p$. We have $\beta_\infty = \frac{\sqrt{p}}{\pi_v} \cdot \frac{1}{\sqrt{p}}$, so $\tilde{\beta}_\infty = \frac{\sqrt{p}}{\pi_v}$. The class of $1/\pi_v \equiv 1 \pmod{3}$ in $\mathbb{Z}_3[\zeta_3]/(1-\zeta_3)^2$ is 1, so $\tilde{c}_{v_3}(1/\pi_v) = 1$. The value of

16.3 Interpretation as the L-function of a Representation

ψ at β is then given by $\tilde{c}_{v_3}(1/\pi_v)\tilde{c}_\infty(\frac{\sqrt{p}}{\pi_v})\|\beta\|^{-1/2} = \frac{\pi_v}{\sqrt{p}}p^{1/2} = \pi_v$. The local factor is therefore

$$\left(1 - \frac{\psi(b(v))}{(Nv)^s}\right)^{-1} = \left(1 - \frac{\pi_v}{p^s}\right)^{-1}.$$

Multiplying together the local factors at v and $\bar{v}$ we get

$$\left(1 - \frac{\pi_v}{p^s}\right)^{-1}\left(1 - \frac{\overline{\pi_v}}{p^s}\right)^{-1} = \left(1 - (\pi_v + \overline{\pi_v})p^{-s} + (\pi_v\overline{\pi_v})p^{-2s}\right)^{-1}$$
$$= \left(1 - a_p p^{-s} + p \cdot p^{-2s}\right)^{-1},$$

where we have again used Lemma 16.2.

Remark 16.5 It is an interesting exercise to check explicitly that—had we chosen a different uniformiser π_v and therefore a different idèle $b(v)$—we would have found the same result. Of course, this is guaranteed by the general theory we have set up.

The upshot is that, taking the product over all places of k different from v_3, the Hecke L-function of ψ gives precisely

$$\prod_{p \neq 3}\left(1 - a_p p^{-s} + p \cdot p^{-2s}\right)^{-1},$$

that is, the L-function of E as defined in Eq. (16.2). By Theorem 15.1 we see in particular that $L(E, s)$ admits analytic continuation to the whole complex plane and satisfies a functional equation. As already mentioned, for non-CM elliptic curves, this statement is known thanks to the modularity theorem, which gives the equality between $L(E, s)$ and the L-function of a suitable automorphic representation (in this case, a modular form of weight 2 for $\Gamma_0(27)$: the interested reader can find the data about this modular form, and its associated L-function, at [3, Newform orbit 27.2.a.a]).

16.3 Interpretation as the L-function of a Representation

In this section we will freely use properties of elliptic curves. All we need can be found in Silverman's classical book [2], see especially Section III.7 for the notion of Tate module.

We will give an interpretation of (16.2) as a "generalised Artin L-function" for a Galois representation with infinite image (or, more precisely, a compatible system of Galois representations: see below). For every integer $n \geq 1$, we let $E[n]$ be the n-torsion of E, that is, the subgroup of $E(\overline{\mathbb{Q}})$ given by

$$E[n] = \{P \in E(\overline{\mathbb{Q}}) : nP = O_E\},$$

where $O_E = [0 : 1 : 0]$ denotes the origin of the elliptic curve addition law. It is well known that $E[n] \cong (\mathbb{Z}/n\mathbb{Z})^2$ as abstract abelian groups, and that furthermore there is a natural Galois action of $\text{Gal}(\overline{\mathbb{Q}}/\mathbb{Q})$ on $E[n]$, giving a representation (over a finite ring)

$$\rho_{E,n} : \text{Gal}(\overline{\mathbb{Q}}/\mathbb{Q}) \to \text{Aut}(E[n]) \cong \text{GL}_2(\mathbb{Z}/n\mathbb{Z}). \tag{16.4}$$

The Galois action can be described as follows. Every Galois automorphism fixes O_E; every other point of $E(\overline{\mathbb{Q}})$ lies in the affine plane (x, y), and the action of $\sigma \in \text{Gal}(\overline{\mathbb{Q}}/\mathbb{Q})$ on $P = (x_P, y_P)$ is given by

$$\sigma(P) = (\sigma(x_P), \sigma(y_P)).$$

Now fix a prime ℓ and choose $n = \ell^k$ (for an integer $k \geq 1$) in Eq. (16.4). Passing to the inverse limit in k, we obtain the ℓ-adic representation

$$\rho_{E,\ell^\infty} : \text{Gal}(\overline{\mathbb{Q}}/\mathbb{Q}) \to \text{Aut}(\varprojlim_k E[\ell^k]) \cong \text{GL}_2(\mathbb{Z}_\ell).$$

The limit $\varprojlim_k E[\ell^k]$ is called the ℓ-adic Tate module of E; it is a free $\mathbb{Z}_\ell$-module of rank 2, usually denoted by $T_\ell(E)$. Finally viewing $\text{GL}_2(\mathbb{Z}_\ell)$ as a subgroup of $\text{GL}_2(\mathbb{Q}_\ell)$ (equivalently, tensoring by $\mathbb{Q}_\ell$) we get a representation of $\text{Gal}(\overline{\mathbb{Q}}/\mathbb{Q})$ over a field of characteristic 0,

$$\rho_{E,\ell^\infty} : \text{Gal}(\overline{\mathbb{Q}}/\mathbb{Q}) \to \text{GL}_2(\mathbb{Q}_\ell).$$

Note that the image of ρ_{E,ℓ^∞} is not finite, so—even though $\text{GL}_2(\mathbb{Q}_\ell)$ can be embedded inside $\text{GL}_2(\mathbb{C})$ by choosing an embedding $\mathbb{Q}_\ell \hookrightarrow \mathbb{C}$—we are not quite in the setting of Artin's L-functions.

We briefly discuss how to extend the notion of decomposition groups, inertia groups and Frobenius elements to the setting of infinite Galois theory. A place v of $\overline{\mathbb{Q}}$ lying over the rational prime p is by definition a compatible collection $\{v_K\}$ of places lying over p, one for each number field K. Compatibility means that for every inclusion of number fields $K \subseteq L$, the restriction of v_L to K coincides with v_K. There is a well-defined notion of decomposition and inertia subgroups at v (denoted by D_v and I_v): they are the inverse limits of the decomposition and inertia groups of v_K over p, as K varies over all number fields. A Frobenius element at v is an element of $\text{Gal}(\overline{\mathbb{Q}}/\mathbb{Q})$ that maps to a Frobenius at v_K for every number field K; it is only well-defined up to the inertia group I_v. As in the finite case, all Frobenius

16.3 Interpretation as the L-function of a Representation

elements corresponding to different places v over p are conjugate to one another (up to the image of inertia).

We can now discuss the Galois representation ρ_{E,ℓ^∞} in greater detail. It is a non-trivial fact, known as the Néron-Ogg-Shafarevich criterion, that since E has good reduction outside the prime 3, for every prime ℓ the representation ρ_{E,ℓ^∞} is unramified outside of ℓ and 3. Concretely, this means the following: if p is a prime different from 3 and ℓ, and I_v is the inertia subgroup of $\text{Gal}(\overline{\mathbb{Q}}/\mathbb{Q})$ corresponding to any place v of $\overline{\mathbb{Q}}$ over p, then $\rho_{E,\ell}(I_v)$ is the trivial group.

Let again $p \neq 3$ be a prime and fix an auxiliary prime $\ell \neq p$. Also fix an arbitrary Frobenius element $\text{Frob}_p \in \text{Gal}(\overline{\mathbb{Q}}/\mathbb{Q})$ at a place v of $\overline{\mathbb{Q}}$ lying over p. Since $\rho_{E,\ell^\infty}(I_v) = \{1\}$, the element $\rho_{E,\ell^\infty}(\text{Frob}_p)$ is well-defined up to conjugacy (the indeterminacy due to inertia disappears after applying ρ_{E,ℓ^∞}). In particular, the characteristic polynomial of $\rho_{E,\ell^\infty}(\text{Frob}_p)$ is well-defined and depends only on p. It can be shown that this characteristic polynomial is independent of the auxiliary prime $\ell \neq p$, and that in fact it is given by $f_p(t) = t^2 - a_p t + p$, where as in the previous sections $a_p = p + 1 - \#E(\mathbb{F}_p)$. Thus, if we instead consider the dual[1] $\rho_{E,\ell^\infty}^\vee$ of the representation $T_\ell(E)$, the normalised characteristic polynomial $\det(\text{Id} - \rho_{E,\ell^\infty}^\vee(\text{Frob}_p)t)$ is $1 - a_p t + p t^2$. The conclusion is that the L-function of E can be written as the product

$$\prod_{p \neq 3} \det(\text{Id} - \rho_{E,\ell^\infty}^\vee(\text{Frob}_p) p^{-s} \mid T_\ell(E))^{-1}, \qquad (16.5)$$

where for each p we choose any $\ell \neq p$ to compute $\rho_{E,\ell^\infty}^\vee(\text{Frob}_p)$. Equation (16.5) is formally very similar to the definition of an Artin L-function, except for the fact that we can choose one of many different representations $\rho_{E,\ell^\infty}^\vee$ to compute the relevant characteristic polynomials of Frobenius. The point here is that the various representations $\rho_{E,\ell^\infty}^\vee$ form what is called a *compatible system*: as already discussed, the characteristic polynomials of $\rho_{E,\ell^\infty}^\vee(\text{Frob}_p)$ do not depend on the auxiliary prime ℓ, provided that ℓ is chosen different from p. This is the starting point of the fascinating theory of L-functions of elliptic curves over number fields.

To conclude, we simply point out that—just as in the case of Artin's L-functions—one should also give an appropriate definition at the ramified places. It turns out that the solution is exactly the same in this more general setting as it was for representations with finite image: instead of considering the action on the full Tate module $T_\ell(E)$, the local factor at a prime p of bad reduction is given by

$$\det(\text{Id} - \rho_{E,\ell^\infty}^\vee(\text{Frob}_p) p^{-s} \mid T_\ell(E)^{\rho_{E,\ell^\infty}(I_p)})^{-1},$$

---

[1] The expert reader will recognise that this dual representation is the natural Galois representation on the first étale cohomology group of $E_{\overline{\mathbb{Q}}}$ with coefficients in $\mathbb{Q}_\ell$.

where I_p is an inertia subgroup at (a place of $\overline{\mathbb{Q}}$ lying over) p. It turns out that, for the case of our elliptic curve E, the local factor at 3 is trivial (that is, equal to 1; this is due to the fact that E has additive reduction at 3), and so the product in Eq. (16.5), where the prime 3 does not appear, is the full L-function of E.

The constructions of this section can be vastly generalised to the notion of *Hasse-Weil zeta functions*, and more generally to the notion of L-functions of geometric origin. The interested reader can find out (much) more for example in Kahn's book [1].

Problems

16.1 Let $E/\mathbb{F}_p$ be an elliptic curve. Justify the equality

$$\zeta(E, s) = Z(E/\mathbb{F}_p, p^{-s})$$

mentioned in Remark 16.2.

Hint. Start by writing

$$\log \zeta(E, s) = - \sum_{x \in E_{(0)}} \log(1 - N(x)^{-s}) = \sum_{x \in E_{(0)}} \sum_{k \geq 1} \frac{N(x)^{-ks}}{k}$$

and then re-arrange suitably.

16.2 Prove Lemma 16.5.

Hint. It may be useful to apply the orthogonality relations of Proposition 6.3.

References

1. Kahn, B.: Zeta and L-functions of Varieties and Motives, London Mathematical Society Lecture Note Series, vol. 462. Cambridge University Press, Cambridge (2020). https://doi.org/10.1017/9781108691536. Translated from the 2018 French original [3839285]
2. Silverman, J.H.: The Arithmetic of Elliptic Curves, Graduate Texts in Mathematics, vol. 106, 2nd edn. Springer, Dordrecht (2009). https://doi.org/10.1007/978-0-387-09494-6
3. The LMFDB Collaboration: The L-functions and modular forms database. https://www.lmfdb.org (2024). Online. Accessed 15 Nov 2024

Index

Symbols
L-function
    Artin, 61
    of an automorphic representation, 12
    Dirichlet, 7, 65
    functoriality, 68
    Hecke, 67
    special value, 89, 91
L-function of an elliptic curve, 251
S-units, 37
p-adic norm, 136
ζ-functions
    global case, 216
    local case, 169
Γ function, 4
ζ function of number field, *see* Dedekind ζ function
p-adic field, 137

A
Abel summation, 22
Adèles, 144, 191
    infinite part, 194
    standard character, 192
Additive fundamental domain, 196
Analytic class number formula, 91, 217, 235
Analytic continuation, 238
    of Dedekind ζ functions, 6, 235
    of Dirichlet L-functions, 7
    of Hecke L-functions, 68, 240
    of the Riemann ζ function, 4, 232
    Tate's global ζ functions, 217
    of Tate's local ζ functions, 174
Artin L-function, *see* L-function, Artin
analytic continuation (conjectural), 73
functoriality, 68
meromorphic continuation, 74
non-vanishing, 76
Artin symbol, 33
Automorphic representation, 12

B
Borel sets, 115
Brauer's induction theorem, 74

C
Character, 15, 129, 160
    conductor, 8
    Dirichlet, 6, 15, 228
    fundamental additive character, 162
    Hecke, 223
    primitive, 8
    quasi-character, 129
    of a representation, 42
    standard adèlic, 192
    of the units of a local field, 166
    unramified, 166, 225
Chebotarev's theorem
    qualitative form, 35
    quantitative form, 95
Class field theory, 224
Class function, 51
Class group, 39
    finiteness, 39
Completed ζ function, 5
Completion, 138
Conductor, 8

Index

D
Dedekind ζ function, 6, 65, 232
    factorisation, 75, 76, 87
Density
    Dirichlet, 86, 95, 103
    logarithmic, 86
    natural, 86
    upper and lower, 87
Different, 139
Dirichlet L-functions, 7, 65
    are Artin L-functions, 67
Dirichlet series, 3
Dirichlet's theorem, 101
    for number fields, 103
    qualitative form, 35
    quantitative form, 88
Dirichlet's unit theorem, 36, 37
Discriminant, 29, 139, 194
Divisor, 205
    principal, 205
Dual group, 81
    of $\mathbb{A}_k/k$, 198
    of the additive group of a local field, 162
    double dual, 82
    functoriality, 82, 130
    topological, 130
    of a topological restricted product, 148
    of the unit group of a local field, 166

E
Euler product, 5, 10, 15
    for Artin L-functions, 61, 64
Euler's Γ function, *see* Γ function

F
Fourier inversion
    on the adèles, 192
Fourier transform, 84, 132, 175, 179, 182, 236
    on $\mathbb{A}_k/k$, 200
    inversion, 84
    inversion formula, 132
    of a product, 154
Frobenius element, 33
    conjugacy class, 34
    in cyclotomic extensions, 34
Functional equation, 4, 9, 232, 234
    for Tate's local ζ function, 171

G
Gauss sum, 187, 246

H
Haar measure, 117
    on the additive group of a local field, 164
    dual, 133
    existence, 118
    on the idèles, 208
    invariance under translation, 124
    left, 117
    on the multiplicative group of a local field, 168
    on a restricted product, 150
    right, 117
    self-dual, 164
    uniqueness, 124
Hecke character, 223
Hecke L-function, 252
    for characters with finite image, 67
    idèlic version, 223

I
Ideal
    fractional, 38
    group of fractional ideals, 38
    inertia degree, 138
    norm, 6
    principal, 39
    ramification index, 138
    unique factorisation, 30
Idèles, 144
    norm, 158
Invariant subspace, 42

J
Jacobi sum, 246

L
Landau's ξ function, 4
Logarithmic embedding, 36

M
Measure
    Haar (*see* Haar measure)
    inner regular, 115
    outer regular, 115
    Radon, 115
    regular, 115
Minkowski's theorem, 32
Multiplicative fundamental domain, 213

Index

N
Number field, 135
  completion, 137
  signature, 35

O
Orthogonality relations
  abelian case, 83
  general case, 53
Ostrowski's theorem, 136

P
Perron's formula, 26, 28
Place, 135
  Archimedean, 136
  complex, 136
  degree, 205
  finite, 135, 136
  inertia degree, 138
  infinite, 136
  non-Archimedean, 135, 136
  ramification index, 138
  real, 136
  ring of integers, 205
  set of all places, 136
Poisson formula, 202
Pontryagin duality, 130
Prime ideal
  Galois action, 32
  inertia degree, 31
  ramification index, 31
  ramified, 31
Prime number theorem, 17
Product formula, 137, 140, 205, 208
Product function, 151

Q
Quasi-character, 129, 146, 175, 178, 181
  equivalence class, 217
  equivalent, 170
  exponent, 167
  of $I_k/k^\times$, 216
  of the units of a local field, 166
  unramified, 166

R
Ramified, *see* Prime ideal, ramified
Regular representation, *see* Representation, regular
Regulator, 37, 211
Representation, 41
  character, 42
  character table, 52
  completely reducible, 46
  contragredient, 44
  dimension, 41
  direct sum, 43
  dual, 44
  faithful, 60
  fixed points, 42
  homomorphism, 41
  induced, 55, 68
  inflation, 55, 68
  irreducible, 42
  isomorphism, 42
  of an abelian group, 46
  regular, 59, 76
  restriction, 54
  subrepresentation, 42
  tensor product, 45
  trivial, 42
Restricted product, 143
  of abstract groups, 143
  of Haar measures, 150
  of rings, 144
  of topological groups, 145
Riemann hypothesis, 27, 102
Riemann-Roch theorem
  geometric form, 206
  Tate's arithmetic version, 203
Riemann's ζ function, 4, 65, 231
Riemann–von Mangoldt exact formula, 26
Riesz-Markov-Kakutani representation theorem, 126
Ring of integers, 6, 138

S
Schur's lemma, 47

T
Tate's global ζ function, 217
  analytic continuation, 217
  functional equation, 217
Tate's local ζ function, 170, 177, 180, 183
  analytic continuation, 174
  functional equation, 171
  non-vanishing, 187
Tauberian theorem, 20
Topological group, 116
  locally compact, 116

U
Uniformiser, 136, 138
Unit theorem, *see* Dirichlet's unit theorem

MIX
Papier aus verantwortungsvollen Quellen
Paper from responsible sources
FSC® C105338

If you have any concerns about our products,
you can contact us on
ProductSafety@springernature.com

In case Publisher is established outside the EU,
the EU authorized representative is:
**Springer Nature Customer Service Center GmbH
Europaplatz 3, 69115 Heidelberg, Germany**

Printed by Libri Plureos GmbH
in Hamburg, Germany